FOREIGN ANIMAL DISEASES

REVISED 2008

SEVENTH EDITION

FOREIGN ANIMAL DISEASES

REVISED 2008

SEVENTH EDITION

Committee on Foreign and Emerging Diseases

of the

United States Animal Health Association

USAHA
PO Box 8805
St. Joseph, MO 64508
Phone: 816-671-1144
Fax: 816-671-1201
email: usaha@usaha.org
Internet site: www.usaha.org

Library of Congress Catalogue Number 2008900990

ISBN 978-0-9659583-4-9

Boca Publications Group, Inc.
2650 N. Military Trail, 240-SZG
Boca Raton, FL 33431
bocagroup@aol.com

Printed in Canada

PREFACE

Educating the veterinary profession about Foreign Animal Diseases has been a long tradition of the U. S. Animal Health Association. The first "Gray Book" edition was published more than half a century ago in 1953, with subsequent editions in 1964, 1975, 1984, 1992 and in 1998.

Traditionally, the task of the reviewing and updating this book, still familiarly known as the "Gray Book" (despite the white cover of recent editions) falls to the Chair and Co-Chair of the USAHA's Foreign and Emerging Disease Committee. We are thus indebted to the U.S. Animal Health Association for the opportunity to assemble this, the 7th edition of *Foreign Animal Diseases*.

There have been vast changes in the world since the last edition was published in 1998. At that time, the World Trade Organization was just three years old and only beginning the tremendous facilitation of international trade that we see today. The last edition was published before Nipah virus in Malaysia, before the massive foot-and-mouth disease outbreak in the United Kingdom, before the advent of the term "agroterror", before SARS had infected any humans, and prior to the possibility of highly pathogenic avian influenza as a human pandemic. Indeed, with so many new pathogens as well as old pathogens surfacing in new and unexpected places, the term "foreign animal disease" is becoming less relevant, even as the threat of foreign animal disease incursions becomes more relevant.

We have utmost respect for and gratitude to the authors of the chapters. Their contributions were timely, articulate, and accurate. This book is rightfully theirs and we are merely organizers and purveyors of their information. We owe special thanks to Visual Information Services at the Plum Island Animal Disease Center, whose staff supplied most of the new photographs in Part IV. In addition, the Animal and Plant Health Inspection Service, Professional Development Staff (PDS), was very generous in allowing us to borrow Dr. Jason Baldwin, who served as an infallible and incredibly diligent copy and content editor. PDS also supplied the funding for the final formatting of the book, including the new cover design.

We also acknowledge our host institutions, the Colleges of Veterinary Medicine at the University of Georgia and Cornell University, for allowing us the time to devote to the editing and assembling of this book. We were each generously given the opportunities to apply our efforts, without any expectations of compensation. Fortunately, the leaders of our respective institutions understand the importance and impact that this volume has on preparing our animal health professionals.

Lastly, we wish to recognize the long-term efforts of Dr. Charles Mebus, in research, diagnosis, and dissemination of information regarding foreign animal diseases. Chuck Mebus has been a mentor to the two of us at various stages in our careers and has always served as a stellar role model, good friend, and a visionary regarding the larger picture of animal health. As it was done for the 6th edition, we re-dedicate this, the 7th edition of *Foreign Animal Disease,* to him.

Corrie Brown, DVM, PhD, DACVP
Chair, Foreign Animal and Emerging Diseases Committee
Josiah Meigs Distinguished Teaching Professor
College of Veterinary Medicine
University of Georgia
Athens, GA, 30602

Alfonso Torres, DVM, MS, PhD
Co-Chair, Foreign Animal and Emerging Diseases Committee
Professor & Associate Dean for Public Policy
College of Veterinary Medicine
Cornell University
Ithaca, NY 14850

FOREWORD

For more than a half century the "Gray Book" has been the key resource for veterinarians, from private practitioners to federal, state and corporate practices and students regarding foreign animal diseases.

The preparation of this book is a tangible example of the remarkable cooperation between all the sectors of the U.S. Animal Health Association. Professionals from academia, U.S. federal and state agencies, and a number of foreign countries have shared their expertise and their time to create this seventh edition.

The Chair and Co-Chair of the USAHA Committee on Foreign and Emerging Diseases, Corrie Brown, DVM, PhD, and Alfonso Torres, DVM, PhD, respectively, have coordinated the compilation of this edition, with USAHA acting as the publishing agency.

On behalf of the USDA, I want to express my appreciation to the leadership of USAHA, Drs. Brown and Torres, and all the authors, reviewers and editors for their selfless contributions. There are no monetary remunerations or royalties for writing this book: its creation grew from the collective understanding of the importance of sustaining a successful history of safeguarding all animal health industries from animal diseases.

Dr. John R. Clifford
Deputy Administrator for Veterinary Services, and
U.S. Chief Veterinary Officer
Animal Plant & Health Inspection Service
U.S. Department of Agriculture
Washington DC

TABLE OF CONTENTS

I

CONTRIBUTORS

Corrie Brown
College of Veterinary Medicine
University of Georgia
Athens, GA 30602-7388
corbrown@uga.edu

Claudio S.L. Barros
Universidade Federal de Santa Maria
Santa Maria, Brazil
claudiosbarros@uol.com.br

Rafael Fighera
Universidade Federal de Santa Maria
Santa Maria, Brazil
anemiaveterinaria@yahoo.com.br

R.O. Gilbert
College of Veterinary Medicine
Cornell University
Ithaca, NY 14853-6401
rog1@cornell.edu

Alan J. Guthrie
Equine Research Centre
Faculty of Veterinary Science
University of Pretoria
Onderstepoort, 0110, Republic of South Africa
alan.guthrie@up.ac.za

Christopher Hamblin
94 South Lane, Ash, Near Aldershot
Hampshire, GU12 6NJ, England
chris.hamblin@ntlworld.com

Christiane Herden
Institut fur Pathologie
Tierarztliche Hochschule Hannover
Hannover, Germany
christiane.herden@tiho-hannover.de

Sharon K. Hietala
University of California-Davis
Davis, CA 95617
skhietala@ucdavis.edu

Daniel J. King
Southeast Poultry Research Laboratory
USDA-ARS
Athens, GA 30605
jack.king@ars.usda.gov

Peter Kirkland
Head, Virology Laboratory
Elizabeth Macarthur Agricultural Institute
Menangle, NSW, Australia
peter.kirkland@agric.nsw.gov.au

Paul Kitching
National Centre for Foreign Animal Disease
Winnipeg, Manitoba, R3E 3M4, Canada
kitchingp@inspection.qc.ca

Steven B. Kleiboeker
Director, Molecular Science and Technology
ViraCor laboratories
1210 NE Windsor Drive
Lee's Summit, MO 64086
skleiboeker@viracor.com

Donald Knowles
USDA-ARS
Pullman, Washington 9914-6630
dknowles@vetmed.wsu.edu

Hong Li
USDA-ARS
Pullman, WA 99164-6630
hli@vetmed.wsu.edu

Susan Little
Department of Pathobiology
Oklahoma State University
Stillwater, OK, 74078-2007
susan.little@okstate.edu

N. James MacLachlan
School of Veterinary Medicine
University of California at Davis
Davis, CA, 95616
njmaclachlan@ucdavis.edu

Terry McElwain
College of Veterinary Medicine
Washington State University
Pullman, WA 99165-2037
tfm@vetmed.wsu.edu

Suman M. Mahan
Pfizer Animal Health
Kalamazoo, MI 49001
suman.mahan@pfizer.com

Peter Merrill
Aquaculture Specialist
USDA-APHIS Import Export
Riverdale, MD 20737
Peter.Merrill@aphis.usda.gov

Jim Mertins
USDA-APHIS-VS-NVSL
Ames, IA 50010
James.W.Mertins@aphis.usda.gov

Samia Metwally
Foreign Animal Disease Diagnostic Laboratory
USDA-APHIS-VS-NVSL
Plum Island, Greenport, NY 11944
samia.a.metwally@aphis.usda.gov

Bethany O'Brien
USDA-APHIS VS
Western Regional Office
Fort Collins, CO
bethany.o'brien@aphis.usda.gov

Doris Olander
USDA VS APHIS
6510 Schroeder Road, Suite 2
Madison, WI 53711
doris.olander@aphis.usda.gov

Donal O'Toole
Wyoming State Laboratory
Laramie, WY 82070
dot@uwyo.edu

John Pasick
National Centre for Foreign Animal Disease
Winnipeg, Manitoba, R3E 3M4, Canada
jpasick@inspection.gc.ca

Jürgen A. Richt
National Animal Disease Center
USDA-ARS
Ames, IA 50010
jricht@nadc.ars.usda.gov

Luis Rodriguez
Plum Island Animal Disease Center
USDA-ARS
Greenport, NY 11944-0848
luis.rodriguez@ars.usda.gov

Fred Rurangirwa
College of Veterinary Medicine
Washington State University
Pullman, WA 99165-2037
ruvuna@vetmed.wsu.edu

Eoin Ryan
Institute of Animal Health,
Pirbright Laboratory
Surrey, UK
eoin.ryan@bbsrc.ac.uk

Jeremiah T. Saliki
College of Veterinary Medicine
University of Georgia
Athens, GA 30602
jsaliki@vet.uga.edu

Tirath S. Sandhu
Cornell University Duck Research Laboratory
Eastport, NY 11941
tss3@cornell.edu

Jack Schlater
USDA-APHIS-VS-NVSL
Ames, IA 50010
Jack.L.Schlater@aphis.usda.gov

Moshe Shalev
Department of Homeland Security
Plum Island Animal Disease Center
Greenport NY 11944-0848
mshalev@gmail.com

David E. Swayne
Southeast Poultry Research Laboratory
USDA-ARS
Athens, GA, 30605
David.Swayne@ars.usda.gov

Belinda Thompson
Animal Health Diagnostic Laboratory
College of Veterinary Medicine
Cornell University
Ithaca, NY 14852
bt42@cornell.edu

John Timoney
Gluck Equine Research Center
University of Kentucky
Lexington, KY 40546-0099
jtimoney@uky.edu

Alfonso Torres
Associate Dean for Public Policy
College of Veterinary Medicine
Cornell University
Ithaca, NY 14852
at97@cornell.edu

Fernando J. Torres-Vélez
College of Veterinary Medicine
University of Georgia
Athens, GA, 30602-7388
ftorres@vet.uga.edu

Thomas E. Walton
5365 N Scottsdale Rd.
Eloy, AZ 85231
vetmedfed@comcast.net

William R. White
USDA-APHIS-VS-NVSL
Foreign Animal Disease Diagnostic Laboratory
Plum Island, Greenport, NY 11944-0848
William.R.White@aphis.usda.gov

Mark M. Williamson
Gribbles Veterinary Pathology
The Gribbles Group,
1868 Dandenong Rd.
Clayton, Victoria, Australia, 3168
mark.williamson@gribbles.com.au

Peter Wohlsein
School of Veterinary Medicine
Hannover, Germany
Peter.Wohlsein@tiho-hannover.de

II

General Considerations

1

PROTECTING THE UNITED STATES FROM FOREIGN ANIMAL DISEASES

The Threat of Foreign Animal Diseases

Decades ago, foreign animal diseases (FADs) were considered in the United States as the purview of the regulatory animal health community only. It was thought that adequate border surveillance and good import controls would keep us secure from having to worry about big outbreaks. Being a large country with oceans on either side and only single neighboring countries to the north and south helped us to feel safe from incursions.

Today, as a result of increases in free market economies and relaxations of restrictions on foreign investment, animals and animal products are moving around the world in unprecedented numbers and at record rates. In 2005, global agricultural exports were US$670B, an 8% increase over the previous year and a 23% increase over five years.

There is a monumental and growing load of traveling fur, feathers, meat, milk and eggs. Consequently, the term "foreign animal disease" is rapidly becoming simultaneously more meaningful and less meaningful. Less meaningful because of the rapid movement of diseases around the world and the increasing likelihood that one of these will be found here (Exotic Newcastle disease and bovine spongiform encephalopathy are two prime examples from recent years). More meaningful for exactly the same reason – more of these disease agents are entering new territories than ever before. Now the incursion of a foreign animal disease is more a probability than a possibility, and could appear just as easily in the middle of the country as at one of our border ports of inspection. A private practitioner who may never venture more than 100 miles from home needs to be aware that a foreign animal disease from a far-flung corner of the world could show up first in the U.S. at his or her doorstep.

To overcome the problem of terminology with "foreign" animal diseases of "domestic" origin, there is increasing world-wide use of the term "Transboundary Animal Diseases" (TADs) developed first by the FAO instead of the use of "foreign animal diseases". Transboundary Animal Diseases are defined as "those that are of significant economic, trade and/or food security importance for a considerable number of countries; which can easily spread to other countries and reach epidemic proportions; and where control/management, including exclusion, requires cooperation between several countries." (Transboundary Animal Diseases: Assessment of socioeconomic impacts and institutional responses. FAO 2004.) (ftp://ftp.fao.org/docrep/fao/meeting/010/ag273e/ag273e.pdf).

International Animal Health Organizations

The world is composed of 195 sovereign states, without any unifying system of international governance. What ties nations together and encourages cooperation is a system of complex interdependence, with primary linkages being commerce, and these linkages expand in size and strength every year. This complex interdependence has been likened to a heavy, wet net draped over a billiard table. As one ball moves, it exerts effects on all the other balls, which are forced to move to varying degrees, as a result of the interdependence. What does this mean for foreign animal diseases? More are likely to happen and also countries are more likely to work together to control spread and prevent their introduction.

This complex interdependence has spawned the development of many new international organizations and spurred the expansion of those previously existing. In the realm of animal health, there are two overarching international organizations working to preserve the collective public good.

The World Organization for Animal Health (OIE)

The OIE is an independent international organization created on January 25th, 1924, with the mandate "to improve animal health worldwide." In May 2003 the organization changed its name from the Office International des Epizooties and became the World Organization for Animal Health but kept its historical acronym OIE. The OIE serves as the reference and standard-setting body for the World Trade Organization (WTO) in all issues involving animal health. The OIE has 169 member countries and territories (as of May 2007) and maintains permanent relations with 35 other international and regional organizations, with regional and sub-regional offices on every continent. The OIE members, through their Chief Veterinary Officers as delegates, constitute the International Committee, which meets once a year and approves resolutions that have been developed with the support of the elected Commissions. These Commissions include the Administrative Commission, four Specialist Technical Commissions (The Terrestrial Animal Health Standards Commission or "Code Commission";

the Scientific Commission for Animal Diseases or "Scientific Commission"; the Biological Standards Commission or "Laboratories Commission"; and the Aquatic Animal Health Standards Commission or "Aquatic Animals Commission") and five Regional Commissions (Africa; Americas; Asia, far East and Oceania; Europe; and Middle East). The day-to-day operations of the OIE are under the responsibility of an elected Director General and a headquarters staff, located in Paris, France.

The OIE maintains a unified list of reportable diseases. For many years the OIE used two lists (A and B) of diseases with different reporting obligations. In January of 2005, the OIE combined the lists into a single entity, containing 130 diseases of interest. A list of these diseases can be found in Table 1 and also at the OIE website (www.oie.int). Four criteria are used for the inclusion of a given animal disease in this list: international spread; zoonotic potential; significant spread within naïve populations; and emerging diseases.

Member countries and territories through their veterinary authorities have certain notification responsibilities to the OIE Central Bureau regarding the presence of a listed disease as follows. (See Terrestrial Animal Health Code, Chapter 1.1.2) (http://www.oie.int/eng/normes/mcode/en_chapitre_1.1.2.htm#article_1.1.2.3.):

- Notification from the Delegate of the country by telegram, fax or e-mail, within 24 hours, of any of the following events:
    - first occurrence of a listed disease and/or infection in a country, a zone or a compartment;
    - re-occurrence of a listed disease and/or infection in a country, a zone or a compartment following a report declaring the outbreak ended;
    - first occurrence of a new strain of a pathogen of an OIE listed disease in a country, a zone or a compartment;
    - a sudden and unexpected increase in the distribution, incidence, morbidity or mortality of a listed disease prevalent within a country, a zone or a compartment;
    - an emerging disease with significant morbidity or mortality, or zoonotic potential;
    - evidence of change in the epidemiology of a listed disease (including host range, pathogenicity, strain), in particular if there is a zoonotic impact.
- Weekly reports by telegram, fax or e-mail subsequent to a notification under point 1 above, to provide further information on the evolution of an incident which justified urgent notification; these reports should continue until the situation has been resolved through either the disease being eradicated or it

becoming endemic so that six-monthly reporting under point 3 will satisfy the obligation of the country to the OIE; in any case, a final report on the incident should be submitted;

- A six-monthly report on the absence or presence, and evolution of diseases listed by the OIE and information of epidemiological significance to other countries;
- An annual report concerning any other information of significance to other countries.

Once a notification is received at the OIE Central Bureau, they in turn notify all OIE-Member countries and territories and post the information on the newly created World Animal Health Information Database (WAHID) on their web site (www.oie.int)

The Food and Agriculture Organization (FAO)

The FAO is an agency of the United Nations headquartered in Rome, Italy, with the mandate of leading international efforts to defeat hunger through raising the levels of nutrition, enhancing agricultural productivity, improving the lives of rural populations and contributing to the growth of the world economy. Regarding animal health, in 1994 FAO created the Emergency Prevention System for Transboundary Animal and Plant Pests and Diseases (EMPRES). The EMPRES Livestock Program is dedicated to promoting the effective containment and control of the most serious livestock diseases as well as newly emerging diseases by the progressive elimination on a regional and global basis through international co-operation involving four critical activities: early warning, early reaction, enabling research, and coordination. EMPRES activities include international efforts for the diagnosis, surveillance, control and/or eradication primarily of (but not exclusively) rinderpest, contagious bovine pleuropneumonia, foot-and-mouth disease, contagious caprine pleuropneumonia, peste des petits ruminants, Rift Valley fever, lumpy skin disease and more recently highly pathogenic avian influenza.

Other International Animal Health Organizations

Significant efforts in promoting the awareness, diagnosis and response to a number of FADs are also conducted by other international organizations in the Americas. They include:

- o Pan-American Health Organization (PAHO), the western hemisphere branch of the World Health Organization for their efforts on the prevention control and eradication of FMD in the Americas through their Pan-American Foot-and-Mouth Disease Center (PANAFTOSA) in Rio de Janeiro, Brazil and their Veterinary Public Health Program.

- The Inter-American Institute for Cooperation on Agriculture (IICA), an agency of the Organization for American States (OAS) with a mandate to provide innovative technical cooperation in many areas in agriculture, including plant and animal health.
- The "Organismo Internacional Regional de Sanidad Agropecuaria" (OIRSA), a regional organization in Mexico and Central America providing assistance in animal health, plant health, food safety, quarantine services and trade facilitation.

Protecting U.S. animal industries

The responsibility for protecting the U.S. animal industries from the effects of FADs falls primarily within the U.S. Department of Agriculture, Animal and Plant Health Inspection Service (APHIS). The mission of APHIS is to protect and improve the health, quality, and marketability of our nation's animals, animal products, and veterinary biologics. The safeguarding of animal health involves at least two major agencies within APHIS: Veterinary Services (VS) and International Services (IS). APHIS - VS is responsible for the development and enforcement of regulations dealing with the importation of animal and animal products that could be vectors of FADs, as well as to provide diagnostic, surveillance and emergency response for the early detection, monitoring, and control/eradication of FADs. APHIS - IS cooperates with the OIE, the FAO and foreign countries to reduce the international spread of animal diseases. The focus is to protect U.S. animal industries by reducing the disease risk through participation in disease-management strategies before animals and poultry are imported into the United States.

Within USDA, research on FADs is conducted at Agricultural Research Service (ARS) laboratories. USDA works in close cooperation with States and Native American Nations regarding surveillance, rapid diagnosis and emergency response against any potential FAD or emerging disease outbreak. Some border control activities, as well as some emergency response actions are now under the jurisdiction of the U.S. Department of Homeland Security. Many other federal agencies are also involved in some aspects of animal health protection and response activities. They include two units within the Department of Health and Human Services: the Centers for Disease Control and Prevention, and the Food and Drug Administration's Center for Veterinary Medicine. In addition, four agencies within the Department of Interior have involvement in these activities.

They include: the U.S. Fish and Wildlife Service, the U.S. Geological Service, the U.S. National Park Service and the Bureau of Indian Affairs. Support roles are also played by the Department of Transportation, the Department of Defense, the Department of Commerce, the Department of Justice, the Department of the Treasury, and the Environmental Protection Agency.

Ultimately the responsibility for the early detection of possible FAD incursion rests primarily on the animal owners and producers, veterinarians in private clinical practice, the animal health organization of each State, many sectors of USDA and other federal agencies, market operators, and animal scientists.

To diagnose FAD outbreaks early, suspicious signs of an FAD must be promptly reported to the State Veterinarian, the APHIS VS Federal Veterinarian, or both. An investigation of the affected herd or flock is immediately conducted by a specially trained FAD diagnostician. On the basis of history, signs, lesions, and species involved, specimens are collected and submitted to the National Veterinary Services Laboratories (NVSL), VS, Ames, IA, or to the Foreign Animal Disease Diagnostic Laboratory (FADDL), which is a branch of the NVSL, but located at Plum Island, NY, to confirm the presence or absence of an FAD.

Emergency Response to an Outbreak

The authorities regarding a response to a suspected animal health event rest on State officials. However, in the case of suspected FADs, the Federal government can, in coordination with the States, declare an emergency situation, triggering a response that is coordinated through the APHIS-VS National Center for Animal Health Emergency Management. The APHIS-VS tasks associated with disease outbreaks and animal emergencies, particularly the routine reporting of FAD investigations, state-specific disease outbreaks or control programs, classic national animal health emergency responses, or natural disasters involving animals, are managed through the Emergency Management Response System (EMRS).

To support the early diagnosis of FADs, an integrated state-federal laboratory partnership, the National Animal Health Laboratory Network, was developed and is designed to provide additional surveillance capacity and also for surge capacity in the event of a FAD outbreak. This network ties many state laboratories in with the NVSL.

The National Response Framework (NRF), managed by the Department of Homeland Security, replaced the National Response Plan, with the mission to establish a comprehensive, national, all-hazards approach to domestic incident responses. It incorporates public and private-sector participation at all levels. The emergency response will be implemented through the use of the Incident Management System, following the guidelines set out in the National Incident Management System (NIMS), which is an integral part of the NRF. The NIMS formalizes the Incident Command System (ICS), developed by the Forest Service in the 1970's, to mobilize resources and people in the management of forest fires. This Incident Command System is composed of five major sections, is highly flexible, with sections growing or shrinking depending on the extent of the outbreak and its complexity. It can involve individuals from a number of different agencies and organizations and is designed to streamline activities, maximize resources, and clarify chains of command. For an FAD outbreak, the ICS might include veterinarians, technicians, disease specialists, and many support personnel, drawn from the military, universities, industry, private practice, as well as federal and state governments. The five major sections of the ICS are: Command, Finance, Logistics, Operations, and Planning (Fig. 1). Duties and responsibilities of the sections are outlined below and also in Fig. 2.

The **Command Section** is led by one *Incident Commander* and this Command Section controls all personnel and equipment, maintains accountability for task accomplishment, and serves as a liaison with outside agencies. The **Planning Section** must create the Incident Action Plan, which defines the response activities and use of all resources. During an FAD outbreak response, the Planning Section will also be concerned with animal welfare, vaccinations, epidemiology, wildlife, laboratory coordination, Geographic Information System (GIS), and disease reporting. The **Operations Section** carries out the responses. Activities and duties here include quarantine, vector control, slaughter and disposal of carcasses. The **Logistics Section** works on supplies and continuing communications, and its functions become more significant in long term or extended operations. The **Finance Section** manages the expenditures required by all sections and participants to respond to a disaster.

Over the past several years, thousands of potential responders to an agricultural event have been trained in ICS. Furthermore, the ICS has been utilized in some FAD incursions in recent years and has proven to be a flexible, efficient, and accommodating vehicle for effective response.

Summary

The last two decades have witnessed massive changes in animal health issues, at global, national, and local levels. The threat of an FAD incursion looms larger than ever, due to the large increases of animal production in developing nations with weak veterinary and public health infrastructures, at a time of increasing dependency on trade (local, regional and international) for their economic well-being. International organizations such as the OIE have assumed greater roles in guiding nations to contribute to and benefit from global animal health. Nationally, our diagnostic and response capabilities have undergone corresponding evolution to allow for a rapid, integrated and efficient system.

Alfonso Torres, DVM, MS, PhD, College of Veterinary Medicine, Cornell University, Ithaca, NY 14850, at97@cornell.edu

- and -

Corrie Brown, DVM, PhD, College of Veterinary Medicine, University of Georgia, Athens, GA, 30602-7388, corbrown@uga.edu

Table 1. OIE Listed Diseases

Multiple species diseases

· Anthrax
· Aujeszky's disease
· Bluetongue
· Brucellosis (*Brucella abortus*)
· Brucellosis (*Brucella melitensis*)
· Brucellosis (*Brucella suis*)
· Crimean Congo haemorrhagic fever
· Echinococcosis/hydatidosis
· Foot and mouth disease
· Heartwater
· Japanese encephalitis
· Leptospirosis
· New world screwworm (*Cochliomyia hominivorax*)
· Old world screwworm (*Chrysomya bezziana*)
· Paratuberculosis
· Q fever
· Rabies
· Rift Valley fever
· Rinderpest
· Trichinellosis
· Tularemia
· Vesicular stomatitis
· West Nile fever

Cattle diseases

· Bovine anaplasmosis
· Bovine babesiosis
· Bovine genital campylobacteriosis
· Bovine spongiform encephalopathy
· Bovine tuberculosis
· Bovine viral diarrhoea
· Contagious bovine pleuropneumonia
· Enzootic bovine leukosis
· Haemorrhagic septicaemia
· Infectious bovine rhinotracheitis/infectious pustular vulvovaginitis
· Lumpky skin disease
· Malignant catarrhal fever
· Theileriosis
· Trichomonosis
· Trypanosomosis (tsetse-transmitted)

Sheep and goat diseases

· Caprine arthritis/encephalitis
· Contagious agalactia
· Contagious caprine pleuropneumonia
· Enzootic abortion of ewes (ovine chlamydiosis)

Equine diseases

· African horse sickness
· Contagious equine metritis
· Dourine
· Equine encephalomyelitis (Eastern)
· Equine encephalomyelitis

· Maedi-visna
· Nairobi sheep disease
· Ovine epididymitis (*Brucella ovis*)
· Peste des petits ruminants
· Salmonellosis *(S. abortusovis)*
· Scrapie
· Sheep pox and goat pox

(Western)
· Equine infectious anaemia
· Equine influenza
· Equine piroplasmosis
· Equine rhinopneumonitis
· Equine viral arteritis
· Glanders
· Surra (*Trypanosoma evansi*)
· Venezuelan equine encephalomyelitis

Swine diseases

· African swine fever
· Classical swine fever
· Nipah virus encephalitis
· Porcine cysticercosis
· Porcine reproductive and respiratory syndrome
· Swine vesicular disease
· Transmissible gastroenteritis

Avian diseases

· Avian chlamydiosis
· Avian infectious bronchitis
· Avian infectious laryngotracheitis
· Avian mycoplasmosis (*M. gallisepticum*)
· Avian mycoplasmosis (*M. synoviae*)
· Duck virus hepatitis
· Fowl cholera
· Fowl typhoid
· Highly pathogenic avian influenza and low pathogenic avian influenza in poultry as per Chapter 2.7.12. of the *Terrestrial Animal Health Code*
· Infectious bursal disease (Gumboro disease)
· Marek's disease
· Newcastle disease
· Pullorum disease
· Turkey rhinotracheitis

Lagomorph diseases

· Myxomatosis
· Rabbit haemorrhagic disease

Bee diseases

· Acarapisosis of honey bees
· American foulbrood of honey bees
· European foulbrood of honey bees

· Small hive beetle infestation (*Aethina tumida*)
· *Tropilaelaps* infestation of honey bees
· Varroosis of honey bees

Fish diseases

· Epizootic haematopoietic necrosis
· Infectious haematopoietic necrosis
· Spring viraemia of carp
· Viral haemorrhagic septicaemia
· Infectious pancreatic necrosis
· Infectious salmon anaemia
· Epizootic ulcerative syndrome
· Bacterial kidney disease (*Renibacterium salmoninarum*)
· Gyrodactylosis (*Gyrodactylus salaris*)
· Red sea bream iridoviral disease

Mollusc diseases

· Infection with *Bonamia ostreae*
· Infection with *Bonamia exitiosa*
· Infection with *Marteilia refringens*
· Infection with *Mikrocytos mackini*
· Infection with *Perkinsus marinus*
· Infection with *Perkinsus olseni*
· Infection with *Xenohaliotis californiensis*

Crustacean diseases

· Taura syndrome
· White spot disease
· Yellowhead disease
· Tetrahedral baculovirosis (*Baculovirus penaei*)
· Spherical baculovirosis (*Penaeus monodon*-type baculovirus)
· Infectious hypodermal and haematopoietic necrosis
· Crayfish plague (*Aphanomyces astaci*)

Other diseases

· Camelpox
· Leishmaniosis

Fig. 1. Five major sections of the Incident Command System.

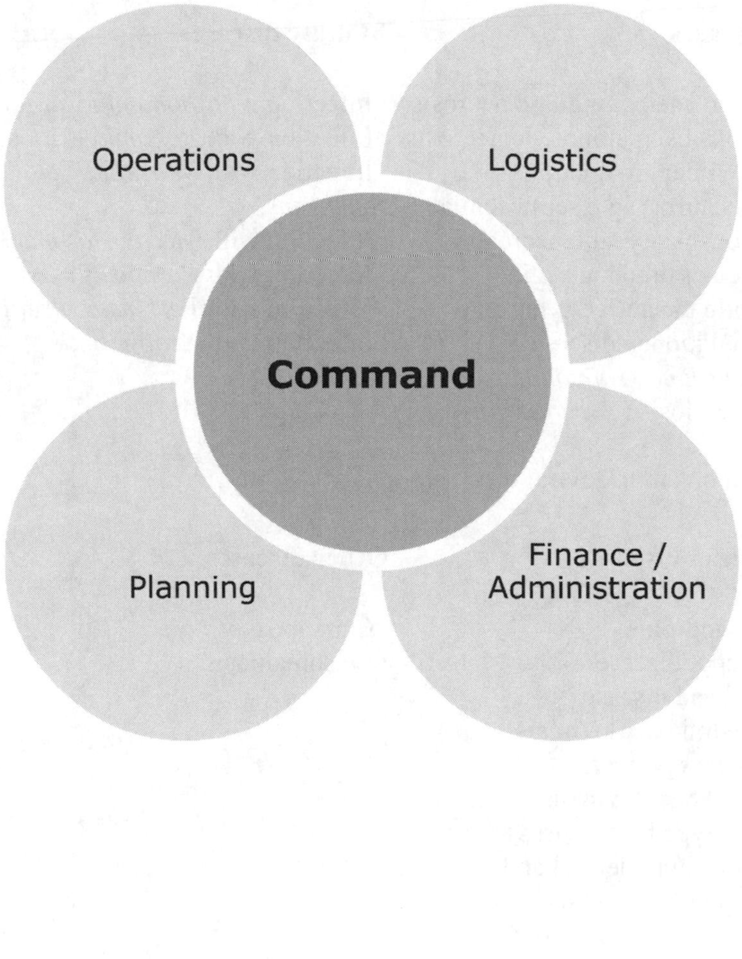

Operations

Logistics

Command

Planning

Finance / Administration

Figure 2. Duties and responsibilities of the five sections of an animal event-associated Incident Command System (ICS).

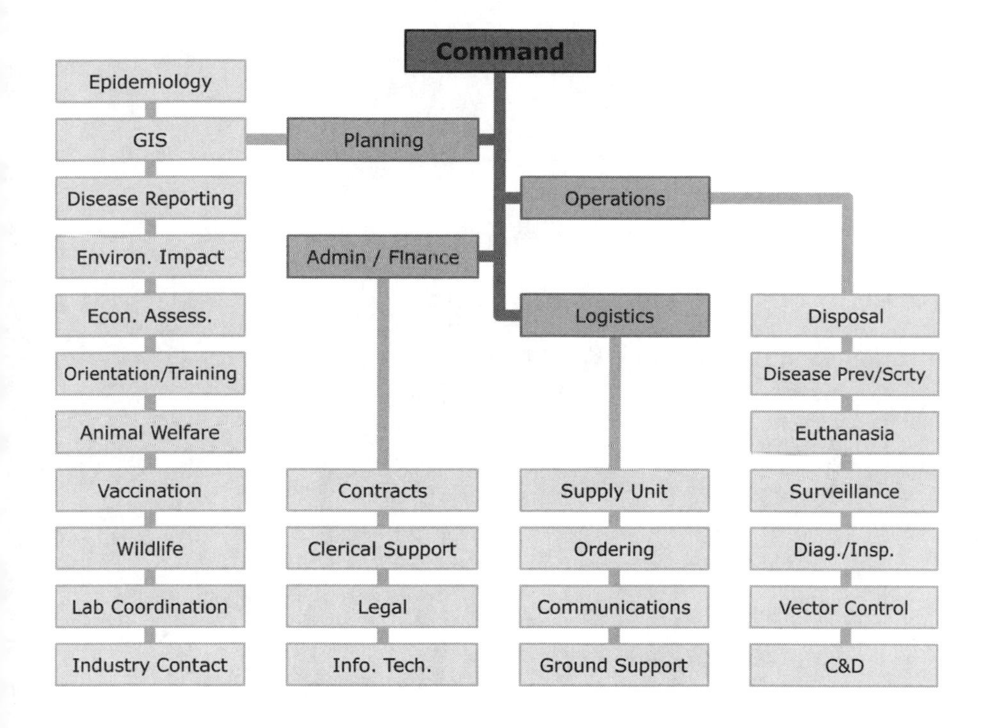

2

DIAGNOSTIC LABORATORY PROCEDURES

Introduction

Diagnostic assays can generally be divided into direct detection assays, those that detect the agent or toxin in case materials, and indirect assays, those which detect the host response as a proxy of exposure to an agent or infectious process. In both instances the accuracy of the diagnostic test procedure is measured by its ability to correctly detect the agent in question, identified as diagnostic sensitivity, and the ability to correctly identify specimens in which evidence of the agent is absent, termed diagnostic specificity. Diagnostic sensitivity and specificity are impacted by several factors inherent in the assay and the host response. In order to understand the diagnostic reliability of an assay or series of assays, it is important to understand the intended use of the assay, or "fitness for purpose", and then evaluate assay performance based on the lowest level of the analyte that can be detected by the assay, referred to as analytic sensitivity, the tendency to cross-react with other agents (analytic specificity), and performance characteristics such as reproducibility, precision, and accuracy for different specimen types and species. The measurement of the different performance characteristics that ultimately document the diagnostic effectiveness of any particular assay is referred to as assay validation.

Direct detection techniques are those that detect the agent, an organism or toxin, in the specimen of interest. Direct detection assays include common laboratory techniques such as microbial culture, virus isolation, toxin identification, gross or microscopic morphologic exam (e.g., parasite identification, electron microscopy), fluorescent antibody staining, immunohistochemistry, and nucleic acid-based technologies such as polymerase chain reaction (PCR).

The most common of the indirect detection methods is serologic testing for detection of agent-specific antibodies, and may include primary binding assays and secondary binding or functional assays. Additional indirect methods of disease diagnosis include techniques that measure cell-mediated responses, such a delayed hypersensitivity testing using skin tests or *in vitro* cell stimulation assays, and clinical chemistry as a measure of the host response to particular disease processes.

Fitness for Purpose

The complex interactions among agent, disease, economic impacts, and public health impacts, combined with the diverse needs for containment or control at the level of individual animals, individual herds, regions, or countries make it impossible for any single assay or laboratory approach to serve all diagnostic needs. Speed, cost, ease of use, diagnostic sensitivity, and diagnostic specificity must be balanced against one another and which of those are prioritized will be dependent on the intended use of the assay (fitness for purpose). In an over-simplified form, assays with high diagnostic sensitivity are selected when the need is agent detection, with the knowledge that high detection sensitivity often comes at the price of lower specificity, i.e., increased risk of false positives. Assays used to diagnose or accurately identify an agent are selected for their high specificity, with the knowledge that high specificity comes at the risk of reduced diagnostic sensitivity.

Assays used for management of individual animals, flocks, or herds are generally designed to offer high specificity. A low rate of false positive results is necessary where an accurate differential diagnosis is critical in management decisions, such as selection of appropriate treatment or therapeutics, containment options, environmental decontamination, decision to cull or destroy, etc. Infectious agents that may transmit in the absence of overt clinical disease are often managed by some form of health certification required prior to breeding, movement, or sale. In contrast to the highly specific assays used in differential diagnosis of clinical disease, assays used in management designed to minimize the risk of transmission of a sub-clinical disease will often sacrifice specificity for increased detection sensitivity.

A key component of regional, national, or international disease surveillance is early detection, requiring assays used for this purpose to be highly sensitive for detection of the agent in question. Surveillance programs may be designed to determine the prevalence or location of particular agents, to demonstrate freedom from a particular agent for trade purposes, or to rapidly identify the transmission of an agent to a new location, species, or ecological niche.

Sampling

Appropriate collection and handling of specimens is a component of the testing process that is often under-appreciated for its critical role in determining the accuracy or reliability of a diagnosis. The site from which a specimen is collected is influenced by the pathogenesis of the disease and the timing of specimen collection relative to onset of disease. If sampling occurs early in the infectious process, sample collection should target the site of agent replication and/or shedding: nasal or pharyngeal swabs for respiratory infections, feces or swabs of the GI tract for enteric infections, fluids or scrapings from epithelial lesions, blood for bacteremia and viremia, etc. Biopsy and necropsy specimens target the appropriate organs or tissues involved, however as clinical signs are rarely specific to a particular agent, laboratories will typically recommend that a full representation of specimens be submitted for the majority of diagnostic work-ups. For a comprehensive list of samples to collect at necropsy for a thorough diagnostic evaluation, please see the following Chapter, Sample Collection.

The best diagnostic specimens are not necessarily limited to affected animals, and in some situations sampling from the environment may prove to be more productive. Environmental sampling may include testing of water, environmental air, bedding and related housing materials, or feed to assess the presence and/or concentration of specific agents or toxins. In some situations where animals are managed as a unit, an epidemiologic or case-control approach to sampling may be used to optimize diagnostic information by using the herd, flock, pen, house, milk-string, etc. as the diagnostic unit of interest.

Preservation of the specimen prior to testing is equally critical and will vary with the sample and proposed testing. Keeping the specimen cool and moist is generally recommended to preserve the tissue and/or agent, to prevent bacterial and fungal overgrowth, and slow enzyme-related degradation, however, extremes of chilling including freezing may be contraindicated where organisms or tissue architecture would be destroyed by ice crystals, or CO_2 from dry ice would alter agent survival. The addition of protein (e.g., gelatin, serum albumin, etc.) and antibiotics in transport media is highly recommended to support sample quality for viral isolation attempts, but must be avoided if isolation of bacteria is indicated. For many bacterial and fungal agents, recovery is significantly improved by placing the specimen in an appropriate selective media immediately after collection. Optimal specimen collection, handling, and preservation are assay-dependent, and it is strongly recommended that the testing laboratory be consulted prior to obtaining diagnostic case material. The transportation of clinical materials, known infectious agents, and toxins has become highly regulated. Information on appropriate containment, packaging, identification, and shipping, as well as permits when required, can be obtained from the International Air Transportation Association (IATA) infectious substances shipping guidelines and the related national and international government agencies (see Chapter 4 in this section, Shipping Diagnostic Specimens).

Direct Detection

The level of laboratory technology used for diagnosis of an infectious disease or intoxication can range from review of the clinical history and gross observation to ultrastructural examination of specimens and agents, and targeted detection of an agent's nucleic acid sequence.

***Pathology, histopathology, immunohistochemistry*:**
Laboratory approaches using pathology and histopathology to evaluate organs and tissues for the presence of characteristic changes or lesions, though specialized and expensive, remain a vital component of veterinary diagnosis. Pathologists trained in the *in vivo* recognition of disease processes are likely to be among the first to recognize and understand the etiology or pathogenesis of any new or emerging disease. Standard histochemical stains used for decades to selectively detect and identify agents or lesions are increasingly being supplemented with more specialized immuno-histochemical approaches. Immunohistochemistry, including immunoperoxidase staining, relies on agent-specific antibody coupled to an enzyme-chromogen marker that allows microscopic visualization of target antigen(s) in individual tissues or cells. Though very specific for detection of the target agent, the procedure can be technically demanding and the results subjective, requiring training and expertise similar to those used in histopathologic evaluations. *In situ* hybridization which uses chemically-labeled complementary strands of nucleic acid to detect either target DNA or RNA in adequately preserved tissue specimens, pairs the pathologist's skills at recognizing histopathologic changes with the laboratory's ability to very precisely locate the agent of interest. The obvious strength of histochemical and *in situ* nucleic acid detection of agents is that detection and localization of the agent in question is interpreted in context of the lesion(s) present.

***Fluorescent antibody staining*:**
The direct detection of virtually any infectious agent is possible in tissues or cells using fluorescent antibody (FA) staining. Specimens are placed on glass slides either in the form of a thin tissue section (4-8μm) or a smear (tissue impression smears, cell smears).The technique relies on the binding of agent-specific antibody to antigen *in situ*. The detecting antibody is usually conjugated to a fluorophore (typically fluorescein isothiocyanate, FITC). When no direct fluorescent conjugate is available, primary antibody can be detected using a fluorescence-conjugated anti-species antibody in a technique termed indirect FA (IFA). The strengths of FA staining are its low cost and speed, allowing for rapid diagnosis within one day. The weaknesses of the technique are the subjective

nature of the reading and the requirement for major equipment including a cryostat and fluorescence microscope.

Hemagglutination and hemadsorption:

Hemagglutination is the ability of some agents, such as influenza viruses, paramyxoviruses, parvoviruses, mycoplasma species, among others, to naturally cross-link red blood cells by binding to a receptor antigen found on the red blood cell surface. Hemadsorption is the ability of viruses, such as African swine fever virus, to cause red blood cells to adsorb onto the surface of virus-infected cells forming rosettes. The hemagglutinating and hemadsorption abilities of several viruses can be used for direct detection and preliminary identification from clinical specimens. While these techniques are useful in establishing that an agent is present in the sample, they are not sensitive or sufficiently specific for definitive etiologic diagnosis. Their main use is in the functional assay of hemagglutination inhibition used for detection of specific antibody.

Isolation and identification:

Most parasites, some bacteria, and a few viruses (e.g., poxviruses) can be detected and identified by microscopic examination of the clinical specimen or extracts thereof. However, most infectious disease agents require amplification in culture to be detected from diagnostic specimens. Isolation of extracellular bacteria is achieved by inoculating preparations of diagnostic specimens onto a variety of inert nutrient media, generally a semi-solid media containing an agar base. Identification of bacterial and fungal isolates can be performed by direct microscopic examination with or without special stains, biochemical reactions, molecular, and immunological techniques. The isolation of obligate intracellular bacteria (e.g., *Chlamydophila*) and viruses is accomplished by inoculating specimens into laboratory animals, embryonated chicken eggs, or appropriate cell cultures.

The growth of most agents in cell culture is characterized by cytopathic effects (CPE) which in some cases can be typical enough to allow presumptive identification or a clue as to which group of viruses is involved. However, some viruses such as bovine viral diarrhea and canine parvovirus can grow in cell culture without causing detectable CPE. In such cases and in other cases where the CPE is atypical, identification techniques such as electron microscopy, fluorescent antibody staining and molecular detection need to be applied for definitive identification. Isolation and identification of pathogens remains the gold standard for most infectious agents, where direct demonstration of a pathogen from the site of a lesion is confirmation of its primary or secondary role in clinical disease. However, care must be taken in interpretation that the recovery of an infectious agent is compatible with the clinical presentation and the pathogenesis associated with the particular agent. Conventional recovery methodologies which can require from days to months dependent on the growth characteristics of the organism, are with increasing frequency being replaced with or supplemented by direct and indirect detection methodologies that are more time and cost-effective and can be more readily standardized.

Molecular Detection:

Molecular diagnostic techniques rely on the natural complementarity of nucleotide base pairs in formation of double-stranded DNA. Short strands of chemically synthesized DNA or RNA, referred to as oligonucleotides or probes, can be used to identify infectious agents by *in vitro* binding (hybridizing) to the complementary DNA/RNA sequence in a diagnostic specimen. Radioactively-labeled probes, though once the standard, have largely been replaced by other marker compounds, such as avidin-biotin, peroxidase, or chemiluminescent molecules. A nucleic acid target sequence can be detected directly by probe binding in the diagnostic specimen when the DNA target is sufficiently abundant. More commonly, however, the target DNA or RNA is detected following PCR-based amplification of the nucleic acid target.

Diagnostic PCR functions by amplifying a selected and generally very specific target or segment of the agent DNA that can be used to accurately identify and differentiate the agent from other closely-related microbes. The PCR method is used to efficiently increase DNA concentration in a diagnostic specimen to detectable levels in a controlled logarithmic fashion. For detection of an RNA sequence, the PCR reaction first requires that cDNA (copy DNA) is generated using reverse transcriptase enzymes (RT-PCR). When a DNA molecule is heated sufficiently, the DNA double helix is denatured into 2 single strands. Each strand serves as a template for replication, which occurs *in vitro* by adding free nucleotides and a polymerase enzyme to drive the replication process, generating a second complementary strand of each target sequence. Each heating and cooling cycle effectively doubles the DNA present in the test solution, so that by 30 cycles the reaction yield will be in excess of a million copies of the initial DNA. In order to perform PCR, the exact sequences of the nucleotides that flank the specific DNA target are used to generate PCR primers (*in vitro* generated nucleotide sequences used to prime or identify the beginning and end of the specific target sequence for each round of replication).

Once amplified the DNA target can be identified by a variety of techniques; including by molecular weight, by binding to a complementary enzyme-labeled nucleic acid sequence (a probe) in techniques variously known as PCR-hybridization, PCR probes, PCR dot-blots, among others. The analytic detection limits of PCR assays are often reported in the range of 1 to 10 copies of DNA or cDNA prior to amplification steps.

The PCR assays are generally divided into **standard PCR**, in which the amplification cycles are completed before the detection steps occur, and **real-time PCR** in which the detection probe is allowed to bind during each cycle of amplification. Real-time PCR has the distinct advantage of speed, in that 30-40 rounds of amplification and detection can be completed in minutes to hours, the PCR system is closed so that no further laboratory manipulation of the amplified DNA is required. The real-time PCR approach can be developed in a quantitative format and is generally abbreviated qPCR, regardless of whether the assay is truly quantitative, in order not to confuse with the RT PCR which is the accepted abbreviation for reverse transcriptase PCR. A modification of the standard PCR reaction, **nested PCR**, is sometimes used to enhance the sensitivity and specificity of the PCR reaction. For nested PCR, an initial PCR reaction is used to amplify a selected DNA fragment, which in turn is used as template for a second round of PCR which targets one or more unique fragments of DNA within the initially-amplified region. The two-step process has the advantage of increasing detection sensitivity, detection specificity, and of diluting potential PCR inhibitors found in many clinical specimens. The significant disadvantage of the two-step process is the high risk of laboratory contamination associated with handling amplified DNA in the laboratory.

Polymerase chain reactions can also be developed in multiplex formats, which allow more than one target to be detected from a single reaction. As **multiplex PCR** approaches continue to be developed and optimized for diagnostic use, the advantages, including cost and time efficiency, are still being balanced against the technical problems associated with loss of detection sensitivity. **Genechips** or **microarrays** consist of thousands of oligonucleotides bound in a specific pattern to a solid surface, typically a glass slide or silica chip. Nucleic acids from the diagnostic specimen compete with fluorescently labeled competitor oligonucleotides for binding to the chip. The pattern of nucleic acid binding is measured and analyzed using fluorescence detectors and computer software. Microarray technology has not seen wide use in veterinary diagnostic laboratories, largely due to the cost of equipment. It is however expected that once established, the technology will be cost-effective on an individual animal basis and examples of diagnostic microarrays have shown promise for detection of antibiotic resistance in food safety applications and rapid classification of emerging pathogens, as occurred with the initial identification of the SARS virus.

There are a variety of molecular techniques, often paired with PCR amplification, that can be used for genotypic characterization or subtyping of microbial agents. Genome-level characterization is used in taxonomic classification, differential detection of virulent strains, identification of genetic sources of antimicrobic resistance, location of virulence factors, recognition of vaccine escape mutants, epidemiologic investigation of disease outbreaks, and identification of inter-species transmission of specific pathogens.

Restriction fragment length polymorphism (RFLP) is a technique that uses one or more well characterized restriction enzymes to digest DNA, which may have been amplified by PCR prior to analysis. A characteristic profile or pattern produced by the different sized DNA fragments remaining after enzyme digestion is visualized by gel electrophoresis. The methodology is useful for sub-typing of closely-related infectious agents (e.g., canine distemper virus and phocine distemper virus) based on genetic sequence variation within specific genes.

Random amplified polymorphic DNA (RAPD) analysis is a technique used in detecting genetic variation and for strain typing. Rather than amplifying a specific region of a genome, RAPD relies on random amplification of genomic DNA using short arbitrary sequences as PCR primers. The advantage of the RAPD approach is that it does not require prior knowledge of an agent's DNA sequence for specific primer design and can be applied to very small amounts of template DNA. Nucleic acid sequence determination followed by computational analysis involving national and international genetic sequence databases, is used to identify and compare the exact nucleic acid sequence of a gene fragment, a gene, or a complete genome.

Sequence analysis can be used diagnostically for forensic investigation, to investigate disease outbreaks, to track evolutionary changes in rapidly mutable micro-organisms, or for precise phenotype or genotype analysis of animals or organisms.

Non-PCR based nucleic acid amplification techniques, referred to as **isothermal amplification** methods, include Nucleic Acid Sequence-Based Amplification (NASBA), rolling circle, and direct signal amplification. The techniques have yet to be widely developed for animal health, but have been applied to detection of human pathogens and show promise for future automation.

Molecular diagnostic techniques are powerful diagnostic tools, particularly for use with agents that are difficult to recover by standard microbiologic or virus isolation techniques (e.g., Malignant Catarrhal Fever), or where maintaining and amplifying an agent for assay reagent purposes may pose unwanted biosecurity risks (e.g., routine surveillance testing for foreign animal diseases). The inherent sensitivity of nucleic acid amplification-based technologies requires a full understanding of the technology, as well as assay design, in order to accurately interpret the findings. Because the presence of DNA or RNA can theoretically be detected at very low levels, it is critical that PCR-based diagnostic results be interpreted in the context of clinical history and agent pathogenesis. It is equally critical that the laboratory and user understand the limitations of genome-based assays including the impacts of natural evolution of an infectious agent. Particularly with RNA viruses, naturally occurring genetic mutation can be sufficient to alter a PCR target such that a previously reliable assay has little or no diagnostic use.

Indirect Detection

Assays used to detect antibody are divided among the primary and secondary binding assays. Primary binding assays detect the antibody protein directly, and include enzyme-linked immunosorbent assays (ELISA), indirect fluorescent antibody (IFA), and radioimmunoassays (RIA). Primary binding assays can be designed specifically to recognize a particular sub-class of immunoglobulin, such as IgM versus IgG. When so designed these assays are effective for use in differentiating acute infection from chronic infection, and in some situations vaccine responses. By comparison, the secondary-binding assays detect antibody using a functional measure, such as agglutination, precipitation, complement fixation, or neutralization. The functional activity measured by secondary binding assays is often, but not always, a biologically important activity of that antibody in the host defense process. Because there is a direct relationship between biologic function, sub-class of immunoglobulin detected, and timing of an infectious process, secondary binding assays can often be selectively used to differentiate early from late immune response (active infection versus prior exposure), and in some cases can correlate with level of protection in a particular animal. The principal criticism of secondary binding assays is that the results require visual interpretation, and so excluding the very few exceptions where automation is employed, secondary binding assays are biased by operator subjectivity.

***Precipitation*:**
The precipitation assays take advantage of the fact that selected sub-classes of immunoglobulin will form insoluble complexes when antigen and antibody occur in proportions at or near equivalence. Precipitating antibody can be detected by allowing test antibody and reference antigen to come into contact in a semi-solid matrix, such as agar gel, and then visually observing the stabilized precipitation. Examples of precipitation-based assays include Ouchterlony or double diffusion where the antigen and antibody diffuse toward each other from wells cut into the gel (e.g., the Coggins test for detecting antibody to Equine Infectious Anemia virus), radial immunodiffusion where the gel is impregnated with antibody and the test antigen diffuses from a well cut into the gel, or immunoelectrophoresis where an electric charge applied to the gel is used to separate complex mixtures of proteins prior to allowing the standard antigen-antibody precipitation reaction to occur. Temperature and size (molecular weight) significantly affect the rate of diffusion, and will impact the position and shape of precipitation reactions. Interpretation of results requires knowledge of these factors.

***Agglutination, microagglutination*:**
Agglutination assays take advantage of the ability of bivalent (e.g., IgG) and pentavalent (e.g., IgM) antibody molecules to cross-link non-soluble antigens to form stable complexes that are visible as clumps in a liquid matrix. The agglutination reaction may be visible to the naked eye, such as occurs when bacterial cells (e.g., *Brucella*, *Mycoplasma*) are used as the target antigen, or may be visible only with magnification, as occurs with microagglutination assays (e.g., *Leptospira* spp.). By chemically binding target antigens onto latex beads, the agglutination assays can be extended to detect antibodies to specific sub-sets of antigens or to targets that would otherwise be too small to form visible antigen-antibody complexes, such as viruses or viral antigens. The agglutination assays are often used as diagnostically sensitive but non-specific screening tools, based on the ability to detect early immune responses associated with the IgM antibody subclass. The key drawback of agglutination reactions is their potential for false positive agglutination due to cross-reactivity with related antigens, and sensitivity to sample quality issues including bacterial contaminants in the sample being tested.

Hemagglutination inhibition (HI):

When agent-specific antibody binds to a hemagglutinating agent the ability to agglutinate red blood cells is blocked. Hemagglutination inhibition assays take advantage of the ability of laboratory personnel to visually evaluate both hemagglutination and the inhibition of the reaction in microtiter wells or test tubes. Some antibodies do not hemagglutinate effectively, and can be detected using an indirect hemagglutination test which adds a second antibody to bind to the antibody on the red blood cell. An example of the indirect approach is the Coombs test. The strength of hemagglutination inhibition assays is their good sensitivity and specificity. The major weakness is that HI techniques are relatively cumbersome and results can be variable dependent on reagent pH, temperature, incubation times, and the species or source or erythrocytes used in the assay.

Complement fixation:

Complement fixation takes advantage of the functional ability of some antigen-antibody complexes to bind complement. The assay uses complement-mediated lysis of sensitized red blood cells in a test tube or microtiter well to indicate the presence of unbound complement, which in turn indicates the absence of the test antibody. When antibody is present in the test sample, the antigen-antibody complexes that form bind the reagent complement, and prevent lysis of the red blood cells. Complement fixing antibodies often arise and decay early in an immune response, allowing the CF assay to be used to differentially detect an early or active infection versus residual antibody from prior exposure (e.g., *Brucella abortus*, Vesicular Stomatitis Virus). The assay is however prone to laboratory error and subjectivity due to its innate complexity and dependence on multiple biologic systems working at optimal reagent concentration, time, and temperature.

Neutralization assays:

When specific antibody in a serum sample binds to live virus, the ability of the virus to infect cells is diminished or abrogated. This technique is used to measure antibody levels (titers) by reacting serial dilutions of a serum sample with a fixed amount of virus and inoculating the serum-virus mixtures onto cell cultures. After incubating for 2-4 days, the antibody titer of the serum is read as the highest dilution of the serum that inhibited virus infectivity. The same principle can be applied for detection of toxins in serum samples. In this case, serum dilutions are mixed with a constant amount of specific anti-toxin antibody and the highest dilution of the serum to neutralize toxin activity is determined. Neutralization assays are highly specific, have good sensitivity, and are generally good measures of the biologic activity of the associated antibody. The drawbacks to neutralization assays are the extended incubation times (often several days), the requirement for live cells, embryonated eggs of specific age, or susceptible animals, the need to propagate reference viruses, and the subjectivity in reading the neutralization results.

***Indirect Fluorescent Antibody*:**

The indirect fluorescent antibody (IFA) method, as previously described for antigen detection, can also serve as a tool for detection and titration of specific antibody. Patient serum is allowed to react with antigen bound to a solid phase, typically agent-infected cells fixed to a microscope slide. Unbound patient antibody is washed free, and a fluorescence-labeled anti-species antibody is used to detect the antigen-antibody complex that has formed. The IFA methodology can be both sensitive and specific, but like the other serologic assays that require subjective reading of results, the IFA technique can vary from laboratory to laboratory based on the quality of the reagents and the technical expertise available.

***ELISA*:**

Enzyme-linked immunosorbent assays (ELISA) can occur in a variety of formats including test tubes or micro-wells (strip ELISA, 96-well plate ELISA), on micro-beads (e.g., Particle Cell Fluorescence Immunoassay, Liquid Array), or as lateral-flow assays on membrane filters. ELISA technology is widely used for detection of both antigen and antibody, and can be designed in a number of formats, the most common of which are direct ELISA and competitive ELISA. In general, ELISA technology requires that the test antigen is bound to an immobile phase (the well, bead, or membrane), and is allowed to react with the test antibody (in the example of an antibody-detection ELISA). Once antigen-antibody complexes have formed, they are detected by allowing a second enzyme-conjugated anti-species antibody to bind prior to exposure to the enzyme-substrate containing an indicator dye. The reaction of the enzyme and substrate is measured, either as optical density or as a visual color change. The interpretation of ELISA assays can variously be based on optical density (OD), on a ratio comparing the sample OD to a negative control or background sample OD (S/N or signal to noise), change in OD over time (kinetic ELISA), competition for binding of the target antigen (competitive or cELISA), or related permutations. ELISA-based tests cover a range of technology from highly automated, high-throughput robotic platforms to single use manual dip-stick type assays. The advantages of ELISA-based technologies are the speed (minutes to hours), and the ability to standardize assays so reproducible results are obtained by a wide range of users.

Interpretation of Diagnostic Assay Results

Interpretation at individual animal level:

Interpretation of diagnostic assays requires comparison of the test result from an individual animal to an identified standard or "norm". For individual animals, the results of a diagnostic test may be compared to the population using diagnostic sensitivity and specificity to calculate the positive and negative predictive value of the result. Alternately, the animal may be compared to itself over the course of infection by testing an acute and a convalescent serum sample for evidence of a significant change in the antibody response, known as seroconversion. Seroconversion is typically defined as a 2 to 4-fold change in titer over a specified time interval, often 2-4 weeks, to measure the immune response to acute infection. Though theoretically sound, detection of seroconversion is not always possible or practical, it is strictly linked to the onset of infection rather than observation of clinical signs (depending on the disease, seroconversion to an infectious agent may occur before or after clinical signs are detected), vaccination history and resulting sub-classes of immunoglobulin present in the animal, and the ability of the assay used to measure differences in antibody concentration and/or sub-classes of immunoglobulin stimulated. The relative response to different antigens in the same animal may also be used to provide diagnostic information for an individual. In some situations, vaccine strains of particular agents are sufficiently different from circulating field strains, such as with influenza viruses, that relative response in a single animal to the two antigens can provide useful diagnostic interpretation.

Interpretation at herd level:

In situations where animals are managed as a group, an epidemiologic approach can be used to aid in interpretation of diagnostic results. In case-control sampling schemes, animals showing clinical signs are matched to herdmates that are free of clinical signs but maintained under the same management. The test results, for example specific antibody responses or titer levels, are statistically compared between the affected and not affected populations to assess agent association and disease risk using odds ratios or similar epidemiologic analyses.

Guide to the Literature

1. BROWN, C. 1998. In situ hybridization with riboprobes: An overview for veterinary pathologists. *Vet Pathol* 35:159-167.
2. FOY, C.A. and PARKES, H.C. 2001. Emerging homogeneous DNA-based technologies in the clinical laboratory. *Clinical Chemistry* 47:990-1000.

3. GREINER, M. and GARDNER, I.A. 2000. Epidemiologic issues in the validation of veterinary diagnostic tests. *Prev Vet Med.* 45(1-2):3-22.
4. JEGGO, M.H. 2000. An international approach to laboratory diagnosis of animal diseases. *Ann New York Acad Sci.* 916:213–221.
5. SALIKI, J.T. 2000. The role of diagnostic laboratories in disease control. *Annals New York Acad Sci*. 916:134-8.
6. SMITH, R.D. 2005. *Veterinary Clinical Epidemiology*, 3rd Edition. Boca Raton, FL.: CRC Press.
7. THURMOND, M.D., HIETALA, S.K. and BLANCHARD, P.C. 1997. Herd-based diagnosis of *Neospora caninum*-induced endemic and epidemic abortion in cows and evidence for congenital and postnatal transmission. *J Vet Diag Invest.* 9(1):44-9.
8. VILJOEN, G.J., NEL, L.H. and CROWTHER, J.R. 2005. *Molecular diagnostic PCR handbook.* Dordrecht: Springer.

Sharon K. Hietala, PhD, University of California-Davis, Davis, CA 95617, skhietala@ucdavis.edu

- and -

Jeremiah T. Saliki, DVM, PhD, University of Georgia, Athens, GA 30602, jsaliki@uga.edu

3

SAMPLE COLLECTION

When conducting a postmortem on an animal suspected of having a foreign animal disease, careful attention to collection and preservation of tissues is essential. This will ensure that the laboratory has the best possible material to enable an accurate diagnosis. Declaration of a foreign animal disease incursion can only be done after laboratory confirmation. So, ultimately rapid response and containment depends on careful sample collection and preservation.

Fresh tissues should be collected into small labeled plastic bags and kept refrigerated. Samples for fixation should be collected directly into 10% buffered formalin. Below is a comprehensive list of tissues for collection.

Guide to Post-Mortem Diagnostic Sample Collection

Integuments	Tissue/Organ	Fresh	Fixed	Notes (*optional if abnormal)
	Skin	X	X	
	Mammary gland	X	X	Look for vesicles or erosions
	Scrotum/Prepuce	*	*	
	Hooves	*		Look for vesicles or erosions
	Anus	*	*	
	Vagina	*	*	
Thoracic cavity				
	Pericardium		*	
	Heart	*	X	
	Pleura		*	
	Trachea	*	X	
	Lungs	X	X	Sample all lobes
	Esophagus	*	X	

	Thoracic fluids	*		
	Mediastinal and Bronchial Lymph Nodes	X	X	
Abdominal cavity				
	Omentum		X	
	Stomach or Abomasum	*	X	* Stomach content (fresh)
	Omasum		X	Ruminants only
	Reticulum		X	Ruminants only
	Rumen		X	Ruminants only
	Pancreas		X	
	Liver	X	X	
	Gall bladder		X	
	Spleen	X	X	
	Ovaries	*	X	
	Testicles	*	X	
	Uterus	*	X	
	Kidneys	X	X	
	Bladder & ureters	*	X	
	Duodenum	*	X	
	Jejunum	X#	X	# Tied segment with contents
	Ileum	*	X	
	Cecum	*	X	
	Ileo-cecal valve	*	*	
	Colon	*	X	
	Rectum	*	X	
	Inguinal Lymph Nodes	X	X	
	Mesenteric Lymph Nodes	X	X	
	Peyer's patches	*	X	
	Abdominal fluids	*		
	Spinal cord	*	X	

Musculoskeletal system				
	Joints (membranes)	*	X	At least 3
	Tendon sheaths	*	X	
	Muscle	*	X	
	Bone		*	
Head				
	Nasal cavity	*		Look for vesicles or erosions
	Oral cavity	*		Look for vesicles or erosions
	Oropharynx	*		
	Larynx	*		
	Tonsils	X	X	
	Teeth		*	
	Tongue	*	X	Look for vesicles or erosions
	Cerebrum	X#	X	# Half fresh for rabies
	Cerebellum	X#	X	# Cross section for Rabies
	Brain stem	X#	X	# Obex
	Eyes	*	X	
	Thyroid	*	X	

Alfonso Torres, DVM, MS, PhD, Associate Dean for Public Policy, College of Veterinary Medicine, Cornell University, Ithaca, NY, 14852, at97@cornell.edu

4

SHIPPING DIAGNOSTIC SPECIMENS

Safe shipment of diagnostic specimens and/or of infectious agent cultures is guided by international regulations for the transport of infectious materials by any mode of transport. These regulations are based upon the Recommendations of the United Nations Committee of Experts on the Transport of Dangerous Goods. The U.S. Department of Transportation (DOT) and International Air Transportation Association (IATA) have brought these regulations into alignment and simplified the category assignments to those recommended by the United Nations.

As these regulations may change with time, it is important that the submitting veterinarian consult with the shipper and the receiving veterinary diagnostic laboratory on the latest packaging and shipping regulations. You must pass a certification exam in order to be qualified to ship dangerous goods.

The following summaries were prepared by regulations in place effective January 1, 2006, available at:

http://www.iata.org/nr/rdonlyres/88834d9f-8ea2-42a0-8da6-2bed8cd2e744/0/sampleissg7thed.pdf

Patient specimens are defined as follows:

"Patient specimen means human or animal material collected directly from humans or animals and transported for research, diagnosis, investigational activities, or disease treatment or prevention. Patient specimen includes excreta, secreta, blood and its components, tissue and tissue swabs, body parts, and specimens in transport media (e.g., swabs, culture media, and blood culture bottles)."

Note: Cultures are not included in this definition, only inoculated transport media, if used.

Unregulated Samples

Many diagnostic patient specimens that are not expected to contain infectious agents (like blood, serum, or urine samples from healthy animals, hair samples for genetic testing, etc), can be sent in a package simply labeled ***"Exempt Patient Specimen."*** A pre-printed label or clear handwritten labeling is acceptable. No shipping paperwork is required. Samples must still be packaged in appropriate leak-proof triple packaging capable of protecting the contents. Absorbent material capable of absorbing the entire liquid contents must be included in the secondary container. When sending unregulated liquids by air transport, use containers rated to an internal pressure of 95 kPa.

Regulated samples

Materials which may or are known to contain infectious substances must be appropriately categorized and labeled. There are two categories of regulated samples or materials: Category A and Category B.

CATEGORY A SUBSTANCES

Category A includes substances transported in a form that, when exposure to it occurs, are capable of causing permanent disability, life-threatening or fatal disease in otherwise healthy humans or animals. Most veterinarians will not have a need to routinely ship specimens included in this category. These shipments require specially rated triple packaging, labeling, and documentation. Unless you are trained in routine handling of these kinds of samples, you should seek assistance before shipping anything in this category.

In general, suspicion of infection with these pathogens in animals requires reporting to the State Veterinarian or the USDA Area Veterinarian in Charge (AVIC). If you are suspicious of a zoonotic infectious disease with any of these organisms, it is appropriate to notify in addition your local or State Public Health Department or State Public Health Veterinarian, and seek assistance with sample packaging and shipping.

Note: Some carriers like UPS and the U.S. Post Office will not accept these packages.

Category A Substances include two groups:
UN 2814 - Infectious substances affecting humans and animals, and
UN 2900 - Infectious substances affecting animals only

UN 2814 - Infectious substances affecting humans and animals

Bacillus anthracis (cultures only)
Brucella abortus (cultures only)
Brucella melitensis (cultures only)
Brucella suis (cultures only)
Burkholderia mallei—Pseudomonas mallei—Glanders (cultures only)
Burkholderia pseudomallei—Pseudomonas pseudomallei (cultures only)
Chlamydohila psittaci—avian strains (cultures only)
Clostridium botulinum (cultures only)
Coccidioides immitis (cultures only)
Coxiella burnetti (cultures only
Crimean Congo hemorrhagic fever virus
Dengue virus (cultures only)
Eastern equine encephalitis virus (cultures only)
Escherichia coli, verotoxigenic (cultures only)
Ebola virus
Flexal virus
Francisella tularensis (cultures only)
Guanarito virus
Hantaan virus
Hantaviruses causing hemorrhagic fever with renal syndrome
Hendra virus
Herpes B virus (cultures only)
Human immunodeficiency virus (cultures only)
Highly pathogenic avian influenza virus (cultures only)
Japanese Encephalitis virus (cultures only)
Junin virus
Kyasanur forest disease virus
Lassa virus
Machupo virus
Marburg virus
Monkeypox virus
Mycobacterium tuberculosis (cultures only)
Nipah virus
Omsk hemorrhagic fever virus
Poliovirus (cultures only)
Rabies and other lyssaviruses (cultures only)

Rickettsia prowazekii (cultures only)
Rickettsia rickettsia (cultures only)
Rift Valley fever virus (cultures only)
Russian spring summer encephalitis virus (cultures only)
Sabia virus
Shigella dysenteriae type I (cultures only)
Tick-borne encephalitis virus (cultures only)
Variola virus
Venezuelan equine encephalitis virus (cultures only)
Vesicular stomatitis virus (cultures only)
West Nile virus (cultures only)
Yellow fever virus (cultures only)
Yersinia pestis (cultures only)

UN 2900 - Infectious substances affecting animals only
African swine fever virus (cultures only)
Avian paramyxovirus Type 1—Velogenic Newcastle disease virus (cultures only)
Classical swine fever virus (cultures only)
Foot-and-mouth disease virus (cultures only)
Lumpy skin disease virus (cultures only
Mycoplasma mycoides—Contagious bovine pleuropneumonia (cultures only)
Peste des petits ruminants virus (cultures only)
Rinderpest virus (cultures only)
Sheep pox virus (cultures only)
Goat pox virus (cultures only)
Swine vesicular disease virus (cultures only)

CATEGORY B SUBSTANCES

Most veterinary diagnostic specimens fall within this category. Category B includes substances that may be or are known to be infectious but do not meet the criteria for inclusion in Category A (see above). Most potentially infectious diagnostic specimens submitted for veterinary diagnostics would fall into this category, such as fecal samples, other tissues, fluids and excreta, and post-mortem tissue samples for bacteriological, parasitological or viral work-up.

These samples should be packaged in leak-proof triple packaging. Absorbent material capable of absorbing the entire liquid contents must be included in the secondary container. If sent by air, specimens must be put in a primary or secondary container capable of withstanding aircraft cargo hold pressure changes. These containers are rated to an internal pressure of 95 kPa, and are available in a variety of sizes of rigid containers and also as rated, leak-proof pouches.

The package must be labeled ***"Biological Substance, Category B"*** and also bear a UN 3373 diamond label. These labels are available through a variety of sources (see Label 1). There is a maximum size and weight limit for these packages which includes 1 Liter or 4 kg (solids) maximum per primary container, and 4 Liters or 4 kg total maximum for the package. The sender and recipient's name, address, and phone number must be legibly printed on the package.

Note: The quantity limits described above do not apply to body parts, organs or whole bodies. Special provisions of the IATA and DOT regulations allows for an exception to the weight and size limits for these samples. On the weigh bill accompanying the shipment, note***: "Special Provision A82 (Title 49 CFR 172.102) or A81 (IATA) to exceed volume and weight limit. The quantity limits do not apply to animal body parts, whole organs or whole bodies known to contain or suspected of containing an infectious substance, UN 3373, Biological specimen, Category B."***

Many commercial vendors provide shipping materials and labels that meet all the above described requirements and regulations.

Cultures

Prudent, professional judgment must be exercised in selecting the appropriate category for shipment of cultures and laboratory isolates, due to the high concentration of pathogens present. If exposure to a culture of a pathogen otherwise classified in Category B may result in moderate to severe disease, it is still expected that the isolate be shipped as Category A, UN2814 or UN2900. Special permits may be required for the shipment of laboratory isolates between facilities. The regulations are described in Title 9 CFR, Chapter 1, Part 121.

For more information regarding possession, use, and transfer of biological agents, select agents, and toxins, see: http://www.aphis.usda.gov/animal_health/

For permit applications APHIS Form 2005 see:
http://www.aphis.usda.gov/animal_health/permits/vet_bio_permits.shtml

Dry Ice

In general, avoid sending diagnostic specimens on dry ice. This is particularly important for agents that are sensitive to low pH conditions (like foot-and-mouth disease virus-containing samples or cultures), due to the formation of carbolic acid from the gaseous CO_2 in the package. In addition, there are many restrictions in the type of carriers that accept dry ice shipments.

Shipments by Air Containing 10% Formalin

Special labeling and documentation is required when shipments will or might go via aircraft, and contain more than 30 ml of 10% formalin. Package must bear a "Class 9" hazardous materials diamond-shaped label and the words ***"Aviation regulated liquid, n.o.s (10% formalin) UN3334"*** (Label 2). The package must be accompanied by a Shipper's Declaration for Dangerous Goods which provides the name, address, and phone number of the shipper and consignee, the proper shipping name, class, UN number, quantity, and description of packing, and the statement ***"I hereby declare that the contents of this consignment are fully and accurately described above by the proper shipping name, and are classified, packaged, marked and labeled/ placarded, and are in all respects in proper condition for transport according to applicable international and national governmental regulations."*** It must have a statement with an emergency phone number of the shipper, and it must bear the signature of the person responsible for shipping the package.

The original and 2 copies (color copies may be required) are placed in an unsealed pouch on the outside of the package. Some carriers may require prior arrangements or the use of their own version of the form. UPS will not accept these packages.

Note: Well-fixed trimmed tissue samples could be sent to the laboratory in sealed plastic pouches with a small amount of formalin to prevent dryness. This way there is no need to declare the shipment of formalin (below limit of 30ml).

Note to Importers and Exporters

Packages containing diagnostic specimens to be submitted to international reference laboratories require additional permits and shipping labels. The foreign laboratory must be consulted in advance to obtain an import permit for clearing customs upon arrival. Also there may be restrictions imposed by the U.S. Government in the shipment of control substances to certain countries of the world. It is recommended that if diagnostic specimens or cultures need to be sent to international reference laboratories, that the USDA or the CDC be consulted for guidance. Additional information on international shipments can be found at:

USDA biological agents and toxins importation, possession, use and transfer
http://www.aphis.usda.gov/animal_health/vet_biologics/vb_import_export_products.shtml

CDC etiologic agent import permit program
http://www.cdc.gov/od/eaipp/

U.S. Department of Commerce, Bureau of Industry and Security (Export permits)
http://www.bis.doc.gov/index.htm

U.S. Food and Drug Administration Import Permits
http://www.fda.gov/ora/import/

U.S. Fish and Wildlife Service Import and Export permits
http://www.fws.gov/permits/ImportExport/ImportExport.shtml

ILLUSTRATIONS ON FOLLOWING PAGES

Alfonso Torres, DVM, MS, PhD, Associate Dean for Public Policy, and Belinda Thompson, DVM, Senior Extension Associate, Animal Health Diagnostic Laboratory, College of Veterinary Medicine, Cornell University, Ithaca, NY, 14852, at97@cornell.edu, bt42@cornell.edu

Label 1: Biological Specimens

Category B: Shipping Label for Biological Substances

The label (right) should print with the proper dimension of an Infectious Substance label (minimum dimensions: 50 mm on a side, and the proper shipping name, "Biological Substance, Category B" must be in letters at least 6mm high). Affix the label to the package by covering with clear plastic tape, so that moisture will not cause printer ink to run.

Label 2: Formalin Shipments

Shipping Label for 10% Buffered Formalin, Class 9

This label is required for shipments containing in excess of 30ml 10% buffered formalin. The label below should print with the proper dimension of a class 9 hazardous goods label (minimum dimensions: 100mm on a side). Cut around the label and affix it to a vertical side of the package (not top or bottom), oriented as shown. Cover with clear plastic tape, to avoid moisture causing the printer ink to run.

Shipper's Declaration of Dangerous Goods for 10% Formalin

Below is an example of a fully completed Shipper's Declaration of Dangerous Goods for a shipment which includes 10% formalin in a quantity exceeding 30ml, and containing no other dangerous goods. The original and two copies (color copies may be required) are placed in an unsealed pouch on the outside of the package.

SAMPLE DECLARATION

SHIPPER'S DECLARATION FOR DANGEROUS GOODS

Dr. Ben Johnson
Johnson Veterinary Hosp.
327 Johnson Blvd.
Johnsonville, NY 11111

Air Waybill No. 45892587

Page 1 of 1 Pages

Shippers Reference Number *(optional)*

Consignee Receiving Dept. 607 253-3900
Animal Health Diagnostic Center
College of Veterinary Medicine, Cornell Univ.
Upper Tower Rd.
Ithaca, NY 14853

Two completed and signed copies of this Declaration must be handed to the operator

TRANSPORT DETAILS

This shipment is within the limitations prescribed for: *(delete non-applicable)*

PASSENGER AND CARGO AIRCRAFT | CARGO AIRCRAFT XXXX

Airport of Departure

Airport of Destination

WARNING
Failure to comply in all respects with the applicable Dangerous Goods Regulations may be in breach of the applicable law, subject to legal penalties. This Declaration must not, in any circumstances, be completed and/or signed by a consolidator, a forwarder or an IATA cargo agent.

Shipment type: *(delete non-applicable)*

NON-RADIOACTIVE | RADIOACTIVE (XXXXXXX)

NATURE AND QUANTITY OF DANGEROUS GOODS

Dangerous Goods Identification							
UN or ID Number	Proper Shipping Name	Class or Division	Subsidiary risk	Packing Group	Quantity and type of Packing	Packing Instruction	Authorization
UN 3334	Aviation regulated liquid, n.o.s (10% neutral buffered formalin)	9			80 ml in a plastic screw-top vial. All packaged in an outer fiberboard box.	906	

Additional Handling Information

Emergency Contact Number: 607 555-1212

I hereby declare that the contents of this consignment are fully and accurately described above by the proper shipping name, and are classified, packaged, marked and labeled/placarded, and are in all respects in proper condition for transport according to applicable international and national governmental regulations.

Name/Title of Signatory
Amy Brown / Vet Tech

Place and Date
Johnsonville, NY 11111 10/6/06

Signature *(see warning above)* Amy Brown

Diagram 1: Packing and Shipping Category B Diagnostic Samples

Step 1: Place solid or liquid specimen in a leak-proof primary containers, and label contents

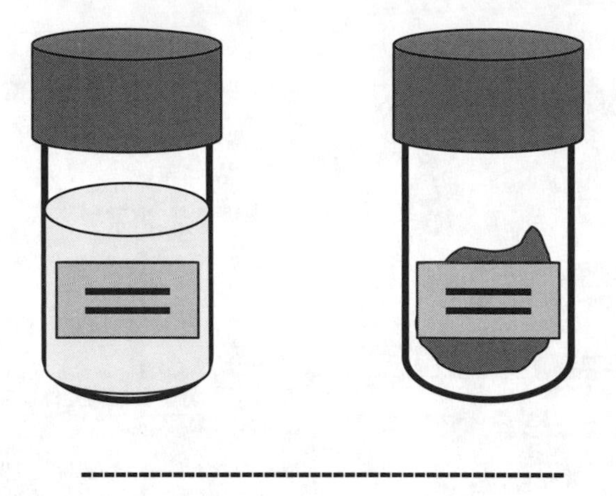

Step 2: Place primary containers inside a flexible (sealable plastic bag) or rigid secondary containers. Place absorbing material inside of secondary container

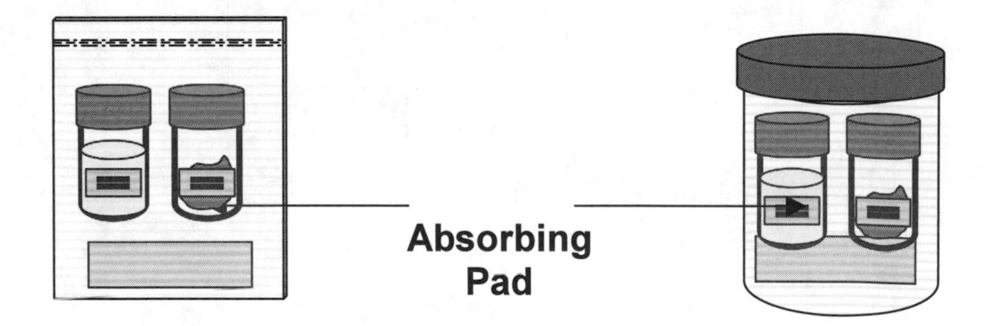

Step 3: Place secondary container in approved insulated water-proof flexible or rigid insulated shipping container. Place ice packs inside of shipping container

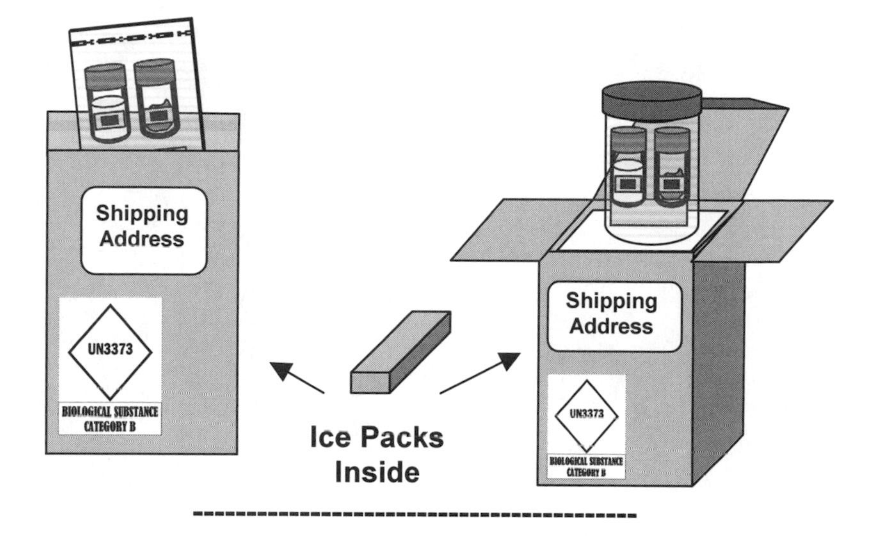

Step 4: Place case history and submission forms in sealed plastic envelope inside of shipping container. Seal and ship

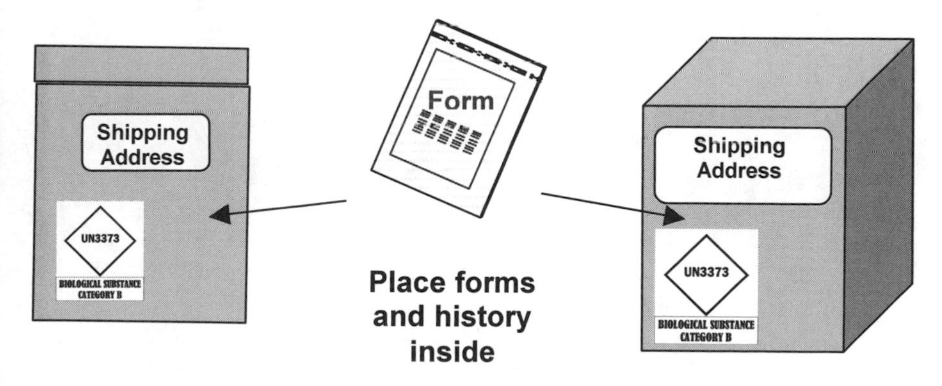

5

ANIMAL PATHOGEN DISINFECTANTS

INTRODUCTION

A crucial part of any FAD control scheme is cleaning and disinfection. Preliminary cleaning entails mechanical brushing of surfaces with a detergent solution and occurs prior to the application of chemical disinfectant. Removal of organic matter will greatly aid the process of chemical disinfection. Disinfection decreases the load of pathogenic organisms that may remain after the cleaning.

Products used to kill microorganisms in various settings (such as veterinary clinics, farms, equipment, and vehicles) are generally called disinfectants, although the technically correct but infrequently used term is "antimicrobial pesticides." The United States Environmental Protection Agency (EPA) regulates these products and their distribution and sale can only be done after registration of the product, which requires the generation of human and environmental safety data and antimicrobial efficacy data.

According to the Food and Agriculture Organization (FAO), soaps and disinfectants are broadly categorized into several groups, including:

- Soaps and detergents
- Oxiding agents
- Alkalis
- Acids
- Aldehydes

Soaps and detergents serve to remove dirt, grease and organic material, facilitating the subsequent process of decontamination. They are cleaning agents. Hot water, brushing and scrubbing will all enhance the action of soaps and detergents. The surfactant action of many soaps and detergents is effective for those viruses that are enveloped, as the lipid nature of the envelope is disrupted by the surfactant. Some detergents contain phenolics or quarternary ammonium compounds which are effective antibacterials but have minimal effectiveness for nonenveloped viruses.

Oxidizing agents, alkalis, acids, and aldehydes are all considered "disinfectants."

Oxidizing agents include sodium hypochlorite (household bleach), calcium hypochlorite (lime) and the commonly used commercially available agent Virkon S. These are the most widely used disinfectants, but the efficacy decreases in the presence of organic matter, which is why preliminary washing is so important. Oxidizing agents inactive viruses by damaging any proteins that have disulfide bonds on the surface of the virus.

Alkalis include sodium hydroxide (caustic soda) and sodium carbonate anhydrous (washing soda). They are virucidal, even under heavy burdens of organic matter. The action of alkalis may be to dissociate ribosomes due to conditions of high pH.

Acids are especially helpful for some viruses, especially foot-and-mouth disease.

Aldehydes include glutaraldehyde, formalin, and formaldehyde gas. These are more expensive, making them unsuitable for large-scale decontamination. Also there are adverse human health consequences associated with formaldehyde gas. Formalin, which is used to fix tissues for pathologic examination, will rapidly inactivate all infectious agents in the immersed specimen.

Federal oversight and regulations

The U.S. government regulates the sale, distribution, use, and disposal of disinfectants. The Federal Insecticide Fungicide and Rodenticide Act (FIFRA), first passed in 1947, defines "antimicrobial pesticide" as a product intended to disinfect, sanitize, reduce, or mitigate the growth or development of microbiological organisms, or protect inanimate objects, industrial processes or systems, surfaces, water, or other chemical substances from contamination, fouling, or deterioration caused by bacteria, viruses, fungi, protozoa, algae or slime. A pest, as defined by FIFRA, means any insect, rodent, nematode, fungus, weed, or any other bacteria, or other microorganisms (except viruses, bacteria, or other microorganisms on or in living animals). A pesticide, as defined by FIFRA, includes any substance or mixture of substances intended for preventing, destroying, repelling, or mitigating pests on inanimate objects and surfaces.

FIFRA authorizes the Environmental Protection Agency to regulate three characteristics of pesticides (including antimicrobial products), i.e., chemical composition, labeling, and packaging. EPA registers pesticide products based on data showing that there will be no unreasonable adverse effects to humans or the environment if the pesticide is used according to the directions on the label. In addition, FIFRA grants EPA the authority to require certification of pesticide users, or ask for additional submission of data.

The EPA classifies disinfectants according to the extent of the submitted efficacy data, i.e., limited, general or broad-spectrum, and hospital. Different categories of antimicrobial pesticide registration actions include new chemicals, new uses of a registered chemical, registration of a product that is identical or substantially similar to an existing product, manufacturing use products, and registration of re-packaged products.

Federal law allows the Agencies to publish regulations. *Code of Federal Regulations,* Title 40, Subchapter E, Parts 152 through 180, contains regulations dealing with the disinfection programs. These regulations describe registration, review, labeling, packaging, data requirements, policies, good laboratory practice standards, State registrations, registration cancellations, emergency exemptions, establishment registration, enforcement, records, worker protection, applicator certification, experimental use permits, State enforcement, hearings, and tolerances.

There is an emergency exemption possibility for using a disinfectant that is not registered. Section 18 of FIFRA authorizes EPA to allow States and Federal agencies to use a pesticide (disinfectant) for an unregistered use for a limited time if EPA determines that emergency conditions exist. This might be the case in the event of a foreign animal disease. A crisis exemption is a type of emergency exemption issued by a State or Federal agency if a need is so immediate that the Agency cannot wait for EPA to approve the request and allows for unregistered disinfectant use for up to 15 days. It is in violation of Federal law for States, Agencies, companies, individuals, etc., to use disinfectants exempted under a Section 18 if these entities are not listed on that specific Section 18 exemption. The use of a registered disinfectant off-label (e.g., at non-label concentrations or on non-listed use sites) or the use of a non-EPA registered generic chemical compound as a disinfectant (e.g., sodium hypochlorite, sodium hydroxide, sodium carbonate, acetic acid, citric acid, etc.) by any entity not specifically listed on the Section 18 exemption is unlawful. A copy of the corresponding Section 18 exemption must be available at the location where the disinfectant is used.

Selection of a disinfectant for a foreign animal disease outbreak

Table 1 contains a list of registered products that may be used for many of the FAD agents.

It is important to only use a disinfectant registered by the U.S. EPA. Some States may require an additional State product registration. The product should be used only according to label directions. The pathogen for which the disinfectant is used must be indicated on the product label. Only use a disinfectant in locations and on surface types indicated on the product label. Disinfectants should only be used on surfaces which are free of organic matter. Surfaces which have been previously cleaned with a detergent should be thoroughly rinsed with clean water to remove residual detergent and allowed to dry prior to disinfectant application. Consideration should be given to the corrosive action of some disinfectants on the surface to be treated. Metallic salts present in hard water used to dilute disinfectants may reduce product efficacy. Disinfectants should be allowed to remain wet on treated surfaces for the time period indicated on the label. Most disinfectants work better at warmer rather than at cooler temperatures. Some disinfectants may taint the treated surfaces. Some disinfectants may have residual activity that would harm animals returned to the treated premises. Disinfectants have a maximum shelf life and should be stored under label indicated conditions. Disinfectant costs should be considered.

Possible causes of disinfection failure

The disinfectant selected must be efficacious for the specific pathogen to be controlled. The disinfectant should not be over-diluted during preparation or application. Organic material remaining on treated surfaces may act as a physical barrier preventing the disinfectant from reaching the pathogen and may combine chemically with the disinfectant to inactivate it. Insufficient amounts of disinfectant applied may not thoroughly cover and penetrate all surfaces. Inadequate temperature and humidity at the time of application may reduce effectiveness. Inadequate rinsing to remove cleaning product may neutralize the disinfectant. Inadequate temperature may limit a disinfectant's efficacy. Inadequate contact time of the disinfectant may limit a disinfectant's efficacy.

For further information about disinfection in the event of an FAD outbreak, contact Dr. Nathan Birnbaum, Senior Staff Veterinarian, USDA-APHIS-VS National Center for Animal Health Emergency Management, Riverdale, MD, Nathan.G.Birnbaum@aphis.usda.gov.

For further information about disinfection registration, contact the United States Environment Protection Agency, Office of Pesticide Programs, Antimicrobial Division, Mail Code 7510-P, 1200 Pennsylvania Ave., NW, Washington D.C. 20460-0001, E-mail: opp-web-comments@epa.gov, Main telephone number: 703.308.6411.

Bethany O'Brien, DVM, USDA-APHIS-VS, Western Regional Office, Fort Collins, CO, bethany.o'brien@aphis.usda.gov

Table 1. Disinfectants for use in FAD outbreaks

EPA-APPROVED PESTICIDES (DISINFECTANTS) FOR USE AGAINST HIGHLY PATHOGENIC DISEASES				
DISEASE	PRODUCT	EPA REG. NUMBER	MANUFACTURER NAME AND CONTACT[1]	ACTIVE INGREDIENT(S)
African Horse Sickness				
	none registered[2]	--	--	--
African Swine Fever				
	Low Ph Phenolic 256	211-62	Central Solutions, Inc.	o-Phenylphenol 2-Benzyl-4-chlorophenol
	Pheno Cen Germicidal Detergent	211-25	Central Solutions, Inc.	o-Phenylphenol, potassium salt p-tert-Amylphenol, potassium salt Potassium 2-benzyl-4-chlorophenate
	Klor-Kleen	71847-2	Medentech Ltd.	Sodium dichloro-s-triazinetrione
	Virkon S	71654-6	DuPont Chemical Solutions Enterprise	Sodium chloride Potassium peroxymonosulfate
Akabane				
	none registered	--	--	--
Avian Influenza[3]				
	Odo-Ban Ready-To-Use	66243-1	Clean Control Corporation	Alkyl dimethyl benzyl ammonium chloride
	Odo-Ban	66243-2	Clean Control Corporation	
	Johnson's Forward Cleaner	70627-10	JohnsonDiversity, Inc.	

[1] See internet for complete commercial information contacts.
[2] "Registered" refers to a FIFRA Section 3 registration.
[3] Products listed for use against avian influenza were not generated from the NPIRS database. This list originated with EPA's "Registered Antimicrobial Products with Label Claims for Avian (Bird) Flu Disinfectants" list dated July 13, 2007, located at: http://www.epa.gov/pesticides/factsheets/avian_flu_products.htm.

EPA-APPROVED PESTICIDES (DISINFECTANTS) FOR USE AGAINST HIGHLY PATHOGENIC DISEASES				
DISEASE	PRODUCT	EPA REG. NUMBER	MANUFACTURER NAME AND CONTACT[1]	ACTIVE INGREDIENT(S)
	Johnson's Blue Chip Germicidal Cleaner for Hospitals	70627-15	JohnsonDiversity, Inc.	
	BTC 2125 M 10% Solution	1839-86	Stepan Company	Alkyl dimethyl benzyl ammonium chloride Alkyl dimethyl ethylbcnzyl ammonium chloride
	NP 4.5 (D&F) Detergent/Disinfectant	1839-95	Stepan Company	
	Scented 10% BTC 2125M Disinfectant	1839-154	Stepan Company	
	BTC 2125 M 20% Solution	1839-155	Stepan Company	
	Quat 44	3838-36	Essential Industries, Inc.	
	Quat Rinse	3838-37	Essential Industries, Inc.	
	Spray Nine	6659-3	Spray Nine Corporation	
	Marquat 256	10324-56	Mason Chemical Company	
	Marquat 128	10324-58	Mason Chemical Company	
	Marquat 64	10324-59	Mason Chemical Company	
	Maquat 10	10324-63	Mason Chemical Company	
	Maquat 20-M	10324-94	Mason Chemical Company	
	Maquat 50DS	10324-96	Mason Chemical Company	
	Maquat 10 FQPA	10324-99	Mason Chemical Company	
	Maquat 256 EBC	10324-118	Mason Chemical Company	
	Maquat 128 EBC	10324-119	Mason Chemical Company	

EPA-APPROVED PESTICIDES (DISINFECTANTS) FOR USE AGAINST HIGHLY PATHOGENIC DISEASES				
DISEASE	PRODUCT	EPA REG. NUMBER	MANUFACTURER NAME AND CONTACT[1]	ACTIVE INGREDIENT(S)
	Maquat 64 EBC	10324-120	Mason Chemical Company	
	Maquat MQ2525M-14	10324-142	Mason Chemical Company	
	Maquat 10-B	10324-143	Mason Chemical Company	
	Maquat FP	10324-145	Mason Chemical Company	
	Maquat 256 PD	10324-164	Mason Chemical Company	
	D-125	61178-1	Microgen, Inc.	
	Public Places	61178-2	Microgen, Inc.	
	Public Places Towelette	61178-4	Microgen, Inc.	
	CCX-151	61178-5	Microgen, Inc.	
	D-128	61178-6	Microgen, Inc.	
	PJW-622	67619-9	Clorox Professional Products Company	
	Opticide-3	70144-1	Micro-Scientific Industries	
	Opticide-3 Wipes	70144-2	Micro-Scientific Industries	
	Disinfectant DC 100	70627-2	JohnsonDiversity, Inc.	
	Sterilex Ultra Disinfectant Cleaner	63761-8	Sterilex Corporation	Alkyl dimethyl benzyl ammonium chloride Alkyl dimethyl ethylbenzyl ammonium chloride Hydrogen peroxide
	HI-Tor Plus Germicidal Detergent	303-91	Huntington Professional Products	Alkyl dimethyl benzyl ammonium chloride Didecyl dimethyl ammonium chloride
	Marquat 256-NHQ	10324-141	Mason Chemical Company	
	Maquat 2420 Citrus	10324-162	Mason Chemical Company	

EPA-APPROVED PESTICIDES (DISINFECTANTS) FOR USE AGAINST HIGHLY PATHOGENIC DISEASES				
DISEASE	PRODUCT	EPA REG. NUMBER	MANUFACTURER NAME AND CONTACT[1]	ACTIVE INGREDIENT(S)
	Formulation HS-652Q	47371-6	H&S Chemical Division c/o Lonza, Inc.	
	Formulation HS-821Q	47371-7	H&S Chemical Division c/o Lonza, Inc.	
	HS-867Q	47371-36	H&S Chemical Division c/o Lonza, Inc.	
	HS-267Q Germicidal Cleaner and Deodorant	47371-37	H&S Chemical Division c/o Lonza, Inc.	
	Formulation HH-652 Q	47371-141	H&S Chemical Division c/o Lonza, Inc.	
	Virex II/128	70627-21	JohnsonDiversity, Inc.	
	Virex II Ready To Use	70627-22	JohnsonDiversity, Inc.	
	Virex II 64	70627-23	JohnsonDiversity, Inc.	
	Virex 11/256	70627-24	JohnsonDiversity, Inc.	
	Virocide	71355-1	CID Lines, NV/SA	Alkyl dimethyl benzyl ammonium chloride Didecyl dimethyl ammonium chloride Glutaraldehyde
	Ucarcide 14 Antimicrobial	464-700	The Dow Chemical Company	Alkyl dimethyl benzyl ammonium chloride Glutaraldehyde
	Ucarcide 42 Antimicrobial	464-702	The Dow Chemical Company	
	Ucarsan 442 Sanitizer	464-715	The Dow Chemical Company	
	Ucarsan 414 Sanitizer	464-716	The Dow Chemical Company	

EPA-APPROVED PESTICIDES (DISINFECTANTS) FOR USE AGAINST HIGHLY PATHOGENIC DISEASES				
DISEASE	PRODUCT	EPA REG. NUMBER	MANUFACTURER NAME AND CONTACT[1]	ACTIVE INGREDIENT(S)
	Synergize	66171-7	Preserve International	
	Maxima 128	106-72	Mason Chemical Company	Alkyl dimethyl benzyl ammonium chloride Octyl decyl dimethyl ammonium chloride Didecyl dimethyl ammonium chloride Dioctyl dimethyl ammonium chloride
	Maxima 256	106-73	Mason Chemical Company	
	Broadspec 256	106-79	Brulin & Company, Inc.	
	Maxima RTU	106-81	Brulin & Company, Inc.	
	Q5.5-5.5NPB-2.5HW	211-50	Central Solutions	
	Sanox II	11600-4	Conklin Co., Inc.	
	7.5% BTC 885 Disinfectant/Sanitizer	1839-173	Stepan Company	
	Quik Control	66243-3	Clean Control Corporation	
	Bardac 205M-7.5B	6836-70	Lonza, Inc.	
	Lonza Formulation Y-59	6836-71	Lonza, Inc.	
	Lonza Formulation S-21	6836-75	Lonza, Inc.	
	Lonza Formulation S-18	6836-77	Lonza, Inc.	
	Lonza Formulation R-82	6836-78	Lonza, Inc.	
	Lonza Formulation S-18F	6836-136	Lonza, Inc.	
	Lonza Formulation R-82F	6836-139	Lonza, Inc.	

EPA-APPROVED PESTICIDES (DISINFECTANTS) FOR USE AGAINST HIGHLY PATHOGENIC DISEASES				
DISEASE	PRODUCT	EPA REG. NUMBER	MANUFACTURER NAME AND CONTACT[1]	ACTIVE INGREDIENT(S)
	Lonza Formulation S-21F	6836-140	Lonza, Inc.	
	Lonza Formulation DC-103	6836-152	Lonza, Inc.	
	Bardac 205M-50	6836-233	Lonza, Inc.	
	Bardac 205M-10	6836-266	Lonza, Inc.	
	Bardac 205M-1.30	6836-277	Lonza, Inc.	
	Bardac (R) 205M-14.08	6836-278	Lonza, Inc.	
	Bardac 205M-2.6	6836-302	Lonza, Inc.	
	Bardac 205M-5.2	6836-303	Lonza, Inc.	
	Microban QGC	70263-6	Microban Systems, Inc.	
	Microban Professional	70263-8	Microban Systems, Inc.	
	Maquat MQ651-AS	10324-67	Mason Chemical Company	
	Maquat 615-HD	10324-72	Mason Chemical Company	
	Maquat 5.5-M	10324-80	Mason Chemical Company	
	Maquat 7.5-M	10324-81	Mason Chemical Company	
	Maquat 86-M	10324-85	Mason Chemical Company	
	Maquat 750-M	10324-115	Mason Chemical Company	
	Maquat 710-M	10324-117	Mason Chemical Company	
	Maquat A	10324-131	Mason Chemical Company	
	DC & R	134-65	HACCO	Alkyl dimethyl

EPA-APPROVED PESTICIDES (DISINFECTANTS) FOR USE AGAINST HIGHLY PATHOGENIC DISEASES				
DISEASE	PRODUCT	EPA REG. NUMBER	MANUFACTURER NAME AND CONTACT[1]	ACTIVE INGREDIENT(S)
	Disinfectant			benzyl ammonium chloride Formaldehyde 2-(Hydroxymethyl)-2-nitro-1,3-propanediol
	Biosol	777-72	Reckitt Benckiser	Alkyl dimethyl benzyl ammonium sacchaarinate Ethanol
	Husky 806 H/D/N	8155-23	Canberra Corporation	Didecyl dimethyl ammonium chloride
	Pheno Cen Germicidal Detergent	211-25	Central Solutions, Inc.	o-Phenylphenol, potassium salt p-tert-Amylphenol, potassium salt Potassium 2-benzyl-4-chlorophenate
	Pheno-Cen Spray Disinfectant/Deodorant	211-32	Central Solutions, Inc.	o-Phenylphenol Ethyl alcohol
	Low Ph Phenolic 256	211-62	Central Solutions, Inc.	o-Phenylphenol 2-Benzyl-4-chlorophenol
	Phenocide 256	6836-252	Lonza, Inc.	
	Phenocide 128	6836-253	Lonza, Inc.	
	Phenolic Disinfectant HG	70627-6	JohnsonDiversity, Inc.	
	Tek-Trol Disinfectant Cleaner Concentrate	3862-177	ABC Compounding Co.	o-Phenylphenol 2-Benzyl-4-chlorophenol 4-tert-Amylphenol
	Advantage 256 Cleaner Disinfectant Deodorant	66171-1	Preserve International	
	LPH Master Product	1043-91	Steris Corporation	o-Phenylphenol 4-tert-Amylphenol

EPA-APPROVED PESTICIDES (DISINFECTANTS) FOR USE AGAINST HIGHLY PATHOGENIC DISEASES				
DISEASE	**PRODUCT**	**EPA REG. NUMBER**	**MANUFACTURER NAME AND CONTACT[1]**	**ACTIVE INGREDIENT(S)**
	Sporicidin Brand Disinfectant Solution	8383-3	Sporicidin International	Phenol Sodium phenate
	Ucarsan Sanitizer 420	464-689	The Dow Chemical Company	Glutaraldehyde
	Ucarsan Sanitizer 4128	464-696	The Dow Chemical Company	
	Accel TB	74559-1	Virox Technologies	
	Virkon	71654-7	DuPont Chemical Solutions	Phenol Peroxyacetic acid
	Oxonia Active	1677-129	Ecolab, Inc.	Hydrogen peroxide Peroxyacetic acid
	OxySept LDI	1677-203	Ecolab, Inc.	
	Peridox	81073-1	Clean Earth Technologies, LLC	
	Vortexx	1677-158	Ecolab, Inc.	Hydrogen peroxide Peroxyacetic acid Octanic acid
	DisinFx	74331-2	SteriFx Inc.	Citric acid Hydrochlorine acid Phosphoric acid
	Dyne-O-Might	66171-6	Preserve International	Iodine
	Virkon S	71654-6	DuPont Chemical Solutions	Sodium chloride Potassium peroxymonosulfate
	Klor-Kleen	71847-2	Medentech, Ltd.	Sodium dichloro-s-triazinetrione
	Clorox	5813-1	Clorox Company	Sodium hypochlorite
	Dispatch Hospital Cleaner with Bleach	56392-7	Caltech Industries	
	Dispatch Hospital Cleaner Disinfectant Towels with Bleach	56392-8	Caltech Industries	

EPA-APPROVED PESTICIDES (DISINFECTANTS) FOR USE AGAINST HIGHLY PATHOGENIC DISEASES				
DISEASE	PRODUCT	EPA REG. NUMBER	MANUFACTURER NAME AND CONTACT[1]	ACTIVE INGREDIENT(S)
	CPPC Ultra Bleach 2	67619-8	Clorox Professional Services Company	
	CPPC Storm	67619-13	Clorox Professional Services Company	
	Aseptrol S10-Tabs	70060-19	Engelhard Corporation	Sodium chlorite Sodium dichloroisocyanurate dihydrate
Bovine Babesiosis				
	none registered	--	--	--
Bluetongue				
	none registered	--	--	--
Borna Disease				
	none registered	--	--	--
Bovine Ephemeral Fever				
	none registered	--	--	--
Bovine Parafilariasis				
	none registered	--	--	--
Bovine Spongiform Encephalopathy				
	none registered	--	--	--
Classical Swine Fever ("Hog Cholera")				
	Pheno Cen Germicidal Detergent	211-25	Central Solutions, Inc.	o-Phenylphenol, potassium salt p-tert-Amylphenol, potassium salt Potassium 2-benzyl-4-chlorophenate
	Pheno-Cen Spray Disinfectant/Deodorant	211-32	Central Solutions, Inc.	o-Phenylphenol Ethyl alcohol
	Tri-Cen	211-36	Central Solutions, Inc.	p-tert-Amylphenol, sodium salt Sodium 2-benzyl-4-chlorophenate Sodium o-phenylphenate
	Q5.5-5.5NPB-2.5HW	211-50	Central Solutions, Inc.	Alkyl dimethyl benzyl ammonium chloride Didecyl dimethyl

EPA-APPROVED PESTICIDES (DISINFECTANTS) FOR USE AGAINST HIGHLY PATHOGENIC DISEASES				
DISEASE	PRODUCT	EPA REG. NUMBER	MANUFACTURER NAME AND CONTACT[1]	ACTIVE INGREDIENT(S)
				ammonium chloride Octyl decyl dimethyl ammonium chloride Dioctyl dimethyl ammonium chloride
	Low Ph Phenolic 256	211-62	Central Solutions, Inc.	o-Phenylphenol 2-Benxyl-4-chlorophenol
	Ucarsan Sanitizer 420	464-689	The Dow Chemical Company	Glutaral
	Ucarsan Sanitizer 4128	464-696		
	1-Stroke Environ	1043-26	Steris Corporation	2-Benzyl-4-chlorophenol o-Phenylphenol 4-tert-Amylphenol
	Fort Dodge Nolvasan Solution	1117-30	Fort Dodgc Animal Health Division of Wyeth	Chlorhexidine diacetate
	Fort Dodge Nolvasan S	1117-48		
	Virkon S	71654-6	DuPont Chemical Solutions Enterprise	Sodium chloride Potassium peroxymonosulfate
	Klor-Kleen	71847-2	Medentech Ltd.	Sodium dichloro-s-triazinetrione
Contagious Agalactia of Sheep and Goat				
	none registered	--	--	--
Contagious Bovine Pleuropneumonia				
	none registered	--	--	--
Contagious Caprine Pleuropneumonia				
	none registered	--	--	--
Contagious Equine Metritis				
	none registered	--	--	--
Dourine				
	none registered	--	--	--
Duck Virus Hepatitis				
	none registered	--	--	--
Epizootic Lymphangitis				
	none registered	--	--	--
Equine Encephalosis				

EPA-APPROVED PESTICIDES (DISINFECTANTS) FOR USE AGAINST HIGHLY PATHOGENIC DISEASES				
DISEASE	PRODUCT	EPA REG. NUMBER	MANUFACTURER NAME AND CONTACT[1]	ACTIVE INGREDIENT(S)
	none registered	--	--	--
Foot-and-Mouth Disease[4]				
	Low PH Phenolic 256	211-62	Central Solutions, Inc.	2-Benzyl-4-chlorophenol o-Phenylphenol
	Oxonia Active	1677-129	Ecolab Inc.	Ethaneperoxoic acid Hydrogen peroxide
	Oxysept LDI	1677-203		
	Lonza DC 101	6836-86	Lonza, Inc.	Alkyl dimethyl benzyl ammonium chloride 1-Decanaminium, N-decyl-N,N-dimethyl-, chloride 1-Decanaminium, N,N-dimethyl-N-octyl-, chloride 1-Octanaminium, N,N-dimethyl-N-octyl-, chloride
	Aseptrol S10-TAB	70060-19	BASF Catalysts, LLC	Sodium chlorite Sodium dichloroisocyanurate dehydrate
	Aseptrol FC-Tab	70060-30		
	Virkon S	71654-6	DuPont Chemical Solutions Enterprise	Sodium chloride Potassium peroxymonosulfate
Getah Virus Disease				
	none registered	--	--	--
Glanders				
	none registered	--	--	--
Heartwater				
	none registered	--	--	--
Hendra Virus Disease				
	none registered	--	--	--
Hemorrhagic Septicemia				
	none registered	--	--	--
Infectious Salmon Anemia				
	Virkon S[5]	71654-6	DuPont Chemical	Sodium chloride

[4] Products listed for use against foot and mouth disease were not generated from the NPIRS database. This list originated from EPA's 8/8/07 FMD table provided via e-mail to APHIS.

EPA-APPROVED PESTICIDES (DISINFECTANTS) FOR USE AGAINST HIGHLY PATHOGENIC DISEASES				
DISEASE	PRODUCT	EPA REG. NUMBER	MANUFACTURER NAME AND CONTACT[1]	ACTIVE INGREDIENT(S)
			Solutions Enterprise	Potassium peroxymonosulfate
Japanese Encephalitis				
	none registered	--	--	--
Jembrana Disease				
	none registered	--	--	--
Louping Ill				
	none registered	--	--	--
Lumpy Skin Disease				
	none registered	--	--	--
Malignant Catarrhal Fever				
	none registered	--	--	--
Nairobi Sheep Disease				
	none registered	--	--	--
Newcastle Disease				
	Vesphene II SE	1043-87	Steris Corporation	o-Phenylphenol 4-tert-Amylphenol
	LPH Master Product	1043-91		
	Vesta-Syde Interim Instrument Decontamination Solution	1043-114		
	Process Vesphene II ST	1043-115		
	Amerse II	1043-117		
	Beaucoup Germicidal Detergent	303-223	Huntington Professional Products	o-Phenylphenol 4-tert-Amylphenol 2-Benzyl-4-chlorophenol
	Matar II	303-225		
	1-Stroke Environ	1043-26	Steris Corporation	
	Tek-Trol Disinfectant Cleaner Concentrate	3862-177	ABC Compounding Co, Inc.	
	Bio-Phene Liquid Disinfectant	71654-17	DuPont Chemical Solutions Enterprise	
	Phenocide 256	6836-252	Lonza, Inc.	o-Phenylphenol 2-Benzyl-4-chlorophenol
	Phenocide 128	6836-253	Lonza, Inc.	
	Phenolic Disinfectant HG	70627-6	JohnsonDiversity, Inc.	
	Mikro-Quat	1677-21	Ecolab Inc.	Alkyl dimethyl

[5] Product not listed in NPIRS as approved for use against ISAV, but the use does appear on Federally approved label.

EPA-APPROVED PESTICIDES (DISINFECTANTS) FOR USE AGAINST HIGHLY PATHOGENIC DISEASES				
DISEASE	PRODUCT	EPA REG. NUMBER	MANUFACTURER NAME AND CONTACT[1]	ACTIVE INGREDIENT(S)
	Odo-Ban Ready-To-Use	66243-1	Clean Control Corp	benzyl ammonium chloride
	Odo-Ban	66243-2	Clean Control Corp	
	Johnson Blue Chip Germicidal Cleaner for Hospitals	70627-15	JohnsonDiversity, Inc.	
	Grenadier	1769-259	NCH Corp	Alkyl dimethyl benzyl ammonium chloride Alkyl dimethyl ethylbenzyl ammonium chloride
	BTC 2125M 20% Solution	1839-155	Stepan Company	
	Maquat 10	10324-63	Mason Chemical Co.	
	Maquat 20-M	10324-94	Mason Chemical Co.	
	Maquat 50DS	10324-96	Mason Chemical Co.	
	Maquat 10-PD	10324-99	Mason Chemical Co.	
	Maquat 256 EBC	10324-118	Mason Chemical Co.	
	Maquat 128 EBC	10324-119	Mason Chemical Co.	
	Maquat 64 EBC	10324-120	Mason Chemical Co.	
	Maquat MQ2525M-CPV	10324-140	Mason Chemical Co.	
	Maquat MQ2525M-14	10324-142	Mason Chemical Co.	
	Maquat 10-B	10324-143	Mason Chemical Co.	
	Maquat FP	10324-145	Mason Chemical Co.	
	Maquat 256 PD	10324-164	Mason Chemical Co.	
	D-125	61178-1	Microgen, Inc.	
	Public Places	61178-2	Microgen, Inc.	
	Public Places Towelette	61178-4	Microgen, Inc.	
	CCX-151	61178-5	Microgen, Inc.	
	Bioguard 453	71654-16	DuPont Chemical Solutions Enterprise	
	Gemstone	5813-91	Clorox c/o PS&RC	Alkyl dimethyl benzyl ammonium chloride Didecyl dimethyl
	Hospital Disinfectant Cleaner	7546-27	U.S. Chemical Corp.	

EPA-APPROVED PESTICIDES (DISINFECTANTS) FOR USE AGAINST HIGHLY PATHOGENIC DISEASES				
DISEASE	PRODUCT	EPA REG. NUMBER	MANUFACTURER NAME AND CONTACT[1]	ACTIVE INGREDIENT(S)
	Maquat 2420-Citrus	10324-162	Mason Chemical Co.	ammonium chloride
	Formulation HS-652Q	47371-6	H&S Chemicals Division c/o Lonza Inc.	
	Formulation HS-821Q	47371-7	H&S Chemicals Division c/o Lonza Inc.	
	FMB 1210-5 Quat	47371-27	H&S Chemicals Division c/o Lonza Inc.	
	HL-867 Q	47371-36	H&S Chemicals Division c/o Lonza Inc.	
	HS-267Q Germicidal Cleaner and Disinfectant	47371-37	H&S Chemicals Division c/o Lonza Inc.	
	FMB 1210-8 Quat Concentrated Germicide	47371-42	H&S Chemicals Division c/o Lonza Inc.	
	Formulation HS-1210 Disinfectant/Sanitizer (3.85%)	47371-147	H&S Chemicals Division c/o Lonza Inc.	
	Formulation HS-1210 Disinfectant/Sanitizer (50%)	47371-164	H&S Chemicals Division c/o Lonza Inc.	
	Formulation HS-1210 Disinfectant/Sanitizer (14.08%)	47371-180	H&S Chemicals Division c/o Lonza Inc.	
	Virex II/128	70627-21	JohnsonDiversity, Inc.	
	Virex II Ready to Use	70627-22	JohnsonDiversity, Inc.	
	Virex II 64	70627-23	JohnsonDiversity, Inc.	
	Virex II/256	70627-24	JohnsonDiversity, Inc.	
	Biosentry 904	71654-19	DuPont Chemical Solutions	Alkyl dimethyl benzyl ammonium chloride

EPA-APPROVED PESTICIDES (DISINFECTANTS) FOR USE AGAINST HIGHLY PATHOGENIC DISEASES				
DISEASE	PRODUCT	EPA REG. NUMBER	MANUFACTURER NAME AND CONTACT[1]	ACTIVE INGREDIENT(S)
				Didecyl dimethyl ammonium chloride Bis(tributyltin) oxide
	Process NPD	1043-90	Steris Corporation	Alkyl dimethyl benzyl ammonium chloride Didecyl dimethyl ammonium chloride Octyl decyl dimethyl ammonium chloride Dioctyl dimethyl ammonium chloride
	Bardac 205M-7.5B	6836-70	Lonza, Inc.	
	Lonza Formulation S-21	6836-75	Lonza, Inc.	
	Lonza Formulation S-18	6836-77	Lonza, Inc.	
	Lonza Formulation R-82	6836-78	Lonza, Inc.	
	Lonza Formulation S-18F	6836-136	Lonza, Inc.	
	Lonza Formulation R-82F	6836-139	Lonza, Inc.	
	Lonza Formulation S-21F	6836-140	Lonza, Inc.	
	Lonza Formulation DC-103	6836-152	Lonza, Inc.	
	Bardac 205M-50	6836-233	Lonza, Inc.	
	Bardac 205M-10	6836-266	Lonza, Inc.	
	Bardac 205M-1.30	6836-277	Lonza, Inc.	
	Bardac (R) 205M-14.08	6836-278	Lonza, Inc.	
	Bardac 205M RTU	6836-289	Lonza, Inc.	
	Bardac 205M-2.6	6836-302	Lonza, Inc.	
	Bardac 205M-5.2	6836-303	Lonza, Inc.	
	Bardac 205M-23	6836-305	Lonza, Inc.	
	Maquat MQ615-AS	10324-67	Mason Chemical Co.	
	Maquat 615-HD	10324-72	Mason Chemical Co.	
	Maquat 5.5-M	10324-80	Mason Chemical Co.	
	Maquat 7.5-M	10324-81	Mason Chemical Co.	
	Maquat 86-M	10324-85	Mason Chemical Co.	
	Maquat 750-M	10324-115	Mason Chemical Co.	

EPA-APPROVED PESTICIDES (DISINFECTANTS) FOR USE AGAINST HIGHLY PATHOGENIC DISEASES				
DISEASE	PRODUCT	EPA REG. NUMBER	MANUFACTURER NAME AND CONTACT[1]	ACTIVE INGREDIENT(S)
	Maquat 710-M	10324-117	Mason Chemical Co.	
	Maquat A	10324-131	Mason Chemical Co.	
	KP 3510	65072-6	Chemstation International	
	Quick Control	66243-3	Clean Control Corp.	
	Microban QGC	70263-6	Microban Systems Inc.	
	Microban Professional Strength Multi-Purpose Antibacterial Cleaner	70263-8	Microban Systems Inc.	
	DC & R Disinfectant	134-65	HACCO of Neogen Corp.	Alkyl dimethyl benzyl ammonium chloride 2-(Hydroxymethyl)-2-nitro-1,3-propanediol Formaldehyde
	Fort Dodge Nolvasan Solution	1117-30	Fort Dodge Animal Health Division of Wyeth	Chlorhexidine diacetate
	Nolvasan S	1117-48	Fort Dodge Animal Health Division of Wyeth	
	Ucarsan Sanitizer 420	464-689	The Dow Chemical Company	Glutaral
	Ucarsan Sanitizer 4128	464-696	The Dow Chemical Company	
	Mikroklene	1677-22	Ecolab, Inc.	Phosphoric acid Butoxypolypropoxypolyethoxyethanol - iodine complex
	Mikroklene DF	1677-58	Ecolab, Inc.	
	Oxonia Active	1677-129	Ecolab, Inc.	Peroxyacetic acid Hydrogen peroxide
	Oxysept LDI	1677-203	Ecolab, Inc.	
	Virkon S	71654-6	DuPont Chemical Solutions Enterprise	Sodium chloride Potassium peroxymonosulfate
	Klor-Kleen	71847-2	Medentech Ltd.	Sodium dichloro-s-

EPA-APPROVED PESTICIDES (DISINFECTANTS) FOR USE AGAINST HIGHLY PATHOGENIC DISEASES				
DISEASE	PRODUCT	EPA REG. NUMBER	MANUFACTURER NAME AND CONTACT[1]	ACTIVE INGREDIENT(S)
				triazinetrione
New World Screwworm ("Screwworm")				
	Champion Insecticide Spray	498-188	Chase Products Co.	Permethrin
	Black Jack Multipurpose 0.5% Insecticide	8848-73	Safeguard Chemical Corp.	
	Sunbugger Flea & Mite Spray	11474-95	Sungro Chemicals, Inc.	
	CT Residual Spray	47000-100	Chem-Tech LTD	
	Permethrin Insecticide Spray	61483-71	KMG-Bernuth, Inc.	
	Permanone Multi-Use Insecticide Spray	73049-301	Valent Biosciences Corp.	
	Co-Ral Coumaphos Flowable Insecticide	11556-98	Bayer Healthcare, LLC	Coumaphos
	Co-Ral Fly and Tick Spray	11556-115	Bayer Healthcare, LLC	
Nipah Virus Disease				
	none registered	--	--	--
Peste Des Petits Ruminants				
	none registered	--	--	--
Rabbit Calicivirus Disease				
	none registered	--	--	--
Rift Valley Fever				
	none registered	--	--	--
Rinderpest				
	none registered	--	--	--
Sheep and Goat Pox				
	none registered	--	--	--
Spring Viremia of Carp				
	none registered	--	--	--
Swine Vesicular Disease				
	none registered	--	--	--
Trypanosomosis				
	none registered	--	--	--
Theileriosis				

EPA-APPROVED PESTICIDES (DISINFECTANTS) FOR USE AGAINST HIGHLY PATHOGENIC DISEASES				
DISEASE	PRODUCT	EPA REG. NUMBER	MANUFACTURER NAME AND CONTACT[1]	ACTIVE INGREDIENT(S)
	none registered	--	--	--
Venezuelan Equine Encephalomyelitis				
	none registered	--	--	--
Vesicular Exanthema of Swine				
	Alcide Brand LD 10:1.1 Base	1677-217	Ecolab, Inc.	Sodium Chlorite
	Virkon S	71654-6	DuPont Chemical Solutions Enterprise	Sodium Chloride Potassium Peroxymonosulfate
	Klor-Kleen	71847-2	Medentech, Ltd.	Sodium Dichloro-s-triazinetrione
Vesicular Stomatitis				
	Alcide Exspor 4:1:1 - BASE	1677-216	Ecolab, Inc.	Sodium chlorite
	Alcide Brand LD 10:1.1 BASE	1677-217	Ecolab, Inc.	Sodium chlorite
	D-125	61178-1	Microgen, Inc.	Alkyl dimethyl benzyl ammonium chloride Alkyl dimethyl ethylbenzyl ammonium chloride
	Virkon S	71654-6	DuPont Chemical Solutions Enterprise	Sodium chloride Potassium peroxymonosulfate
	Bio-Phene Liquid Disinfectant	71654-17	DuPont Chemical Solutions Enterprise	2-Benzyl-4-chlorophenel 4-ter-Amylphenol o-Phenylphenol
	Biosentry 904	71654-19	DuPont Chemical Solutions Enterprise	Alkyl dimethyl benzyl ammonium chloride Bis(tributyltin) oxide Didecyl dimethyl ammonium chloride Alkyl dimethyl benzyl ammonium chloride
Wesselsbron Disease				
	none registered	--	--	--

6

MASS CULLING OF ANIMALS FOLLOWING OUTBREAKS OF EXOTIC DISEASES

INTRODUCTION

Foreign animal diseases pose a considerable threat to the livestock and poultry in the United States. Outbreaks of these diseases, often termed exotic or transboundary diseases, may threaten agricultural production, incur significant economic losses, and threaten human health. Control strategies to eradicate outbreaks of exotic diseases may include vaccination, mass surveillance with culling of infected animals, control of insect vectors, quarantine and control of animal movement, biosecurity, and other measures.

Because these methods alone may not stop outbreaks of some exotic diseases, mass culling of susceptible animals may be necessary to control and eradicate disease. In addition to culling infected premises, killing may be necessary of animals in contiguous premises, or those at risk of becoming infected, according to the circumstances unique to each outbreak.

In addition to economic, epidemiologic, and political considerations of animal and human health, decisions to implement mass culling of infected or potentially infected animals, must take into account social and ethical concerns.

In this chapter, we describe general principles of mass culling and methods of killing animals to eradicate foreign animal diseases in agricultural settings.

Follow the guidelines below when performing Mass Culling:

Plan.

Plan ahead to ensure that all of the following are considered: procedures to be used, equipment and materials, staff numbers and their skills, safety and well-being of personnel, carcass disposal, measures of infection control and biosecurity, public relations, and related operations.

Kill animals on the affected site.

Animals or poultry in the diseased or at-risk population should be killed on-site where practical. This reduces the risk of the possible spread of the infectious agent into the environment. The handling and movement of animals should be minimized.

Maintain biosecurity.
The culling operation must take all reasonable steps to maintain biosecurity, including the preferential use of non-invasive methods of killing where necessary, personal protective equipment, and disinfectants. Infected animals should be killed first, followed by contact animals and then the remaining animals.

Kill animals as quickly as possible.
Following the decision to kill animals during an outbreak, animals should be killed as quickly as possible; the interval from diagnosis to culling is a key indicator of the efficiency of the disease-eradication operation. Normal husbandry should be maintained until the animals are killed.

Kill animals humanely.
Killing animals on affected premises should be practical, efficient, and humane. It should induce loss of consciousness and death without causing pain, distress, anxiety, or apprehension. Young animals should be killed before older ones.

Employ skilled operators.
All personnel involved in killing animals should be trained, skilled, and competent to carry out the cull, in order to ensure humane killing and their own safety.

Ensure the health and safety of personnel.
Protect personnel from physical hazards during culling, e.g., from the animals, the environment, zoonotic agents, and the methods they use to kill animals. Treat personnel exposed to zoonotic infections with prophylactic medications, including antibiotics and vaccines as required. Provide adequate rest and emotional support to operators, and monitor their health after the completion of the culling operation.

Dispose carcasses safely in a disease-specific manner that meets environmental concerns.
...For information, see the following chapter, Carcass Management.

MASS CULLING METHODS

All euthanasia methods should induce loss of consciousness followed by cardiac and/or respiratory arrest that leads to complete loss of brain function. The underlying mechanisms include hypoxia, depression of neurons necessary for life function, and physical disruption of the brain. We list below the main methods of euthanasia for eradicating outbreaks of foreign animal diseases, each with its specific requirements, advantages and disadvantages.

1. Free bullet

Method:

Free bullet shooting is appropriate for cattle, sheep, goats, horses and pigs. It is recommended for animals that are difficult to handle or restrain. The ideal point of shooting varies between species. Its outcome depends on the degree of brain damage caused by the bullet. Ideally the bullet should enter the cranium and destroy the brainstem which controls breathing and the cardiovascular system, thus killing the animal. Insufficient brain damage may lead to suffering and possible recovery.

There are specific landmarks for the optimal target area in each of the species. It is important to adhere to these landmarks in order to avoid incomplete killing and animal suffering, and to minimize risks to human safety.

Equipment:

Skilled and licensed operators should use appropriate firearms and ammunition to kill animals outdoors on soft ground, in order to avoid ricochets. Rifles, shotguns and handguns, including single-shot humane killers, may be used for this method. There are specific requirements for cartridges, calibre and type of bullet for each animal species.

Advantages:

Free bullet shooting is an excellent method for animals agitated, or those animals that cannot be restrained, e.g., those in open fields.

Disadvantages:

The free bullet method poses considerable risks to human safety. Also, it may not kill the animals, may cause leakage of body fluids which may present a biosecurity risk, and may preclude evaluation of the brain that has been damaged by the shooting.

2. Penetrating captive bolt

Method:

Captive bolt guns, powered by gunpowder or compressed air, may be used to kill cattle, horses, pigs, poultry, and sheep. The impact of the bolt causes concussion and trauma to the cerebral hemisphere and brainstem. The captive bolt should be aimed on the skull, perpendicular to the frontal bone, in a position to penetrate the cortex and mid-brain of the animal. Although the physical damage to the brain caused by penetration of the bolt may result in death, pithing or bleeding out should be performed as soon as possible after the shot to ensure the death of the animal. Adequate restraint is crucial to ensure accurate delivery of the captive bolt. Sedation, therefore, may be desirable to improve accuracy and reduce animal stress induced by this procedure.

Equipment:

Different sizes of captive bolt guns are used for different species.

Advantages:

This method is safer for operators than the free bullet method, and reduces the need to move animals.

Disadvantages:

Captive bolts may not penetrate the brains of older pigs because of their thick skulls. Misfiring and inaccurate placements of the captive bolt gun may compromise animal welfare. The captive bolt guns must be maintained, cleaned, and several must be used in order to reduce overheating. Poor gun maintenance and misfiring, and inaccurate gun positioning and orientation may result in poor animal welfare. Moreover, destruction of brain tissue may preclude diagnosis of some diseases.

3. Electrocution

Method:

Electrocution applied by alternating electrical current causes an immediate loss of consciousness through depression of neuronal activity of the brain, or ventricular fibrillation. Signs of effective electrical stunning are extended limbs, opisthotonos, downward rotation of the eyeballs, and tonic spasm changing to clonic spasm, with eventual muscle flaccidity.

The electrical current may be applied in a number of ways that vary with the situation, species, and age of the animal. A single application of current is appropriate for small farm animals such as calves, pigs, sheep, and goats -

electrodes are applied head to back or head to body, positions that span both brain and heart.

Killing large animals by electrocution requires two steps: a) electrical stunning, in which alternating electrical current is applied to the head to induce unconsciousness, followed by b) electrical current applied across the chest to stop the heart. This method requires adequate restraint of the animals, considerable operator training and skill, and safety precautions.

In poultry, a single application of current to the head induces unconsciousness, but this must be followed by bleeding, cervical dislocation, or decapitation. Another method to kill poultry by electrocution involves an electrified water bath stunner. Shackled poultry are drawn through a water bath that has sufficient electrical current. The length of the water bath should be enough to ensure that birds are subjected to the charge for a minimum of 10 seconds of current application.

Equipment:
An electrical supply and electrodes to delivery electrical current are required. Portable water bath stunners are available for killing poultry.

Advantages:
When administered properly, electrocution provides a method of euthanasia without any exposure of tissues or body fluids.

Disadvantages:
Electrocution euthanasia requires considerable operator knowledge. Misplacement of electrodes can cause incomplete stunning, severe pain, and danger to humans in the area. Restraint is essential, especially for the two-stage method. Applying to individual animals is physically demanding and operator fatigue can occur. Electro-immobilization can hide any signs of consciousness, so it is important to monitor to ensure death has occurred. Electrical stunning is not recommended for piglets, lambs, and kids, as they require prolonged current application across the heart.

4. Gases

Method:
Killing with gas involves exposing animals to a mixture of gas that induces unconsciousness and subsequently death by causing hypoxia. This method is most useful for poultry, piglets, and neonatal sheep and goats. The following gases and gas combinations can be used:

Carbon dioxide at high concentrations (80-90%) will induce unconsciousness within 30 seconds. However, this concentration causes irritation or respiratory membranes and distress for the animals, with hyperventilation and agitation.

Carbon dioxide at 30% is not irritating and if mixed with an inert gas, such as argon or nitrogen, will kill animals in 7 minutes.

Inert gases such as xenon, krypton or argon, all have anesthetic properties and can be used to fill a chamber into which animals are introduced.

Carbon monoxide (CO) combines with hemoglobin to cause anoxia. A 1% concentration of CO is enough to be lethal. CO can be obtained commercially. If used from a combustion engine, the gas has to be filtered to remove impurities because these impurities will cause respiratory distress.

Gas may be introduced into a chamber and then the animals placed in the chamber, or the gas may be introduced into the animal housing (i.e., poultry house). The chamber of the poultry house should be sealed until all animals are dead.

Equipment:
Equipment varies according to the situation but always entails a suitable chamber, or if an entire house is gassed, means of sealing the house. Gas should be preferably in a compressed form.

Advantages:
Using gases is non-invasive, and does not require restraining the animals and so minimizes stress. There is no tissue or blood exposure.

Disadvantages:
Most of the gases have some hazardous aspects for humans. Carbon monoxide is especially dangerous. There must be excellent ventilation systems to disseminate the gas after administration. Some of the gases and gas combinations can be expensive or difficult to acquire.

5. Injectable Chemicals

***Method*:**
Chemicals injected intravenously to the animals cause death by depressing their central nervous system. Barbituric acid derivatives are the drugs most commonly used for euthanasia of small and large animals. Barbiturates depress the central

nervous system in descending order, beginning with the cerebral cortex, with loss of consciousness progressing to deep anesthesia. With an overdose, deep anesthesia progresses to apnea, by depressing the respiratory center, which is followed by cardiac arrest. All barbituric acid derivatives used for anesthesia are acceptable for euthanasia when administered intravenously. This method is suitable for all species, including large animals, but is most useful when only a few animals are to be killed. Barbiturates may be the method of choice for killing pregnant animals, because they cross the placental barrier.

Equipment:
Animals must be restrained adequately to permit intravenous administration of the euthanasia solutions. The other pieces of "equipment" required are syringes, needles, and the drug to be injected.

Advantages:
Intravenous injection of euthanasia drugs is a time-tested method for rapid and humane of killing animals.

Disadvantages:
This technique must be done by trained personnel. Because the drugs used for intravenous euthanasia are classified as controlled substances, their distribution and use limit their use to persons with, or those directly under, persons licensed to se them.

6. Cervical Dislocation

Method:
Either stretching the neck to cause cervical dislocation or mechanically crushing the neck by pliers or burdizzos will result in death from asphyxiation or cerebral anoxia.

Equipment:
No equipment is required for manual cervical dislocation. It may be useful to have a barrel or box, in which to put the birds immediately after this procedure, as there may be considerable reflex movements for several seconds up to a minute. When crushing necks, pliers or burdizzos are required.

Advantages:
Cervical dislocation is the least expensive method of euthanatizing chickens, and works well for small groups of birds.

Disadvantages:
Death by cerebral anoxia takes several seconds, during which poultry may experience pain or distress. This euthanasia method may be very time consuming when used in a large poultry house. It is unknown whether cervical dislocation causes immediate unconsciousness.

CONCLUSION

This paper is a brief overview of the major methods of euthanasia of animals during outbreaks of foreign animal diseases. Euthanasia of large numbers of animals to eradicate such outbreaks requires consideration of economic, social, ethical, and political issues, animal welfare, human safety, and biosecurity. Understanding the methods of euthanasia most appropriate for the animal species and the agents that infect them during foreign animal disease outbreaks, will facilitate preparedness for, response to, and eradication of such diseases Additional information on this topic is available in the excellent references listed below.

REFERENCES

1. Opinion of the Scientific Panel on Animal Health and welfare on a request from the Commission related to welfare aspects of the main systems of stunning and killing the main commercial species of animals, 2004. *The EFSA Journal* 45:1-29.nn
2. GALVIN, J.W., BLOKHUIS, H., CHIMBOMBI, M.C., JONG, D. and WOTTON, S. 2005. Killing of animals for disease control purposes. *Rev. sci. tech. Off. Int. epiz.*, 24(2):711-722, 2005.
3. AVMA Guidelines on Euthanasia. JAVMA News. JAVMA. September 15, 2007. http://www.avma.org/issues/animal_welfare/euthanasia.pdf .
4. Guidelines for the Killing of animals for Disease Control Purposes. Appendix 3.7.6. Article 3.7.6.1. In Terrestrial Animal Health Code. 2007. http://www.oie.int/eng/normes/mcode/en_chapitre_3.7.6.htm

Moshe Shalev, MSc, VMD, DACLAM, Department of Homeland Security, Plum Island Animal Disease Center, Greenport NY 11944-0848, mshalev@gmail.com

- and-

Eoin Ryan, BVSc, IAH Pirbright Laboratory, Surrey, UK, eoin.ryan@bbsrc.ac.uk

- and-

Corrie Brown, DVM, PhD, College of Veterinary Medicine, University of Georgia, Athens, GA, 30602-7388, corbrown@vet.uga.edu

7

CARCASS MANAGEMENT

Carcass management is often a necessary part of animal disease or mortality management activities. Carcasses may be generated by direct disease impacts on livestock (animals dying) or by the regulatory response to a Foreign Animal Disease (FAD) incursion.

The objective of all carcass management activities, whether for routine disease management or FAD events, is to dispose of the carcasses in a cost-efficient manner that protects animal, human, and environmental health. Carcass disposal typically involves multiple stakeholders that are both regulatory and non-regulatory. In recent years this topic has become increasingly complex and controversial and is now an issue that combines science, public policy, and public relations.

There is no one-size-fits-all carcass management strategy. In some cases, even a single carcass may pose a significant disposal challenge, as is the case for a cow affected by bovine spongiform encephalopathy. In other cases the disposal of relatively large numbers or volume of carcasses and associated materials can be readily accomplished with systems currently in place. Many factors should be considered when preparing a carcass management plan. The following reflects some, but not all of the factors to be considered:

- The number of carcasses and geographic distribution
- The amount of associated materials (litter, manure, milk, eggs, etc.)
- The anticipated duration of the event and the rate of carcass generation
- Regional agricultural, demographic, and geographic patterns
- Weather conditions – short- and long-term forecasts
- The availability of different carcass disposal methods
- The capability of the disposal method to meet biosecurity requirements
- The availability of resources to support disposal operations
- The immediate and long-term costs of different carcass disposal methods
- Public and political concern, awareness, and acceptance
- Media interest or potential interest
- Public health risks that may result from disposal activities
- Short- and long-term economic consequences of disposal methods under consideration
- Environmental and other regulatory concerns (local, state, and federal)

This chapter is written specifically to address carcass management issues for an FAD event. In these cases, in addition to the factors listed above, there are other considerations that are disease-specific, including:

- Mode(s) of transmission (airborne, vectors, etc.)
- The transmissibility of the disease (low or high)
- The zoonotic potential (perceived or real)
- The resistance of the agent to inactivation
- The persistence of the agent in carcasses and in the environment
- Trade and economic consequences of the FAD involved

For FAD events, an early decision, with respect to local/on-farm disposal or regionalization of disposal, will typically be required. From a biosecurity risk-management perspective, any disposal method should be as close to the FAD location as possible. However, many of the disposal methods discussed below are not mobile or are not available at a specific location or in a region. Therefore, a regionalized approach may become necessary.

Whenever possible it is best to implement a single, appropriate disposal method for the entire FAD control operation. This facilitates the development and use of standard operating procedures. Often, however, in cases involving mass numbers of mammalian species, multiple methods may be needed in order to address the rapid escalation in the number of carcasses, widespread or uneven geographic distribution or the local/regional availability of transportation and disposal resources.

There are seven categories of disposal methods discussed in this chapter, some with a number of variations. These are: burial, landfill, incineration/combustion, rendering, composting, alkaline hydrolysis, and slaughter. Each method has its own disease and non-disease specific advantages and disadvantages. Each is covered below.

1 Burial

a) Definition: Internment of one to thousands of carcasses in an undeveloped or semi-developed excavation.
b) Treatment to inactivate pathogens: None are actively used but long term passive inactivation occurs for most pathogens.
c) Types: On-farm burial, off-farm burial (frequently for carcasses from more than one premises), simple trench-type excavation or engineered burial with limited environmental controls.
d) Advantages: Initial cost is low; small to moderately large numbers/volume of carcasses can be disposed of over a short time

period, typically utilizing locally available equipment. On-farm or local burial minimizes biosecurity risks by limiting the movement of the carcasses.

e) Disadvantages: Environmental factors such as high water table, porous soils, bedrock, and the proximity to sensitive waters, habitats or groundwater sources all limit the utility and acceptance of burial by many regulatory agencies and members of the general public. Additionally, landowners (who may not be the animal owners) may be unwilling to accept this option. Burial may also face significant regulatory hurdles and delays and potentially expensive remediation costs after the FAD event. Burial is technically difficult to manage in an environmentally sound manner for a large number or volume of carcasses that accumulate rapidly.

f) Best use: Disposal of a small to moderately large number or volume of carcasses in a location that does not pose significant risk to animal, human, or environmental health.

g) Not recommended when: Burial is generally considered unacceptable if the carcasses poses any significant risk to animal, human, or environmental health due to either the proximity of sensitive targets or the presence of toxins or pathogens that persist in the environment.

h) Other: Post-burial remediation may be required to address ground water contamination or other concerns.

2 Landfill

a) Definition: Internment of carcasses in a permitted and engineered site that has an impermeable liner, appropriate gas capture mechanisms, and a leachate management system (LMS). A LMS is a drainage network within a landfill liner that collects the liquids generated by the decomposing waste, which are then moved on to an onsite treatment system.

b) Treatment to inactivate pathogens: No active treatment methods are used, but longer term passive inactivation does take place for most pathogens due to exposure to the harsh biochemical conditions found in the landfill environment.

c) Types: Typically this entails an existing municipal solid waste facility or purpose-built disposal sites for carcass disposal with many of the characteristics of modern landfills.

d) Advantages: The initial cost is low to moderate, and small to moderately large numbers of carcasses can be disposed of over a short time period. Landfills may be locally available and this can minimize biosecurity risks by keeping carcasses within the region of the FAD event. It is possible to establish good controls over disposal activities at the site. Historically, landfills have been considered an acceptable form of carcass disposal.

e) Disadvantages: Transportation of carcasses is required, increasing the logistical complexity of the disposal operation. Landfills are owned by public or private entities, and whether publicly or privately held, many landfills may refuse to accept carcasses that are generated by a disease control operation. At a minimum, special arrangements and approvals have to be made before a landfill will accept carcasses. These arrangements can include additional fees exceeding the normal tonnage fees. One of the critical considerations made by a landfill is the amount of leachate generated by a large numbers of carcasses. The volume of liquid may exceed the capabilities of the existing LMS in the landfill. Finally, there may be local opposition to the transportation of carcasses, odor, and nuisance (e.g., extended working hours). For purpose-built facilities constructed for the FAD event there will be a lag-time for design and construction. Additionally, weather may impact the cost, and local opposition is not unusual. For many pathogens, pest control is a critical ownership issue.

f) Best use of landfill: Disposal of a small to moderately large number or volume of carcasses.

g) Landfill not recommended: For carcasses that are highly contaminated with agents or chemicals that persist in the environment for long periods of time and the numbers of carcasses that will exceed the capacity of the leachate handling system and the stability of the site.

h) Other: Specific pest control measures, based on the types of carcasses and the pathogen, should be implemented.

Many landfills send some or all of the leachate generated at the site to waste water treatment plants. These plants process the leachate along with all the other waste water and the solids are typically used for local land application. Any perceived or real risk to this process can result in the waste water treatment plant refusing to accept the landfill leachate. This potential inability to dispose of leachate is a major concern for landfills and can lead to municipalities declining to accept additional carcasses.

3 Incineration/Combustion

a) Definition: Incineration encompasses many methods of disposal by thermal combustion of carcasses to ash. The methods range from crude to highly controlled.

b) Treatment to inactivate pathogens: Disposal by complete incineration (no unburned portions remaining) is an active process that will reliably and consistently destroy all FAD agents except the transmissible spongiform encephalopathies.
c) Types: Open air pyres, air curtain destructors (ACDs), and fixed animal cremation, medical waste or dedicated animal incinerators with afterburner systems.
d) Advantages: Pathogen destruction is excellent. Pyre and ACD can handle a small to moderately large number or volume of carcasses and can often be done on-farm. Fixed incinerators of all types offer controlled high-temperature destruction and minimal ash management problems.
e) Disadvantages: Pyres and ACDs require large fuel inputs to achieve complete incineration of the carcasses (particularly for cattle). Additional permits from state and local regulatory authorities are typically needed. Both pyres and ACDs are sensitive to weather conditions and carcass types. The combined mass of carcasses and fuel for a successful pyre or ACD disposal operation results in significant ash volumes that must be managed. Additionally, the visible and graphic nature of pyres and ACDs (burning carcasses and columns of dark smoke) may attract unwanted attention and contribute to public opposition to disease eradication efforts. Most fixed incinerator units have a limited capacity, require transportation, and are costly to operate, therefore they will generally be most useful for small numbers of carcasses that need verifiable high temperature destruction.
f) Best use of incineration: Pyre or ACD: situations in which local disposal is critical (anthrax) for a small to moderate number of intact carcasses to burn. Fixed incinerator: situations where verifiable high temperature destruction is required.
g) Incineration not recommended: No method of incineration is recommended for a large number or volume of carcasses, particularly cattle which have high water content.
h) Other: Pyres for large numbers of livestock require large amounts of space and fuel, as well as significant construction and labor inputs. Additional criteria for pyres selection should include proximity to dwelling places, prevailing winds, and ground and surface water. Sites should be evaluated for the ability to handle heavy equipment and traffic.

4 Rendering

a) Definition: The process of heating of animal carcasses or parts to temperatures that result in the production of fats, proteins, and water.

b) Treatment to inactivate pathogens: Heat treatment reliably destroys all pathogens except TSEs and heat resistant spores (such as *Bacillus anthracis*).
c) Types: Continuous and batch methods.
d) Advantages: Significant mass reduction, excellent capacity and throughput exist in many facilities, potential utility value of end products (fat and proteins can be burned for energy recovery if they cannot be sold in normal channels), existing transportation and handling capacity.
e) Disadvantages: Rendering plants typically are set up for a particular product stream such as poultry, byproducts of the packing industry, or dead stock. One type of plant may not be equipped to handle materials of a different type. Additionally, rendering plants will not generally accept materials from sheep, goats, and cervid species. Rendered product generated from carcass management activities may not be marketable in the normal channels and may require the use of alternate disposal methods such as landfilling or incineration.
f) Best use of rendering: Rendering can be used for a broad range of situations in which carcasses need to be disposed of, including disease and non-disease associated mortality as well as animals destroyed because of regulatory activities.
g) Rendering not recommended: Any carcasses or materials that have or may be associated with a TSE, toxin, or any materials that may leave an undesirable byproduct.
h) Other: Rendering has two components: transportation and disposal. The fleets managed by the rendering industry are an excellent means of transporting carcasses, whether to rendering or another disposal option such as offsite composting or landfill.

5 Composting

a) Definition: The process by which products of animal origin undergo biological decomposition in the presence of oxygen.
b) Treatment to inactivate pathogens: When properly done, composting will achieve temperatures of 130°-140°F or higher for sustained periods. These temperatures sustained for 1-2 days generally inactivate viruses and most pathologic bacteria. Some exceptions include *Bacillus anthracis, Mycobacterium tuberculosis*, and the TSEs.
c) Advantages: Unlike other fixed facility methods of disposal, compost sites can be established at or in close proximity to the site where carcasses are being generated. The materials utilized can often be locally sourced and the end product can be applied as a soil amendment.

d) Disadvantages: Large scale composting requires considerable skill and knowledge on the part of the manager. Significant labor for initiation and maintenance may be necessary for large composting operations. Under some circumstances the disposal of end product by land application may be refused and alternate methods of disposal of the end product may be required.
e) Best use of composting: Composting can be used as a sole method for many animal disease and mortality events. Composting is best done at a single or limited number of sites where there is adequate oversight and management.
f) Composting not recommended: Composting is not useful for materials that have or may be associated with a TSE, toxin, or any materials that may leave an undesirable byproduct.

6 Alkaline Hydrolysis

a) Definition: A process of heat, pressure, and sodium or potassium hydroxide treatment to break down organic material to produce peptides, amino acids, and soaps. Alkaline hydrolysis units are commonly known as "tissue digesters."
b) Treatment to inactivate pathogens: Heat, pressure, and high pH all combine to result in the destruction of all pathogens, including TSEs.
c) Advantages: A highly controlled process that produces a sterile end product.
d) Disadvantages: Cost, limited capacity, and potential difficulty of disposing of end product. In North America disposal of BSE positive carcasses by alkaline hydrolysis may not be possible due to community and local refusal to accept the end product.
e) Best use of alkaline hydrolysis: Excellent method of disposal for small quantities of carcasses or tissues, including those that are TSE positive.
f) Alkaline hydrolysis not recommended: Large scale carcass disposal.
g) Other: There are only limited numbers of alkaline hydrolysis units in North America.

7 Slaughter

Unlike other methods, slaughter is essentially a euthanasia method or preprocessing step. Ultimately the carcasses will need to be disposed of in one of three ways:

1. Human consumption
2. Non-human consumption
3. Other disposal methods such as rendering

a) Definition: The killing or butchering of livestock, generally for food.

b) Treatment to inactivate pathogens: None - other than the natural processes that occur post mortem.
c) Advantages: Large plants can handle many animals and can process carcasses in such a manner that allows preparation for immediate or delayed disposal. This is especially valuable when the depopulation rate exceeds the capacity of the disposal systems. Slaughter facilities have existing hygiene and biosecurity systems and protocols can be modified to address specific situations or conditions.
d) Disadvantages: Plants may be reluctant to participate if normal business is ongoing and may be disrupted or if there is a perceived threat of negative
business impacts (even for non-human consumption).
e) Best use of slaughter: Mass euthanasia and preprocessing.
f) Slaughter not recommended: When transporting live animals poses a disease transmission risk.
g) Other: Temporary slaughter facilities can be established in the zone of active disease.

SUMMARY

There are many variables present when confronted with a FAD event. This variability dictates that carcass management is not an activity that can follow a formula. Carcass management involves many government and non-government stakeholders as well as a range of technical experts. The objective is to dispose of the carcasses in a biosecure manner that protects animal, human, and environmental health and that is cost effective. This technical and regulatory complexity requires a team approach that makes use of a number of experts.

Doris Olander, DVM, MS,
USDA-VS-APHIS, 6510 Schroeder Road, Suite 2, Madison, WI 53711,
doris.olander@aphis.usda.gov

III

DISEASES

1

AFRICAN HORSE SICKNESS

1. NAME

African horse sickness (Perdesiekte, Pestis Equorum, La Peste Equina)

2. DEFINITION

African horse sickness (AHS) is an infectious but non-contagious insect-transmitted viral disease with extremely high mortality in horses and mules. The disease is manifested by pyrexia, inappetence, and clinical signs and lesions compatible with impaired respiratory and circulatory functions. Clinical signs are characterized by edema of subcutaneous, intermuscular, and lung tissues; transudation into the body cavities; and, by hemorrhages, particularly of the serosal surfaces.

3. ETIOLOGY

African horse sickness virus (AHSV) is classified in the genus *Orbivirus* in the family *Reoviridae*. It is similar in morphology, and shares many properties with other orbiviruses such as bluetongue and equine encephalosis viruses. Virions are approximately 70nm in diameter and have an icosahedral symmetry. They contain 10 double-stranded RNA genome segments encapsulated within a double-layered capsid made up of 32 capsomeres, each comprising seven structural proteins.

Nine serotypes of AHSV are known, of which the last was isolated in 1960. Antisera which neutralize homologous virus may partially cross-neutralize heterologous serotypes.

The virus is relatively heat stable as the infectivity of citrated plasma containing AHS virus is not inactivated by heating at 55-75°C for 10 minutes. Virus derived from cell cultures in a medium containing calf serum is stable for 3 months at 4° C. The virus can be stored for at least 6 months at 4° C in saline containing 10% serum. Putrefaction does not destroy the virus; putrid blood may retain its infectivity for more than 2 years. Virus can be recovered for 12 months from washed infected erythrocytes stored at 4° C. The optimal pH for virus survival is 7.0-8.5; the virus is sensitive to acid pH values but is relatively resistant to pH changes on the alkaline side of neutrality. It is resistant to ether and other lipid solvents.

4. HOST RANGE

a. Domestic animals

African horse sickness primarily affects equids. Horses are most susceptible to the disease but mules are less so. Donkeys and zebras are very resistant, most infections being subclinical.

Dogs are the only other species that contract a highly fatal form of the disease, with infection occurring after ingestion of AHS-infected horse meat. It is doubtful that dogs play any role in the epizootiology of AHS, as *Culicoides* spp. do not readily feed on them. Camels have been reported to become inapparently infected with AHS virus, but the role, if any, of this species in the epizootiology of the disease is currently unknown.

b. Wild animals

Zebras are highly resistant to the disease and only show a mild fever following artificial infection. Although antibody to AHS has been found in sera collected from African elephants and black and white rhinoceros, these species play no role in the epizootiology of AHS.

c. Humans

Humans cannot become infected with field strains of AHS virus, either through contact with infected animals or from working in laboratories. However, it has been reported that certain neurotropic vaccine strains may cause encephalitis and retinitis in humans following transnasal infection under special circumstances.

5. EPIDEMIOLOGY

a. Transmission

The AHSV is transmitted by midges (*Culicoides* spp.). The most significant vectors seem to be *Culicoides imicola* and *C. bolitinos*. Other species, such as *C. variipennis*, which is common in many parts of the United States, and *C. brevitarsis,* which is common in Australia, should also be considered as potential vectors. The virus is transmitted biologically by midges, and these insects are most active just after sunset and at sunrise.

Biting flies (e.g. *Stomoxys, Tabanus*) may transmit AHS mechanically. Mosquitoes have also been implicated as biological vectors. The role of these insects in the epizootiology of the disease is regarded as minimal compared with that played by the *Culicoides* species.

b. Incubation period

The incubation period following natural infection is believed to be 3-9 days. In experimental cases of AHS the incubation period usually varies between 5-7 days, but it may be as short as 3 days.

c. Morbidity

In susceptible horses, morbidity and mortality ranges between 70-95%. In mules and donkeys the morbidity is lower.

d. Mortality

Mortality is about 50% in mules and 5-10% in European and Asian donkeys, but is not observed among African donkeys and zebra.

6. CLINICAL SIGNS

Four clinical forms of AHS are recognized.

The 'Dunkop' or 'Pulmonary' Form

This is the peracute form of the disease, from which recovery is exceptional. This form is characterized by very marked and rapidly progressive respiratory failure and the respiratory rate may exceed 50 breaths per minute. The animal tends to stand with its forelegs spread apart, its head extended, and the nostrils dilated. Expiration is frequently forced with the presence of abdominal heave lines. Profuse sweating is common, and paroxysmal coughing may be observed terminally often with frothy, serofibrinous fluid exuding from the nostrils. The onset of dyspnea is usually very sudden and death occurs within 30 minutes to a few hours of its appearance.

The 'Dikkop' or 'Cardiac' Form

At first the supraorbital fossae fill as the underlying adipose tissue becomes edematous and pushed forward, raising the skin well above the level of the zygomatic arch. This can later extend to the eyelids, lips, cheeks, tongue, intermandibular space, and laryngeal region. Subcutaneous edema may extend a variable distance down the neck towards the chest, often obliterating the jugular groove. However, ventral edema and edema of the lower limbs are not observed. Occasionally, signs of colic may develop. Terminally, petechial hemorrhages develop in the conjunctivae and on the ventral surface of the tongue.

The 'Mixed' Form

This form is a mixture of the pulmonary and cardiac forms. It is seen at necropsy in the majority of fatal cases of AHS in horses and mules. Initial pulmonary signs of a mild nature that do not progress are followed by edematous swellings and effusions, and death results from cardiac failure. In the majority of cases, however, the subclinical cardiac form is suddenly followed by marked dyspnea and other signs typical of the pulmonary form.

Horse Sickness Fever

This is the mildest form and is frequently overlooked in natural outbreaks. Apart from the febrile reaction, other clinical signs are rare and inconspicuous. The conjunctivae may be slightly congested, the pulse rate may be increased, and a certain degree of anorexia and depression may be present. This form of the disease is usually observed in donkeys and zebra or in immune horses infected with a heterologous serotype of AHS virus.

7. POST- MORTEM LESIONS

a. Gross

The lesions observed at necropsy depend on the clinical form of disease. In the pulmonary form the most characteristic changes are edema of the lungs or hydrothorax. In peracute cases, extensive alveolar edema and mottled hyperemia of the lungs are seen, whereas in cases with a somewhat more protracted course extensive interstitial edema is present, but hyperemia is less evident. Occasionally the lungs may appear relatively normal, but the thoracic cavity may contain as much as 8 L of fluid. Other less commonly observed lesions are periaortic and peritracheal edematous infiltration, diffuse or patchy hyperemia of the glandular fundus of the stomach, hyperemia and petechial hemorrhages in the mucosa and serosa of the small and large intestines, subcapsular hemorrhages in the spleen, and congestion of the renal cortex. Most of the lymph nodes are enlarged and edematous, especially those in the thoracic and abdominal cavities. Cardiac lesions are usually not conspicuous, but epicardial and endocardial petechial hemorrhages are sometimes evident.

In the cardiac form the prominent lesion is edema characterized by a yellow, gelatinous infiltration in the subcutaneous and intermuscular fascia primarily of the head, neck, and shoulders. Occasionally, the lesion may also involve the brisket, ventral abdomen and rump. Moderate to severe hydropericardium is a common feature, and there are extensive petechial and ecchymotic hemorrhages on the epicardium and endocardium, particularly of the left ventricle. Lung edema is not present, or is mild, and hydrothorax is rare. Gastrointestinal tract lesions are similar to those found in the pulmonary form, except that submucosal edema of the cecum, large colon and rectum is more pronounced.

In the mixed form the lesions are a combination of those found in the pulmonary and cardiac forms.

b. Key microscopic

Microscopic features of vascular adventitial edema are subtle and easily overlooked. Without a good suspicion grossly that the disease is African horse sickness, it is unlikely that the diagnosis will be made on histopathology.

8. IMMUNE RESPONSE

a. Natural infection

Animals that recover from natural infection with AHS develop a solid life-long immunity against the homologous serotype and may develop a partial immunity against heterologous serotypes. Foals produced by immune dams acquire a passive colostral immunity that may protect them for 3-6 months.

b. Immunization

The work of Alexander and du Toit resulted in the development of mouse-brain attenuated live vaccines that were used successfully for several decades. However, the adaptation of the virus to the brains of adult mice resulted in a neurotropism that occasionally caused encephalitis in horses, mules, and particularly in donkeys. An alternate and safer method of attenuation was achieved by plaque selection in Vero cell cultures. The polyvalent vaccine currently used in South Africa consists of a trivalent and a quadrivalent vaccine that are administered 3 weeks apart.

In enzootic regions, the mortality rate is modified in proportion to the immunity acquired by the equine population as a result of previous vaccination or exposure to natural infection – inactivated or recombinant vaccines may prove to be viable alternatives for the current modified live virus vaccines.

9. DIAGNOSIS

a. Field diagnosis

A field diagnosis is virtually impossible during the early febrile phase of the disease. However, a presumptive diagnosis should be possible once the characteristic clinical signs have developed. The typical necropsy findings support the presumptive clinical diagnosis.

b. Laboratory diagnosis

i. Samples (basic for this disease)

Live animals: blood collected in heparin.
Dead animals: Spleen, lung and lymph nodes. Specimens for virus isolation should be kept on ice during shipment to the laboratory.

ii. Laboratory procedures

Virus isolation and serotype identification remain the gold standard for confirmation of diagnosis.
Polymerase chain reaction can be utilized.
Group or type specific serological tests on paired acute and convalescent phase sera is useful for serosurveillance.

10. PREVENTION AND CONTROL

The introduction of equids incubating AHS is the most important means of introducing the disease into an area or country free of the disease. Zebra and African donkeys that do not develop clinical signs of AHS are particularly dangerous. Equids imported from infected countries should be quarantined in insect-proof facilities prior to export or at the point of entry. At present, there is a minimum 60-day post-arrival quarantine period under vector-protected conditions for horses brought into the U.S. from countries where the disease is endemic.

Once an outbreak of AHS is suspected, it is imperative that control measures be implemented immediately. An area around the outbreak should be declared a controlled area. The movement of all equids within, into, and out of the controlled area should be terminated and movement controls rigidly enforced. All equids should be stabled, at least from dusk till dawn, and sprayed with insect repellents and insecticides. If sufficient stabling facilities are not available, barns can be used. Even if not vector-protected, such housing reduces the risk of infection. Additionally, the rectal temperatures of all equids in the area should be taken regularly. Pyrexia generally precedes overt disease by about 3 days thus allowing the early detection of infected animals. Animals with pyrexia should be housed in vector-protected stables until the etiology of the pyrexia has been established.

Once the diagnosis has been confirmed, vaccination of all susceptible animals with the relevant AHS vaccine should be considered. This decision will be guided largely by the success of measures already taken.

11. GUIDE TO THE LITERATURE

1. COETZER, J.A.W. and GUTHRIE, A.J. 2004. African horse sickness. In: *Infectious Diseases of Livestock*, JAW Coetzer, RC Tustin, eds., Cape Town: Oxford University Press Southern Africa, pp. 1231-1246.
2. MELLOR, P.S. and HAMBLIN, C. 2004. African horse sickness. *Vet. Res*. 35:445-466.
3. OELLERMANN, R.A., ELS, H.J. and ERASMUS, B.J. 1970. Characterization of African horsesickness virus. *Arch. Gesamte Virusforsch*, 29:163-174.
4. SWANEPOEL, R., ERASMUS, B.J., WILLIAMS, B. 1992. Encephalitis and chorioretinitis associated with neurotropic African horsesickness virus infection in laboratory workers: Part III. Virological and serological investigations. *S. Afr. Med. J.* 81:458-461.

See Part IV for photos.

Alan J. Guthrie, BVSc, MMedVet, PhD, Equine Research Centre, Faculty of Veterinary Science, University of Pretoria, Onderstepoort, 0110, Republic of South Africa, alan.guthrie@up.ac.za

2

AFRICAN SWINE FEVER

1. NAME

African swine fever (Peste porcine Africaine; fiebre porcina Africana; maladie de Montgomery; hog cholera)

2. DEFINITION

African swine fever (ASF) is a highly contagious and often fatal disease of domestic swine that can occur in peracute, acute, subacute and chronic forms. Following infection of domestic swine, ASF is characterized by fever, high viremic titers, hemorrhagic lesions and, depending on the strain, high morbidity and mortality.

3. ETIOLOGY

African swine fever is caused by ASF virus (ASFV), which is the sole member of the genus *Asfivirus* within the newly designated Asfarviridae family. *Asfar* is a sigla derived from "African swine fever and related" viruses. The ASFV is an enveloped virus that is 200 nm in diameter with a distinctive virion morphology characterized by a dense 80 nm core. The core is composed of the viral genome with an icosahedral capsid and covered by an internal lipoprotein envelope. Mature virus has an external envelope derived by budding through the cellular membrane. The genome of ASFV consists of double-stranded DNA and is large and relatively complex. There is an indefinite number of distinct ASFV strains.

The ASFV is highly stable and quite resistant to heat, putrefaction and high or low pH. Heat inactivation requires a temperature of 56°C for 70 min or 60°C for 20 min. Inactivation also occurs at pH less than 3.9 or greater than 11.5 in serum-free media. Because ASFV is an enveloped virus, ether and chloroform will eliminate infectivity.

4. HOST RANGE

a. Domestic animals

Of domestic animals, the host range of ASF is limited to swine (*Sus scrofa*) with all breeds and ages of domestic swine similarly affected.

b. Wild animals

A variety of wild swine are susceptible to infection with ASFV. In Africa, ASFV infects warthogs (*Phacochoerus aethiopicus*), bushpigs (*Potamochoerus spp.*), and giant forest hogs (*Hylochoerus meinerizhageni*), all of which are susceptible to infection although clinical signs are typically inapparent. Wild and feral swine in Europe and North America are fully susceptible to ASFV infection and show clinical signs and mortality rates similar to those observed in domestic swine. One important exception is the U.S. collared peccaries (javelina), which are thought to be resistant to infection with ASFV.

The natural tick hosts of ASFV are the argasid ticks *Ornithodoros moubata* (also known as *O. porcinus porcinus* and *O. porcinus domesticus*) which are found throughout sub-Saharan Africa and typically inhabit warthog burrows. ASFV replicates to high titers in these ticks and is transmitted transstadially, transovarially and sexually (male to female). It is likely that most *Ornithodoros* species are competent ASFV vectors. *Ornithodoros marocanus* (previously known as *O. erraticus*) found in the Iberian Peninsula are known to be biological vectors of ASFV, which greatly complicated ASFV eradication efforts in that region. Additionally, *Ornithodoros* ticks found in the Caribbean and the Americas (*O. coriaceus*, *O. turicata*, *O. puertoricensis*, and *O. parkeri*) have been shown to be potential biological vectors and possibly long-term hosts of ASFV.

c. Humans

ASFV has never been shown to infect humans and thus is not considered to be a zoonotic pathogen.

5. EPIDEMIOLOGY

a. Transmission

The primary method of spread into previously ASFV-free countries is thought to be through feeding uncooked or minimally cooked garbage containing ASFV-infected pork products to domestic pigs. ASFV remains infectious for 3–6 months in uncooked products such as sausage, chorizo, filet and dry hams. Once introduced into domestic swine, ASFV is spread by direct or indirect contact. Blood and secretions from infected pigs contain abundant amounts of virus.

In sub-Saharan Africa, ASF is maintained in a sylvatic (wild-life) cycle by infecting both wild Suidae (e.g. warthogs and bushpigs) and the argasid (soft-bodied) ticks *Ornithodoros porcinus porcinus*. It is likely that ASFV may be an insect virus and that mammalian species, especially domestic swine, represent "accidental hosts."

b. Incubation period

The incubation period following inoculation by direct contact with ASFV-infected pigs varies from 5-15 days, but is less than 5 days following infection by a tick bite.

c. Morbidity

In all forms of ASF, morbidity rates will be very high among pigs in contact due to the extremely contagious nature of the virus and high levels of viral shedding particularly in secretions or excretions that contain blood.

d. Mortality

Mortality rates vary widely, from 5% to almost 100%. Mortality rates approach 100% in the early stages of an epidemic, but if the disease becomes endemic, subacute and chronic forms often become more prevalent and mortality rates decrease.

6. CLINICAL SIGNS

Disease caused by ASF can range from peracute to subclinical, chronic, or inapparent, depending on the virulence of the infecting strain, route of exposure and dose. Within the continent of Africa, acute ASF is most commonly observed. However, in recent decades the predominant form of ASF outside Africa has been the subacute or chronic form of disease.

Peracute and acute ASF are caused by highly virulent viral strains. In the peracute form of disease, few if any clinical signs will be observed and the first indication is typically death of the infected pigs. Clinical signs seen in acute ASF consist of fever (40.5°-42° C), increased pulse and respiratory rate, and early leukopenia and thrombocytopenia (48-72 hours). Vomiting, diarrhea (sometimes bloody) and eye discharges may be observed. Reddening of the skin, especially the tips of ears, tail, distal extremities, and ventral aspects of the chest and abdomen will be observed commonly in white pigs. Anorexia, listlessness, cyanosis and incoordination will be observed within 24-48 hours before death, which most commonly occurs 6-13 days post-infection (p.i.). Abortion may occur in pregnant sow. Survivors are thought to be virus carriers for life.

The subacute form of ASF is caused by moderately virulent viral strains. The symptoms are similar though less intense than those described above for acute ASF. The duration of illness is 5-30 days and mortality rates vary widely but are typically lower (e.g. 30-70%). Pigs that succumb will typically do so within 15-45 days p.i. Abortion may be seen in pregnant sows.

In the chronic form of ASF, which is caused by strains of low virulence, the clinical signs are varied and may include weight loss, irregular febrile periods, respiratory signs, necrosis in areas of skin, chronic skin ulcers, and/or arthritis. Clinical signs related to pericarditis, adhesions in the lungs, and swellings over joints may be observed. These signs develop over 2–15 months, and affected pigs have low mortality rates. Pigs suffering from chronic ASF may have recurring episodes of more acute disease, eventually leading to death.

7. POST- MORTEM LESIONS

a. Gross

The main lesions of acute ASF are hemorrhagic and are seen most consistently in the spleen, lymph nodes, kidneys and heart. Disseminated intravascular coagulation and thrombocytopenia contribute to the hemorrhagic syndrome seen with acute ASF. The spleen may be necrotic, congested, friable, and enlarged up to 3 times the normal size. Lymph nodes also are enlarged, hemorrhagic and friable, and this is especially notable in the gastrohepatic and renal lymph nodes. Kidneys will typically have petechial hemorrhages on the cortical and cut surface as well as in the renal pelvis. Perirenal edema and hydropericardium with hemorrhagic fluid is present in some cases. Other lesions that may be seen with acute ASF include congestion of the liver and gall bladder, petechial hemorrhages in the mucosa of the urinary bladder, hydrothorax and petechial hemorrhages of the pleura, lung edema, and gall bladder edema. Intense congestion may also be observed in the meninges and choroid plexus.

Lesions associated with subacute ASF are similar to those seen in acute ASF, but are considerably less severe. The spleen may be enlarged up to 1.5 times its normal size, and lymph nodes may also be enlarged though only mildly hemorrhagic. Slight petechiation may be visible on the kidneys. Chronic ASF is characterized by lesions of the respiratory tract along with fibrinous pleuritis, pleural adhesions, caseous pneumonia and enlarged lymph nodes. Non-septic fibrinous pericarditis and necrotic skin lesions are also common.

b. Key microscopic

Microscopically the lesions of acute ASF are characterized by microthrombosis and damage to the endothelial cells with accumulation of dead cells in the subendothelium. Lymphoid tissue destruction is present mainly in the T cell areas of the organs. In subacute ASF, there may be pulmonary lesions (pleuritis, pneumonia) and lymphoreticular hyperplasia.

8. IMMUNE RESPONSE

a. Natural infection

Pigs infected with virulent strains of ASF often die prior to development of a detectable humoral immune response. Pigs that do not die peracutely or acutely with ASF will typically mount an antibody response and have significant levels of ASF-specific cytotoxic T lymphocytes. Pigs that do not succumb to ASF typically demonstrate solid protection against challenge with homologous but not heterologous viral strains. Repeated exposure to viral strains can establish some level of heterologous protection although it is not complete.

b. Immunization

There is no vaccine for ASF.

9. DIAGNOSIS

a. Field diagnosis

Disease caused by highly virulent ASFV isolates is characterized by extremely high morbidity, with case fatality rates approaching 100%. Necropsy findings of characteristic lesions suggest a diagnosis of ASF. The highly contagious nature of this pathogen will result in rapid spread within a production system. In contrast to disease caused by highly virulent ASFV isolates, it is extremely difficult to clinically diagnose ASF caused by strains of moderate or low virulence; disease caused by these strains is variable and may strongly resemble that caused by a large number of common pathogens.

b. Laboratory diagnosis

i. Samples

Standard necropsy specimens of affected tissues should be analyzed, along with generous amounts of spleen, kidneys, and the distal part of the ileum. The brain must also be submitted. Additionally, the maxillary, mesenteric, gastrohepatic and renal lymph nodes should all be submitted. Whole blood should be collected in EDTA and serum should be drawn from both acute and, if available, recovered or convalescent animals.

ii. Laboratory procedures

Fluorescent antibody staining of tissues and the polymerase chain reaction are both useful to establish a specific diagnosis of ASFV. Additionally, virus isolation of ASFV is possible although it must be performed on peripheral blood mononuclear cells (PBMC) that have been freshly isolated from swine blood or on swine bone marrow. ELISA is used for serology.

10. PREVENTION AND CONTROL

Currently there is no effective vaccine or treatment for ASF. Prevention requires effective trade regulations and good biosecurity. Once established in an area, tick control becomes very important in successful eradication of the disease. Disinfection and inactivation of ASFV can be accomplished by sodium hydroxide (0.8%; 30 min); hypochlorites (2.3% chlorine; 30 min); formalin (0.3%; 30 min); ortho-phenylphenol (3%; 30 min); and by iodine compounds.

11. GUIDE TO THE LITERATURE

1. KING D.P., REID S.M., HUTCHINGS G.H., GRIERSON S.S., WILKINSON P.J., DIXON L.K., BASTOS A.D.S. and DREW T.W. 2003. Development of a TaqMan® PCR assay with internal amplification control for the detection of African swine fever virus. *J. Virol. Methods*, 107:53-61.
2. LUBISI, B.A., BASTOS, A.D., DWARKA, R.M. and VOSLOO, W. 2005. Molecular epidemiology of African swine fever in East Africa. *Arch. Virol.* 150:2439-2452.
3. KLEIBOEKER, S.B. and SCOLES, G.A. 2001. Pathogenesis of African Swine Fever Virus in *Ornithodoros* ticks. *Animal Health Research Reviews.* 2:121-128.
4. KLEIBOEKER, S.B. 2002. Swine Fever: Classical Swine Fever and African Swine Fever. *Vet. Clin. of North Am. Food Animal Pract.* 18:431-451.
5. MEBUS, C.A. 1988. African swine fever. *Adv. Virus Res.* 35:251-269.
6. PENRITH, M.L., THOMSON, G.R., BASTOS, A.D., PHIRI, O.C., LUBISI, B.A., DU PLESSIS, E.C., MACOME, F., PINTO, F., BOTHA, B. and ESTERHUYSEN, J. 2004. An investigation into natural resistance to African swine fever in domestic pigs from and endemic area in southern Africa. *Rev. Sci. Tech.* 23:965-977.

See Part IV for photos.

Steven B. Kleiboeker, DVM, PhD, DACVM, Director, Molecular Science and Technology, ViraCor laboratories, 1210 NE Windsor Drive, Lee's Summit, MO 64086, skleiboeker@viracor.com

3

AKABANE DISEASE

1. NAME

Akabane disease, Congenital arthrogryposis-hydranencephaly (AG/HE) syndrome, curly calf disease, dummy calf disease.

2. DEFINITION

The congenital arthrogryposis-hydranencephaly (AG/HE) syndrome is an infectious disease of the bovine, caprine, and ovine fetus caused by intrauterine infection with Akabane virus and, infrequently, related arboviruses. Disease occurs following transmission of the virus to a susceptible dam by biting midges or sometimes by mosquitoes. The impact on the developing fetus varies depending upon the stage of gestation at which infection occurs. Clinical presentations include abortion, stillbirth, premature birth, and the birth of live animals with a range of deformities including arthrogryposis and hydranencephaly. Signs of encephalitis may also be seen in some animals that have been infected just before or soon after birth. Adult animals are not clinically affected.

3. ETIOLOGY

The etiologic agents of congenital AG/HE syndrome are arboviruses in the *Orthobunyavirus* genus of the family *Bunyaviridae*. Akabane virus is the virus that most frequently is the cause of the AG/HE syndrome. Some of the related viruses have been incriminated following experimental infections but have been rarely if ever associated with disease following natural infection.

4. HOST RANGE

a. Domestic animals

The congenital AG/HE syndrome associated with Akabane virus has only been observed in cattle, sheep, and goats. Although antibody against these viruses has been detected in horses, no clinical evidence of fetal infection has been reported.

b. Wild animals

Infections of wild ruminants do occur and fetal damage must be considered, but this has not been described.

c. Humans

There is no evidence that humans can be infected by Akabane virus.

5. EPIDEMIOLOGY

a. Transmission

The occurrence and distribution of cases of AG/HE are determined by the insect vector that transmits Akabane and related viruses. Cases occur on a well-defined seasonal basis. These viruses are transmitted by hematophagous insects, especially small biting midges of the genus Culicoides. *Culicoides milnei* and *C. imicola* are vectors in Africa; *C. oxystoma* is a vector in Japan; and, *C. brevitarsis* is believed to be the principal vector of Akabane virus in Australia, although *C. wadei* may also play a role in some areas. . Akabane virus has also been isolated from *Anopheles funestus* mosquitoes in Kenya, and from *Aedes vexans* and *Culex tritaeniorhynchus* mosquitoes in Japan. Confirmation of biologic transmission by many of these species is lacking, but epidemiological evidence incriminates them. There is no indication that any of these teratogenic arboviruses are transmitted in any other way than by an insect vector.

b. Incubation period

The onset of viremia with Akabane virus generally occurs 1-6 days after infection and may last 4-6 days before antibodies to the virus are detected. However, the virus may persist for a considerably longer period in the developing fetus while clinical signs are not observed for months until an affected fetus is aborted or reaches term.

c. Morbidity

The morbidity rate is influenced by both the stage of gestation at which the fetus is infected and by the strain of virus.

Calves can be affected from the third month of gestation through to term. There can be a moderately high incidence of HE & AG (the products of infection at 3-4 and 5-6 months of gestation respectively). Losses of 25-30% are common in a herd where animals are infected between 3-6 months of pregnancy, but rates of up to 50% have been reported on occasion. In contrast, the incidence of encephalitis (infection near term) is extremely low (<5%).

Sheep and goats are affected over a narrower range of gestational ages (mainly 28-50 days) with highest losses following infection between 28-36 days of gestation. Small ruminants also tend to have multiple defects that are severe. The variations in virulence have been determined following experimental inoculation of sheep. Some strains of Akabane virus may induce defects in only 15% of fetuses while others have produced disease in up to 80% of infected animals.

d. Mortality

A significant proportion of Akabane virus infected animals are stillborn, especially those with HE. However, due to the high risk of obstetric complications, there is also a high perinatal mortality rate in AG cases. Most live-born affected calves, lambs, or kids die shortly after birth or must be euthanized for humane reasons. A small proportion of mildly affected calves may survive for a long time if they are in an environment where there is a low risk of death from misadventure or where they are intensively managed.

Akabane virus usually has no direct effect on the dam. However, if animals that are delivering AG/HE affected progeny are not closely supervised during parturition there can be maternal deaths (3-5%) or permanent injuries, especially when associated with the delivery of severe cases of AG.

6. CLINICAL SIGNS

a. Cattle

In a region where cattle have been exposed at all stages of gestation, the first cases of disease are usually noticed towards the end of autumn (cases of encephalitis). These animals can show a spectrum of neurological signs ranging from flaccid paralysis to exaggerated movements and excitation. During mid to late winter, cases of arthrogryposis are observed, followed by the delivery (stillborn and live) of cases of hydranencephaly in late winter and spring. Animals with mild arthrogryposis may be mobile but not have the use of a limb. The affected joints are usually fixed in flexion. More severe cases have multiple joints and limbs locked in flexion, often resulting in fatal dystocia which prevents movement by the animal if it is born alive. A full spectrum of abnormalities of the central nervous system can be observed in calves affected with porencephaly and hydranencephaly. Severe HE is common and affected animals that survive parturition exhibit 'dummy calf' behavior (dullness, slow suckling, paralysis and incoordination). They can usually stand, though some require assistance, and are usually blind, deaf and unaware of their surroundings. A high proportion will die from misadventure if not closely supervised. These cases are usually euthanized. Many of these cases will also have an extended gestation as a result of the severe damage to the fetal CNS.

Infection of adult animals usually produces no overt clinical signs, but very sporadic cases of encephalitis have occurred when newborn calves and occasionally older animals became infected with some strains of Akabane virus. Dystocia is common when congenitally deformed calves (especially severe AG cases) are born. Maternal deaths or permanent injuries and infertility may occur.

b. Sheep

In small ruminants, the same range of defects occurs but there is a greater overlap of defects due to the shorter gestation and narrower period of susceptibility. A higher proportion of lambs and kids are stillborn.

7. POST- MORTEM LESIONS

a. Gross

Cattle: Animals affected with Akabane virus may be aborted, stillborn or born alive. An individual fetus or newborn may have arthrogryposis or hydranencephaly or both syndromes. Aborted and stillborn fetuses, even those with severe changes, may initially appear to be normal unless the CNS is carefully examined. Lesions are predominantly associated with damage to the CNS. Arthrogryposis is the most readily observed change without removing the calvarium. Affected joints cannot be straightened even by application of force because of fixation of the joint, most frequently in the flexed position, due to soft tissue damage. There is often atrophy of skeletal muscle on affected limbs.

When large numbers of animals are involved in an outbreak, there will be a gradual progression of the range and severity of both AG and HE. Initially calves will have AG affecting perhaps 1 or 2 joints on a single limb, but later cases may have severe fixation of multiple joints on all limbs and perhaps spinal abnormalities as well. Some of the last cases of AG are also likely to be affected with mild lesions of HE (often with small cystic cavitations in the cerebral cortex). Similarly, the severity of HE also increases during the progression of an outbreak, from cases with porencephaly through to severe HE, with virtual absence of the cerebral hemispheres. Most calves delivered in the last 4-6 weeks of an outbreak will have severe HE. The brain stem appears to be grossly normal, even when there is complete absence of the cerebral hemispheres.

Small ruminants: Torticollis, scoliosis, and brachygnathism are more frequent in small ruminants. Hypoplasia of the lungs, thymus and spinal cord has also been observed in small ruminants. Gross CNS lesions due to Akabane tend to be symmetrical.

b. Key microscopic

Histopathology on Akabane cases with severe HE is unrewarding and of minimal diagnostic value. Large areas of brain will be absent and surrounded by tissues with relatively normal architecture.

In animals where AG is the only presenting sign, microscopic examination of the spinal cord is useful. Cases due to a teratogenic arbovirus such as Akabane will have characteristic changes, including demyelination and loss of motor neurons in the ventral horns of the spinal cord. In some instances these lesions can be quite localized. Finding such changes will assist in the exclusion of non-infectious causes of skeletal muscle abnormalities. Although there is severe atrophy of muscle often with replacement of muscle fibers with fibrous tissue, these lesions are not pathognomonic.

8. IMMUNE RESPONSE

a. Natural infection

Animals that have been infected with Akabane virus after birth produce a rapid and long-lasting antibody response to the virus. After a short (5-7 day) period of viremia, neutralizing antibodies are produced and the virus is cleared. Antibodies are detected by serological tests 14 days after infection. An important diagnostic aspect is that the fetus produces an immune response, particularly in cattle and usually in sheep and goats. When an infected fetus reaches term, antibodies will be found in pre-colostral serum and body fluids.

b. Immunization

Both inactivated and live attenuated vaccines have been used to protect against infection with Akabane virus. Live attenuated vaccines induce immunity to the full spectrum of viral proteins and are expected to prevent infection within 3 weeks of administration of a single dose of vaccine. In contrast, 2 doses of inactivated vaccine administered at an interval of 3-4 weeks are required to provide optimal immunity. A neutralizing antibody response ensues, with a high level of protection against fetal infection about 5-6 weeks after the first dose of vaccine has been given.

9. DIAGNOSIS

a. Field diagnosis

A presumptive field diagnosis of congenital AG/HE due to a teratogenic virus can be made on the basis of the clinical presentation, gross lesions and epidemiology. The sudden onset of aborted, mummified, premature or stillborn fetuses, and newborn calves with arthrogryposis and hydranencephaly should be suggestive. The dam will have had no clinical history of disease. A retrospective study would indicate that the there was potential for biting insect activity during the third to sixth months of pregnancy.

b. Laboratory diagnosis

i. Samples

The most important specimens for confirmation of infection with Akabane virus are those for serology. Fetal or precolostral sera and body fluids (pericardial, pleural or peritoneal, in order of preference) should be collected along with a serum sample from the dam. While maternal serology is not diagnostic, a negative result will conclusively exclude the agents for which testing is being conducted. Conversely, a positive result will be suggestive in areas where the agent is not usually detected. Fresh, chilled samples of placenta, fetal muscle, cerebrospinal fluid and brain and spinal cord should be collected from aborted fetuses that are less than 5 months of gestation. Portions of brain, spinal cord, affected skeletal muscle, spleen, lung, kidney, heart and lymph nodes should be preserved in 10% buffered formalin for histopathology.

If fresh specimens can be delivered to a laboratory within 24-48 hours, they should be kept chilled. If delivery will take longer, quick freeze at low temperatures (at or below -70° C) and do not allow them to thaw during transit.

ii. Laboratory procedures

The principal tests for confirmation of Akabane virus infection are serological tests for detection of antibodies. Both virus neutralization tests and enzyme-linked immunosorbent assays (ELISA) are used to demonstrate antibodies. A positive result in precolostral or fetal serum samples is diagnostic; a positive result in adult animals in an in area that is usually free of Akabane virus can be suggestive but not confirmatory.

Virus isolation in cell culture should be attempted from placenta, fetal muscle, or central nervous tissue from a fetus aborted early in gestation (usually before the end of the fourth month of pregnancy and soon after a fetus has been infected). The polymerase chain reaction (PCR) has also been used to detect viral RNA in samples from the aborted fetus and may give positive results over a longer time span than virus isolation.

10. PREVENTION AND CONTROL

Vaccination using either live attenuated or inactivated vaccine has been used in Japan and Australia to immunize animals in areas at the perimeter of the established vector area. Inactivated vaccination is also used for animals that are to be introduced into the vector area when pregnant. A limitation of the live vaccine is that it is only suitable for immunization of nonpregnant animals. Vaccines must be given well before the intended time of introduction to allow immunity to develop.

For short periods, vector control measures may be effective. These measures include the use of insect repellents and the covering or treatment of vector breeding sites. However, these measures are usually ineffective in preventing fetal infection over a period of more than a few days. The only other option for the movement of naive animals into a recognized vector region is to move nonpregnant animals or pregnant animals when vectors are not active.

11. GUIDE TO THE LITERATURE

1. HARTLEY, W. J., de SARAM, W. G., DELLA-PORTA, A. J., SNOWDON, W. A., and SHEPHERD, N. C. 1977. Pathology of congenital bovine epizootic arthrogryposis and hydranencephaly and its relationship to Akabane virus. *Aust. Vet. J.* 53:319-325.2.
2. JAGOE, S., KIRKLAND, P.D. and HARPER, P.A.W. 1993. An outbreak of Akabane virus induced abnormalities in calves following agistment in an endemic region. *Aust. Vet. J.* 70:56-58.3.
3. KIRKLAND, P.D., BARRY, R.D., HARPER, P.A.W. and ZELSKI, R.Z. 1988. The development of Akabane virus-induced congenital abnormalities in cattle. *Vet. Rec.* 122:582-586.4.
4. PARSONSON, I.M., DELLA-PORTA, A.J. and SNOWDON, W.A. 1981. Akabane virus infection in the pregnant ewe. 2. Pathology of the foetus. *Vet. Microbiol.* 6:209-224. 5.
5. ST GEORGE, T.D. and KIRKLAND, P.D. 2004. Diseases caused by Akabane and related Simbu-group viruses. In: *Infectious Diseases of Livestock*, 2nd ed., JAW Coetzer, RC Tustin, eds., Oxford University Press, Oxford, pp. 1029-1036.

See Part IV for photos.

Peter Kirkland, PhD, Head, Virology Laboratory, Elizabeth Macarthur Agricultural Institute, Menangle, NSW, Australia, peter.kirkland@agric.nsw.gov.au

4

ARTHROPOD LIVESTOCK PESTS AND DISEASE VECTORS

(Note: More detail on screwworms can be found in Chapter 40)

1. NAME, as above

2. DEFINITION

The insects and acari (ticks and mites) that feed on or otherwise adversely influence animal health, as well as the infectious and parasitic diseases many of these arthropods transmit, are some of the most important threats to the livestock industry in the United States. Where they occur, the diseases directly caused by or transmitted by arthropods limit livestock production, compromise the health of wild populations and in some cases, create a serious zoonotic health risk. The diversity of parasitic arthropods worldwide is vast; there are over 850 species of ticks alone, of which more than 80 are known to have some medical importance. Exotic arthropods previously introduced into North America have readily established sustainable populations (Table 1, below), and once established, eradication is difficult if not impossible to achieve, particularly when ample numbers of wildlife or domestic animal hosts are available. Intercepting exotic arthropods at points of entry is essential to protecting the health of domestic and wild animals in the United States.

3. ETIOLOGY

Arthropod livestock pests and disease vectors include flies, lice, ticks, and mites (Table 2, below). Of these, ticks are the most frequently intercepted introduced arthropod not only because ticks are commonly introduced, but also because they are macroscopic and thus readily identified, and because they remain attached to the host animal for several days and thus are evident at inspection. However, over 70 species of arthropods and all the major arthropod groups of concern have been intercepted at points of entry; many of these have been encountered in recent years.

4. HOST RANGE

Although some parasitic arthropods, such as mites, exhibit pronounced species specificity, the majority of insects and ticks are promiscuous feeders and will readily parasitize a variety of wild and domestic animals as well as humans. Equidae, particularly zebras, are a common host for introduction of exotic arthropods and in one review accounted for 69% of arthropod interceptions at USDA quarantine facilities. Some estimates suggest that approximately 5% of cattle legally imported into the U.S. from Mexico are turned away due to infestation with ticks, including *Rhipicephalus* (*Boophilus*) spp. Although numbers vary, this percentage places the potential number of cattle capable of introducing exotic tick species if not intercepted at approximately 40,000-50,000 head each year. Foreign arthropods also may be introduced on wildlife, such as migratory birds, whereas larval flies and other free-living stages of arthropod pests may be introduced in shipments of goods not related to animal agriculture. Finally, natural dispersal by wind currents from adjacent areas allows flies to readily invade new areas. This virtually limitless array of potential hosts and the opportunity for introduction of free-living stages of arthropod pests presents a particular challenge for effective interception efforts. Moreover, the broad host range of most arthropod pests and disease vectors, particularly the ability of many parasitic arthropods to survive on wild vertebrates, also contributes to the threat of establishment of foreign arthropod pests and disease vectors.

Screwworms provide an excellent example of an arthropod pest with a broad host range. For example, in 2000, 2 different introductions into the United States of the New World screwworm, *Cochliomyia hominivorax*, were intercepted by veterinarians. The first entered a quarantine facility in Miami, Florida on a horse from Argentina, and the second was identified by a private practitioner on a cat that had returned with its owner from a military base in Cuba. While screwworms are best known in tropical areas for the devastating livestock disease they cause, the flies, which are obligately parasitic in the larval stage, were endemic throughout the southern United States until eradicated by a USDA-coordinated program of systematic release of sterile male flies. Currently screwworms have been eradicated through North and Central America as far south as Panama, but the flies remain active in South America, creating a risk of reintroduction on any infested animals entering the United States from endemic areas.

Wild animals also harbor arthropod pests and can bring them into the United States through natural migration routes, as may occur with migratory birds, or via introduction of infested wild animals for zoological collections or commercial trade. Cattle egrets banded in the Caribbean have been recovered in the Florida Keys, documenting how readily these birds, and thus the ticks that infest them could move into the United States. Cattle egrets are of particular interest because populations in the Caribbean are known to be infested with *Amblyomma variegatum*, a vector of *Ehrlichia* (formerly *Cowdria*) *ruminantium*, the causative agent of heartwater in ruminants, and because both heartwater and *A. variegatum* are established in several locations in the Caribbean. Arthropods are also introduced on wild animals imported commercially. In one review of arthropods introduced on animals between 1961 and 1993, 14% of interceptions were from a variety of imported species of antelope. In more recent years, commercial trade in wild reptiles, particularly turtles imported as part of the exotic pet trade, has become recognized as an important route for ticks to enter the U.S., and is known to have allowed the introduction of at least 29 exotic tick species, primarily into Florida. In 1997, a breeding population of *Amblyomma marmoreum*, a tick that is also capable of transmitting *E. ruminantium*, was found established on a premises in Florida; a subsequent survey of reptile facilities in Florida found exotic tick species at 29 different locations. When considering the broad host range that most arthropods can exploit, it is not surprising that exotic arthropods have also entered the United States on infested humans. *Amblyomma hebraeum*, a vector of *E. ruminantium* and Africa tick-bite fever, a human disease, has twice been recovered from a person returning from travel to Africa.

5. EPIDEMIOLOGY

The epidemiology of arthropod pests and disease vectors varies according to the species, and sometimes the strain, considered. In general, arthropods are acquired either from the environment or via direct contact with an infested animal. All parasitic arthropods require some time to develop either on the animal or in the environment. *Psoroptes ovis*, the sheep scab mite, infests sheep and cattle worldwide although in the United States infestations most commonly are reported from beef cattle in central and western states. Transmission of mites generally follows direct contact between animals, however fomites also allow transmission. This mite reportedly can survive in the environment for several days. In outbreaks, morbidity rates are high and mortality, particularly in calves, can occur. However, asymptomatic infestations develop in as many as 30% of animals. Infested animals that are not showing any clinical signs are considered an important source of mites in outbreaks.

Screwworm myiasis is acquired when the adult female flies deposit eggs onto a superficial wound; in some cases, eggs are deposited on intact skin. The larvae then hatch from the eggs and burrow into the skin, creating a progressively larger wound as they feed on host tissues and increase in size. After 5-7 days of feeding the larvae emerge from the wound, drop to the ground, and pupate under the soil. Upon emergence from pupation, adult flies intermittently feed on serous fluids from animal wounds and mate, surviving for approximately 2-4 weeks. Morbidity varies according to the density of reproducing flies in a given area, but it is not unusual to find infection rates that approach 100% of neonates in areas where screwworm myiasis is endemic. Treatment of individual animals with a single infestation is readily achieved, but mortality is common in untreated animals, particularly very young animals with heavy infestations. Fatalities, when they occur, are often seen 1-2 weeks after initial infestation due to sepsis and toxemia.

Infestations with the louse fly *Hippobosca longipennis*, commonly referred to as the dog fly, are acquired from the environment when the winged adults emerge from pupation and locate a host, usually a carnivore. The flies feed for several days, mate on the host, and the females then deposit mature larvae into natural crevices in the environment. Female flies then return to the host to continue to feed and mate, intermittently leaving to deposit additional larvae in the environment. Individuals may survive and infest an animal for several months. The timing of development through the pupal stage to the adult fly is variable; depending on environmental conditions, pupation may be completed in a matter of weeks or may require several months. Pupal stages of *H. longipennis*, once established, can survive in the environment even after infested hosts are treated for adult flies. Thus, repeated, sustained treatment of infested hosts and efforts to clean the environment, where possible, may be necessary to eliminate the flies. When infestations occur, most co-housed animals on a premises are infested although the intensity of infestation may vary between animals. Although the flies are irritating, have a painful bite, and can induce anemia, mortality associated with infestation of *H. longipennis* is considered rare.

The exotic ticks of greatest concern for introduction include those in the genera *Ixodes, Amblyomma, Rhipicephalus*, and *Rhipicephalus (Boophilus)*, all of which are in the Family Ixodidae, the hard ticks. Infestations with hard ticks primarily are acquired from questing stages in the environment, although ticks may occasionally transfer directly between animals in close contact. Exotic tick species are of greatest concern because of the pathogens they transmit (Table 2, below). The morbidity and mortality of a given tick-borne disease varies with the disease agent considered. However, hard ticks also induce painful bite wounds at which secondary infections may develop, and the ticks feed on blood, resulting in anemia which may be severe enough to induce fatality, particularly when overwhelming infestations occur on young animals that are otherwise compromised by poor nutrition or concurrent infections. Most animals on an infested premises are likely to harbor some ticks, although certain individuals may be infested with a disproportionate number. In areas with particularly high tick populations, mortality in young animals due to tick infestations is not uncommon.

6. CLINICAL SIGNS

a. *Arthropod pests*

Psoroptes ovis, the sheep scab mite, infests both cattle and sheep and causes severe pruritus and the development of large hyperkeratotic crusts which are loosely adherent to the skin by a viscous, serous fluid. Lesions, which may include alopecia, secondary excoriations, and lichenification, often first appear dorsally on the shoulders and neck of affected animals but quickly spread to involve large areas of skin. With chronic, widespread infestations, anemia, weight loss, emaciation, and development of secondary infections may occur.

The New World screwworm fly, *Cochliomyia hominivorax*, and the Old World screwworm fly, *Chrysomya bezziana*, both infest minor cutaneous wounds such as the umbilicus of neonatal animals or breaks in the skin caused by tick bites, minor lacerations, dehorning, or branding. In early infestations larvae may be difficult to identify. However, as the larvae grow and the wound enlarges, careful visual inspection will reveal fly larvae oriented perpendicular to the skin surface. A malodorous serosanguinous discharge is often present. Larvae may burrow into subcutaneous pockets or into nasal, anal, or vaginal cavities, making detection even more difficult.

The clinical signs associated with *Hippobosca longipennis*, a louse fly of carnivores commonly found on dogs, have not been well characterized. However, this parasite is thought to cause anemia due to blood feeding, and irritation from the painful bites has been documented.

b. *Disease vectors*

The disease agents transmitted by arthropod vectors (Table 2) are more likely to cause overt clinical illness, and thus be the focus of any disease description, than are the vectors themselves. The clinical signs due to individual vector-borne disease agents are outlined in each respective chapter. However, arthropod vectors of disease, such as ticks, mosquitoes, and biting flies, also induce clinical disease in and of themselves, including anemia, pruritus, production loss, dermatitis, pyoderma, and sepsis. Ticks, mosquitoes, and other biting flies can cause blood loss that, in cases of extreme tick infestation or large numbers of blood-feeding flies emerging simultaneously to feed on livestock, may result in clinical anemia and exsanguination. At more moderate infestation levels, the feeding activity of ticks, mosquitoes, and other biting flies is likely to cause a cutaneous hypersensitivity response that results in pruritus. The irritation associated with the hematophagous behavior of these arthropods may result in production loss due to increased time devoted to grooming or, in the case of large, aggressive biting flies such as horse flies, deer flies, stable flies, or horn flies, to attempted avoidance of the vectors. Bite wounds also create a route for establishment of secondary infections which may present as dermatitis associated with localized pyoderma and can, in severe cases, proceed to sepsis. Tick bite wounds are also a common entry point for screwworm fly larvae.

7. POST- MORTEM LESIONS

Because they cause superficial cutaneous infestations, arthropod pests and the lesions they induce may be identified post-mortem the same way they are detected ante-mortem, by careful inspection of the skin lesions. In some cases, such as with screwworm myiasis, a more detailed inspection including dissection of subcutaneous tracts is more readily achieved post-mortem and thus diagnosis may be facilitated by necropsy exploration of skin wounds. Tick infestations and associated lesions are also apparent both ante-mortem and post-mortem. In cases where anemia leading to exsanguination caused by rapidly feeding hematophagous insects such as *Hippobosca longipennis*, mosquitoes, or other biting files has occurred, post-mortem lesions present may include little more than cutaneous hypersensitivity reactions due to biting flies and evidence of blood-loss anemia. In these cases, a thorough understanding of the activity levels of blood-feeding arthropods in a given area may be needed to secure a diagnosis.

8. IMMUNE RESPONSE

Although an immune response to arthropod pests does develop, and, in some cases, limits the arthropod burden in subsequent infestations, the response is generally not protective and animals remain susceptible to infestation throughout their life. Vaccines have been developed for aiding in the control of *Rhipicephalus (Boophilus)* spp. ticks in some areas of the world. Research into arthropod vaccines continues to grow, particularly for those directed against ixodid (hard) ticks and *Psoroptes ovis*. However, immunization against arthropods is not yet widely available. At present, no arthropod vaccines are marketed in the United States.

9. DIAGNOSIS

a. *Field diagnosis*

Field diagnosis of parasitism by arthropod pests or disease vectors relies on recovery and direct identification of the insects or acari either on the animals or in the environment. Some arthropods have distinctive features and can be readily identified by experienced personnel in the field, while others require examination by an entomologist experienced in morphologic identification of international parasitic arthropods. Even when a preliminary identification is achieved in the field, several representative specimens should be submitted to an expert laboratory to confirm the diagnosis.

b. Laboratory diagnosis

Accurate laboratory identification of arthropod pests and disease vectors requires examination of specimens by an entomology expert trained in features unique to a given taxa and capable of accurately consulting pertinent keys. Ideally, when unusual diagnoses are achieved, representative specimens will be deposited in a museum collection so that others may retrieve the specimens and examine them to confirm the identification.

i. Samples

Arthropods to be submitted for identification should be collected from animals or from the environment and immediately placed in 70% ethanol prior to shipping to the diagnostic laboratory. Whenever possible, several specimens in multiple stages of development should be submitted. For example, when acquiring specimens from an animal infested with numerous ticks that are suspected to be exotic, rather than just preserving 1-2 engorged female ticks, the animals should be closely inspected and both male and female ticks, as well as any immature stages present, collected and submitted.

Photographs of the gross lesions associated with an infestation can also be helpful and should be taken and submitted with the specimens if possible.

ii. Laboratory procedures

Arthropods are identified by close morphologic examination by experienced entomologists and comparison to published keys and museum specimens. When a limited number of specimens are available or when specimens are broken or damaged, precluding morphologic identification, molecular identification by PCR and sequencing of characteristic gene targets may be of use. Preservation in 70% ethanol allows both morphologic and subsequent molecular analysis of specimens.

10. PREVENTION AND CONTROL

a. Preventing introductions

With coordinated, sustained effort, arthropod pests and disease vectors can be prevented from entering the United States on infested animals. Interceptions are facilitated by the facts that most arthropod pests and disease vectors of concern are macroscopic and that the cutaneous lesions they cause are grossly apparent. However, continued vigilance is necessary when managing animals arriving from areas where these parasites are endemic, including careful inspection for the presence of ectoparasites, quarantine of adequate duration to allow development of any associated lesions, and routine prophylactic treatment of animals with insecticides and/or acaricides. International cooperation in controlling arthropod pests in areas where they are endemic also augments efforts to prevent introductions. For example, after screwworm was successfully eradicated from the southern United States, cooperative efforts were continued to extend the eradication southward through Mexico to its current line of demarcation in Panama. Although not yet successful, international collaborative efforts have also been pursued to eradicate *Amblyomma variegatum* from several Caribbean nations.

b. Responding to introductions

State and federal veterinary officials are notified immediately upon initial identification of a confirmed or suspected exotic livestock pest or disease vectors. Wildlife hosts exist for many of these arthropods, and some of the species can survive in the environment for several months; a rapid response is necessary to achieve containment. Premises on which an infestation has been identified will likely be quarantined by state or federal animal health officials. All animals present are examined for the presence of disease and observed for subsequent development of clinical disease. Care is taken to examine records and interview involved individuals in order to establish contact animals and other locations the infested animals may have visited, and thus spread the arthropods. All infested animals and premises are treated with effective insecticides or acaricides to eliminate the arthropods and then reevaluated for evidence of infestation. Several successive treatments of both animals and environment may be necessary to completely eradicate the introduced arthropods. The potential for establishment of permanent populations of almost any arthropod pest is possible and of great concern. However, in general, successful eradication is more likely to be achieved for those species of arthropods that are permanent host ectoparasites, and whose survival and spread is closely linked to that of their hosts (e.g. mites, louse flies, screwworm larvae, ticks when first introduced). For arthropods with a more free-living lifestyle, such as mosquitoes, other biting flies, and screwworm adults, populations are generally more difficult to contain and more challenging to eliminate once introduced.

c. Summary

Exotic arthropods are commonly introduced into the United States on imported animals and, less frequently, on wildlife species migrating from areas where these pests are endemic. Once introduced, many of these organisms are able to readily establish sustainable, reproducing populations, thereby posing a long-term threat to livestock, wildlife, other domestic animals, and people. Quarantine of all animals entering the United States and continued surveillance of wildlife and domestic animals for the presence of exotic arthropod pests and disease vectors is critical to prevent introduction and establishment of these organisms and the subsequent veterinary and public health consequences that would ensue. Introductions will continue to occur, requiring continued efforts on the part of federal and state animal health officials and the veterinary profession as a whole to both exclude these arthropods from the United States and to assist in sustained international efforts to control or eradicate them from endemic areas.

Table 1. Examples of exotic arthropod pests and disease vectors that were introduced to and have become established in the continental United States.

Common name	Scientific name	Approximate year of year of introduction
Stable fly	*Stomoxys calcitrans*	1781
Brown dog tick	*Rhipicephalus sanguineus*	early 1800s
Horn fly	*Haematobia irritans*	1887
Iguana tick	*Amblyomma dissimile*	late 1800s
Frog tick	*Amblyomma rotundatum*	1930s
Red fire ant	*Solenopsis invicta*	late 1930s
Face fly	*Musca autumnalis*	1952
hairy maggot blow fly	*Chrysomya rufifacies*	1980
Asian tiger mosquito	*Aedes albopictus*	1985
Honey bee mite	*Varroa destructor*	1987
Oriental latrine fly	*Chrysomya megacephala*	late 1980s

Table 2. Arthropod livestock pests and disease vectors considered at risk for introduction into the United States.

Arthropod	Disease causes/transmits
Flies	
Cochliomyia hominivorax	screwworm myiasis
Chrysomya bezziana	screwworm myiasis
Hippobosca longipennis	irritating bite wounds
Mites	
*Psoroptes ovis**	sheep scab, mange of ruminants
Ticks	
Amblyomma hebraeum	heartwater
Amblyomma variegatum	heartwater, dermatophilosis, Nairobi sheep disease
Rhipicephalus (Boophilus) annulatus	bovine babesiosis, bovine anaplasmosis, benign bovine theileriosis
Rhipicephalus (Boophilus) microplus	bovine babesiosis, bovine anaplasmosis, benign bovine theileriosis, spirochetosis of ruminants and horses
Rhipicephalus appendiculatus	East Coast fever and related pathogens, bovine babesiosis, louping ill, Nairobi sheep disease, Kinsley sheep disease
Ixodes ricinus	bovine babesiosis, bovine anaplasmosis, louping ill, tick-borne fever of ruminants

* occasionally reported from cattle in western states but not reported in sheep in the U.S. since 1970s

11. GUIDE TO THE LITERATURE

1. BRAM, R.A. and GEORGE, J.E. 2000. Introduction of nonindigenous arthropod pests of animals. *J Med Entomol,* 37:1-8.
2. GEORGE, J.E., DAVEY, R.B. and POUND, J.M. 2002. Introduced ticks and tick-borne diseases: the threat and approaches to eradication. *Vet Clin North Am Food Anim Pract*, 18:401-16.
3. MULLEN, G.R. and DURDEN L.A. 2002. *Medical and Veterinary Entomology*. Academic Press, Elsevier Science, London, 597p.
4. BURRIDGE, M.J. and SIMMONS, L.A. 2003. Exotic ticks introduced into the United States on imported reptiles from 1962 to 2001 and their potential roles in international dissemination of diseases. *Vet Parasitol,* 113:289-320.
5. JULIANO, S.A. and LOUNIBOS, L.P. 2005. Ecology of invasive mosquitoes: effects on resident species and on human health. *Ecology Letters*, 8:558-574.
6. BOWMAN, D.D. 2006. Successful and currently ongoing parasite eradication programs. *Vet Parasitol*, 139:293-307.

Susan Little, DVM, PhD, Department of Pathobiology, Oklahoma State University, Stillwater, OK, 74078-2007, susan.little@okstate.edu

5

AVIAN INFLUENZA

1. NAME

Avian influenza, fowl plague

2. DEFINITION

Avian influenza (AI) is a viral infection of birds including chickens, turkeys, guinea fowls, and other avian species. Symptoms vary in severity from asymptomatic infections to mild respiratory and reproductive diseases to an acute, highly fatal systemic disease. Avian influenza viruses are divided into 2 broad pathotypes, low (LP) and high pathogenicity (HP), based on experimental studies in chickens. For this chapter, only notifiable avian influenza (NAI) viruses, as defined by OIE, are discussed:

a) High pathogenicity or highly pathogenic notifiable avian influenza (HPNAI) viruses have an intravenous pathogenicity index (IVPI) in 6-week-old chickens greater than 1.2 or, as an alternative, cause at least 75% mortality in 4- to 8-week-old chickens infected intravenously. H5 and H7 viruses which do not have an IVPI greater than 1.2 or cause less than 75% mortality in an intravenous lethality test should be sequenced. This will determine whether multiple basic amino acids are present at the cleavage site of the hemagglutinin molecule (HA0): if the amino acid motif is similar to that observed for other HPNAI isolates, the isolate being tested should be considered as HPNAI.

b) Low pathogenicity notifiable avian influenza (LPNAI) are all influenza A viruses of H5 and H7 subtype that are not HPNAI viruses.

The terms "high pathogenicity avian influenza (HPAI)" and "high pathogenicity notifiable avian influenza (HPNAI)" are equivalent, while "low pathogenicity notifiable avian influenza (LPNAI)" is a subset within "low pathogenicity avian influenza (LPAI)." HPNAI viruses are responsible for the historically severe systemic disease termed fowl plague.

3. ETIOLOGY

Avian influenza viruses are negative-sense, single-stranded RNA viruses in the family *Orthomyxoviridae*, genus *Influenzavirus A,* and commonly termed "influenza A virus" or "Type A influenza virus." AI viruses are further categorized, based on the 2 surface glycoproteins, into 16 different hemagglutinin (H1-16) and 9 different neuraminidase (N1-9) subtypes. Each AI virus would have a specific hemagglutinin (H) and neuraminidase (N) subtype designation such as H5N1, H9N2, etc. LPAI viruses can be any of the 16H subtypes (H1-H16) while NAI viruses are only of the H5 and H7 subtypes. However, most H5 and H7 NAI viruses are LP and not HP.

Biologically, LPAI and LPNAI are identical except LPNAI viruses (H5 and H7) have the potential to mutate and become HP. Therefore, LPNAI viruses (H5 and H7) are a more serious concern for immediate control than LPAI viruses of H1-4, H6 and H8-16 subtypes. LPAI viruses can infect poultry, usually causing respiratory disease or drops in egg production. LPAI viruses also cause asymptomatic infections in a variety of aquatic birds.

4. HOST RANGE

Most avian species appear to be susceptible to at least some of the NAI viruses. A particular isolate may produce severe disease in turkeys, but not in chickens or any other avian species. Therefore, it would be impossible to generalize on the host range for NAI, for it will likely vary with the isolate. This assumption is supported by reports of farm outbreaks where only a single avian species within several species present on the farm became infected. Mammals do not appear to be involved in the epidemiology of NAI and infections with NAI are rare. However, since 1997, sporadic infections with H5N1 HPNAI viruses have occurred in humans, tigers, leopards, civets, stone martens, domestic cats and dogs, and pigs in Asia, Europe and Africa. In carnivores, the infections were linked to feeding on infected poultry or wild birds; in humans and pigs, infections were linked to close contact with infected poultry.

Since 1997, the H5N1 HPNAI has caused sporadic human infections - principally severe respiratory infections - in several Asian and African countries. During the 2003 H7N7 HPNAI outbreak in the Netherlands, some depopulation crew members and farmers developed self-limiting conjunctivitis. For other NAI viruses, human infections have been rare or non-existent. Fatalities have been limited to 170 cases with Eurasian H5N1 (1997–2006) HPNAI viruses and one case with the Dutch H7N7 (2003) HPNAI virus. Since AI viruses are Type A influenza viruses, the possibility exists that they could be involved in the development, through genetic reassortment, of new mammalian influenza virus A strains including those affecting humans.

5. EPIDEMIOLOGY

a. Transmission

Ecological data indicate various migratory waterfowl, sea birds, and shore birds are the reservoir for all NAI virus genes. In addition, epidemiological and molecular genetic evidence supports the hypothesis that these aquatic birds are generally responsible for introducing the LPNAI viruses into poultry. Once introduced into poultry, the viruses adapt to poultry and are spread from flock-to-flock or village-to-village by human endeavor such as the movement of infected birds, contaminated equipment, shoes, clothing, egg flats, feed trucks, and service crews.

Virus may readily be isolated in large quantities from the feces and respiratory secretions of infected birds. Transmission of the disease can take place through shared and contaminated drinking water. Airborne transmission may occur if birds are in close proximity and with appropriate air movement. Birds are readily infected via inoculation of virus into the conjunctival sac, nares, or the trachea. The HPNAI virus can be recovered from the yolk and albumen of eggs laid by hens at the height of the disease, but not from eggs laid by hens infected with LPNAI virus. Evidence for vertical transmission is lacking, maybe because the HPNAI virus is extremely embryo lethal. However, surface contamination of egg shells by LPNAI virus could be a theoretical means of transmission within a hatching cabinet. The hatching of eggs from an LPNAI infected flock would likely be associated with some risk unless the eggs are sanitized prior to incubation and the resulting chicks are quarantined and tested.

b. Incubation period

The incubation period for an individual bird is usually 1-7 days, depending upon the isolate, the dose of inoculum, the species, and the age of the bird. However, OIE recognizes a 21-day incubation period which takes into account the transmission dynamics of the virus within a population. For LPNAI and HPNAI, the "infectious period" – i.e. the period of time that virus is shed from infected birds - may be a more appropriate concept for control of the disease.

c. Morbidity

With LPNAI, morbidity is variable in birds. The prognosis for chicken or turkey flocks infected with HPNAI is poor. Morbidity rates may be near 100% within 2-12 days after the first signs of illness.

d. Mortality

Mortality in LPNAI varies, but is usually low unless secondary bacterial or viral pathogens are present, or if the birds are housed under poor management conditions. Mortality rates with HPNAI may be near 100% within 2-12 days after the first signs of illness.

6. CLINICAL SIGNS

With LPNAI viruses, the infections may be asymptomatic in poultry (including domestic ducks) with detection only after serological examination. Laying poultry may exhibit 5-30% decrease in egg production with the presence of thin-shelled and misshapen eggs, but production may return to near normal levels following recovery. Turkeys are generally more severely affected by reproductive disease than chickens. With LPNAI, decreases in feed and water consumption are common. However, the most frequent clinical syndrome, as seen in domestic ducks and geese, is respiratory disease with associated signs including respiratory noises (such as snicks, clicks and wheezing), swollen sinuses, matted eyelids, and nasal discharge. Birds may huddle near heaters and have a hunched posture and ruffled feathers.

With HPNAI, the most frequent signalment in chickens is sudden death, many times without clinical signs or gross lesions, and usually death in large numbers of birds. However, infections with HPNAI viruses may result in marked depression, with ruffled feathers, inappetence, decreased water consumption, cessation of egg production, and watery diarrhea 1-2 days before the increased mortality is noted. Respiratory signs are less frequent with HPNAI than LPNAI infections but can be present depending on extent of tracheal involvement. Mucus accumulation can vary. It is not unusual in caged birds, such as layers, for the disease to begin in a localized area of the house. Severely affected birds would be found in only a few cages before the disease would spread to neighboring cages. Death may occur within 24 hours of first signs of disease, frequently within 48 hours for chickens, but it can be delayed for as long as a week with some HPNAI strains or if birds have partial immunity. Some severely affected hens may occasionally recover.

In broiler chickens, the signs of HPNAI disease are frequently less obvious with severe depression, inappetence, and a marked increase in mortality being the first abnormalities observed. Neurologic signs such as torticollis and ataxia may also be seen. The disease in turkeys and other gallinaceous poultry is similar to that seen in layers, but it lasts 2 or 3 days longer and neurological signs may be more frequent.

Infection of domestic ducks and geese by HPNAI viruses has been uncommon and many experimental studies have demonstrated resistance to infection or asymptomatic infection. However, infections and nervous signs may be common with the recent H5N1 HPAI viruses in ducks, geese and swans as well as in pigeons and some wild birds.

7. POST- MORTEM LESIONS

With LPNAI in young birds, catarrhal to purulent sinusitis and rhinitis is common, especially with swollen infraorbital sinuses in turkeys. Serous to catarrhal tracheitis is common, sometimes with air sacculitis and pneumonia, but the lower respiratory lesions are typically seen only when accompanied by secondary bacterial infections. In chicken and turkey layers, the ovary may be hemorrhagic with ova involution and degeneration. The coelomic cavity is frequently filled with yolk from ruptured ova, causing severe airsacculitis and inflammation in birds that survive for 7-10 days.

For HPNAI, birds that die with peracute disease and young birds may not have significant gross lesions other than severe congestion of the musculature and dehydration. In the acute to subacute form, and in mature birds, significant gross lesions are frequently observed. Mature chickens frequently have swollen combs, wattles, and edema surrounding the eyes. The combs are often cyanotic at the tips and may have plasma or blood vesicles on the surface with dark areas of ecchymotic hemorrhage and necrotic foci. The last eggs laid, after onset of illness, are frequently soft and easy to fracture because of decreased calcium deposition. The diarrhea begins as watery bright green and progresses to almost totally white. Edema of the head, if present, is often accompanied by edema of the neck. The conjunctivae are congested and swollen with occasional hemorrhage. The legs, between the hocks and feet, may have areas of diffuse hemorrhage and edema.

With HPNAI, the conjunctivae are severely congested – occasionally with petechiation. The trachea may appear relatively normal except that the lumen contains excessive mucous exudate. It may also be severely involved with hemorrhagic tracheitis similar to that seen with infectious laryngotracheitis. When the bird is opened, pinpoint petechial hemorrhages are frequently observed on the inside of the keel as it is bent back. Very small petechiae may cover the abdominal fat, pericardium, serosal surfaces and peritoneum. Kidneys are swollen, severely congested and may occasionally be plugged with white urate deposits in the tubules. The lungs may be deep red in color from a combination of congestion and hemorrhage and may exude edematous fluid when cut. Hemorrhages may be present on the mucosal surface and in the glands of the proventriculus, particularly at the junction with the ventriculus. The lining of the ventriculus peels easily, frequently revealing hemorrhages, erosions and ulcers.

The intestinal muscosa may have hemorrhagic areas – especially in the lymphoid foci such as the cecal tonsils or Peyer's patches in the small intestine.

In layers with HPNAI, the ovary may be hemorrhagic or degenerated with darkened areas of necrosis. The peritoneal cavity is frequently filled with yolk from ruptured ova, causing severe airsacculitis and inflammation in birds that survive.

8. IMMUNE RESPONSE

a. Natural infection

Natural infection results in development of a cellular and humoral immune response. The protective humoral immunity is primarily directed against the H protein as measured in a hemagglutinin inhibition (HI) test, but is not cross protective between different H subtypes. These antibodies are first detected between 7-10 days after infection and in all birds by 14 days. Antibodies to N protein are subtype specific and measured by neuraminidase inhibition (NI) test, but protection is inferior to humoral immunity to the H protein.

Antibodies against the internal proteins such as the nucleoprotein and matrix protein are measured by the agar gel immunodiffusion (AGID) test or several commercial ELISA tests, but such antibodies are not protective. AGID antibodies can first be detected at 5-7 days after infection in some birds and in all birds by 10 days. Such antibodies indicate infection by Type A influenza virus, irrespective of the H and N subtype of the virus, and these AGID antibodies are very useful in serological screening or monitoring for AI infection. The AGID response is similar regardless of whether it results from virus infection or from immunization by an inactivated whole AI virus vaccine.

b. Immunization

Some vaccine options are possible and all may produce good immunity. Please refer below to Prevention and Control for more details.

9. DIAGNOSIS

a. Field diagnosis

Respiratory disease cases or cases of drops in egg production should be investigated for LPNAI. The combination of typical clinical signs and gross lesions along with detection of influenza type A antigen by commercial antigen capture immunoassay gives a presumptive diagnosis. However, this presumptive diagnosis would be indistinguishable from LPAI caused by H1-4, H6 and H8-16 subtypes. Additional laboratory tests are needed to distinguish LPNAI and LPAI.

HPNAI should be suspected with any flock where sudden deaths follow severe depression, inappetence, and a drastic decline in egg production. The presence of facial edema, swollen and cyanotic combs and wattles, and petechial hemorrhages on serosal surfaces increases the likelihood that the disease is HPNAI. However, an absolute diagnosis is dependent upon the detection and identification of the causative virus. A presumptive diagnosis can be made in flocks with high mortality, typical clinical signs, and gross lesions compatible with HPNAI and detection of Type A influenza antigen in oropharyngeal or tracheal swabs on antigen capture immunoassay. Definitive diagnoses of LPNAI and HPNAI should be followed by more specific tests to identify the AI virus, especially H subtype and pathotype.

b. Laboratory diagnosis

i. Samples

Specimens sent to the laboratory should be accompanied by a history of clinical signs and gross lesions, including any information on recent additions to the flock. Diagnosis depends upon detection of AI viral nucleic acids in real-time reverse transcriptase polymerase chain reaction (RRT-PCR) from tracheal or oropharyngeal swabs; from the isolation and identification of the virus from tracheal or cloacal swabs; or, from feces or internal organs. Specimens should be collected from several birds. It is not unusual for many of the submitted specimens to fail to yield virus.

Swabs are the most convenient way to transfer AI virus from tissues or secretions of the suspect bird to brain and heart infusion broth or other cell culture maintenance medium containing high levels of antibiotics. Dry swabs should be inserted deeply to ensure obtaining ample epithelial tissue. Oropharynx, trachea, lung, spleen, cloaca, and brain should be sampled. If large numbers of dead or live birds are to be sampled, swabs from up to 5 birds can be pooled in the same tube of broth - but do not mix swabs from different sites or tissues. Blood for serum should be collected from several birds in standard serum tubes for serological examination. If the specimens can be delivered to a laboratory within 24 hours, they should be placed on ice or cold packs. If delivery will take longer, quick freeze the specimens on dry ice or liquid nitrogen, but do not allow them to thaw during transit. Freezing and storage at standard freezer temperature (-20°C) is not recommended.

ii. Laboratory procedures

The RRT-PCR test is the preferred diagnostic test for NAI and it gives an answer within 3 hours. In the U.S., swabs from oropharynx or trachea of birds are extracted of RNA and first tested using *Influenzavirus A* matrix gene primers and probes. A positive test indicates an AI virus, and the specimen is further tested with H5 and H7 primers and probes which will determine if an NAI virus is present. To determine if the virus is HPNAI or LPNAI requires either the cloning of the hemagglutinin gene from the specimen with sequencing to determine the H cleavage site, or the isolation and characterization of the virus.

For virus isolation, 9- to 11-day-old embryonating chicken eggs are inoculated with swab or tissue specimens. Avian influenza virus will usually kill embryos within 48-72 hours. If the virus isolated is identified as a Type A influenza virus through the AGID or antigen capture immunoassays, it is then tested using a battery of specific antigens to identify its serological subtype (H and N type) at a reference laboratory. The pathotype is determined by *in vivo* pathogenicity studies in chickens or by sequencing the hemagglutinin cleavage site and comparing this sequence to other LPNAI and HPNAI viruses.

Sera from infected chickens usually produce AGID positive antibody tests as early as 5 days after first detection of the virus or 3-4 days after appearance of clinical signs. AGID testing is best used for LPAI surveillance and requires use of H5 and H7 specific HI tests to identify NAI virus infections. AGID is not a reliable test for detecting AI virus infections in ducks or geese.

10. PREVENTION AND CONTROL

The best strategy for controlling NAI is eradication. This is achieved through comprehensive control strategies with 5 components: inclusion and exclusion biosecurity practices; diagnostics and surveillance; elimination of infected animals; increasing host resistance; and, educating personnel in AI control strategies. The level of incorporation and practice of these 5 components will determine whether the control strategy will be effective at eradicating AI or preventing its introduction.

Vaccines can be used as a single tool in a comprehensive control strategy by increasing host resistance to AI virus infection and decreasing environmental contamination, but additional components are essential to achieve success. The practice of accepted sanitation and biosecurity procedures in the rearing of poultry is of utmost importance. Appropriate biosecurity practices should be applied, including the control of human traffic and the use of quarantine before introduction of birds of unknown disease status into the flock. Cleaning and disinfection procedures are the same as those recommended in the chapter on velogenic Newcastle disease.

Inactivated oil-emulsion vaccines, although fairly expensive, have been demonstrated to be effective in preventing clinical signs and death, increasing resistance to infection, reducing infections, reducing shedding of AI virus from respiratory and intestinal tracts, and preventing contact transmission in a variety of birds. In addition, recombinant fowl poxvirus with H5 AI hemagglutinin gene insert has shown similar protection in chickens. However, no AI vaccine will provide absolute prevention of infections and environmental contamination: biosecurity, surveillance and other management practices must be practiced concurrently for AI control to be successful.

There are 3 drawbacks to current vaccines for controlling NAI: 1) must match the vaccine and field hemagglutinin subtype to get protection, 2) vaccines must be injected, and, 3) identifying infected animals in vaccinated populations requires special procedures.

Several new vaccine technologies have shown promise for mass low-cost immunization, but to date none are available. Detection and elimination of infected animals in a vaccinated population, so-called DIVA strategy, is critical for a successful control program. For any vaccinated population, unvaccinated sentinels can be used for virological or serological detection of infection. Alternatively, with inactivated AI vaccines, vaccinated birds can be serologically examined for infection by demonstration of antibodies to the non-structural 1 (NS1) protein or by using special vaccines that have a different neuraminidase from the field virus, and then detecting anti-neuraminidase antibodies against the field virus. For the recombinant fowl poxvirus-AI-H5 vaccine, detection of AGID antibodies is indicative of infection since the vaccine lacks AI nucleoprotein and matrix protein. Virological surveillance can be accomplished by testing the daily mortality for AI virus by antigen capture immunoassay, RRT-PCR, or by virus isolation.

Several antiviral drugs which target the matrix (e.g. amantadine hydrochloride) or neuraminidase (e.g. oseltamivir phosphate) proteins are available for use in humans. These medications are effective in reducing the severity of influenza Type A in humans. Experimental evidence indicated efficaciousness of amantadine in poultry when the drug was administered in drinking water for reducing disease losses, but drug-resistant viruses quickly emerged, negating the initial beneficial effects. Thus, anti-influenza drugs should not be used in poultry.

11. GUIDE TO THE LITERATURE

1. CAPUA, I. and MUTINELLI, F. 2001. *A Color Atlas and Text on Avian Influenza*, Bologna: Papi Editore.
2. PERKINS, L.E.L. and SWAYNE, D.E. 2003. Comparative susceptibility of selected avian and mammalian species to a Hong Kong-origin H5N1 high-pathogenicity avian influenza virus. *Avian Diseases*, 47:956-967.
3. SPACKMAN, E., SENNE, D.A., MYSER, T.J., BULAGA, L.L., GARBER, L.P., PERDUE, M.L., LOHMAN, K., DAUM, L.T. and SUAREZ, D.L. 2002. Development of a real-time reverse transcriptase PCR assay for type A influenza virus and the avian H5 and H7 hemagglutinin subtypes. *Journal of Clinical Microbiology*, 40:3256-3260.
4. STALLKNECHT, D.E. 1998. Ecology and epidemiology of avian influenza viruses in wild bird populations: waterfowl, shorebirds, pelicans, cormorants, etc., In: *Proceedings of the Fourth International Symposium on Avian Influenza*, DE Swayne, RD Slemons, eds., U.S. Animal Health Association, Virginia, pp. 61-69
5. SWAYNE, D.E. and HALVORSON, D.A. 2003. Influenza. In: *Diseases of Poultry*, 11th ed., YM Saif, HJ Barnes, AM Fadly, JR Glisson, LR McDougald, DE Swayne, eds., Iowa State University Press, Ames, IA, pp. 135-160.
6. SWAYNE, D.E. and PANTIN-JACKWOOD, M. 2006. Pathogenicity of avian influenza viruses in poultry. *Developments in Biologicals*, 124:61-67
7. SWAYNE, D.E., SENNE, D.A. and BEARD, C.W. 1998. Influenza. In: *Isolation and Identification of Avian Pathogens*, 4th ed., DE Swayne, JR Glisson, MW Jackwood, JE Pearson, WM Reed, eds., American Asosication of Avian Pathologists, Kennett Square, Pennsylvania, pp. 150-155.
8. SWAYNE D.E. and SUAREZ, D.L. 2000. Highly pathogenic avian influenza. *Revue Scientifique et Technique Office International des Epizooties*, 19:463-482.

See Part IV for photos.

D.E. Swayne, DVM, PhD, USDA-ARS, Southeast Poultry Research Laboratory, 934 College Station Rd., Athens, GA., 30605, David.Swayne@ars.usda.gov

6

BABESIOSIS

1. NAME

Babesiosis

Bovine babesiosis also known as piroplasmosis, Texas fever, redwater, tick fever

Equine babesiosis

also known as equine piroplasmosis, biliary fever

2. DEFINITION

Babesiosis is an infectious disease caused by a protozoan- an intraerythrocytic tick-borne parasite that infects a wide variety of vertebrate hosts, typically inducing disease characterized by intravascular hemolysis leading to anemia and other variable clinical signs and lesions. This chapter will deal primarily with bovine babesiosis because, from an economic point of view, cattle are most significantly affected by the disease. Equine, ovine, caprine and porcine babesiosis are discussed briefly.

3. ETIOLOGY

Babesiosis is caused by protozoan organisms of the genus *Babesia* which was first described by Babés in Romania as a parasite of bovine red blood cells (RBCs). To date, more than 70 *Babesia* spp. have been recognized based on their morphology, serologic tests, and molecular characteristics. In nature, *Babesia* spp. are transmitted by vector ticks and only exceptionally by other means. A list of *Babesia* spp. that affect livestock, their vector ticks, and dimensions is presented in Table 1.

4. HOST RANGE

Although it is possible for a single *Babesia* species to infect more than one vertebrate host (e.g., *B. microti* affects rodents and humans; *B. divergens* and *B. bovis* affect cattle and humans), *Babesia* spp. are typically host specific.

Bovine babesiosis, also know as piroplasmosis, Texas fever, redwater, and tick fever is a febrile, tick-borne disease of cattle caused by 1 of at least 7 *Babesia* spp. It is generally characterized by extensive intravascular hemolysis leading to depression, anemia, icterus, hemoglobinuria, and, in the case of *B. bovis* infection, neurological signs. The two *Babesia* spp. of most concern as causes of disease in cattle are *B. bigemina* and *B. bovis*, which are transmitted primarily by *Boophilus* ticks. *B. bovis, B. bigemina*, and their vector ticks once occurred in large areas of the U.S. and still occur in Mexico and throughout the tropical and subtropical areas of the Western Hemisphere.

Cattle are the main hosts, but the Asian water buffalo (*Bubalus bubalis*) and African buffalo (*Syncerus caffer*) may also become infected. Outbreaks of *B. bigemina* and *B. bovis* have been described in domestic water buffaloes in Northern Brazil where this animal is an important livestock species. It is possible that other ungulates are infected, but from a practical point of view such hosts are probably not important as reservoirs of infection.

B. bigemina and *B. bovis* are widespread in the cattle population and occur wherever *Boophilus* ticks are encountered, which includes North and South America, parts of Southern Europe, Africa, Asia, and Australia. Babesiosis also occurs in the Caribbean and South Pacific islands. Cattle and the tick hosts provide the major reservoir of infection.

Babesia divergens appears to be a serious pathogen for cattle in the United Kingdom and northern Europe, where it is spread by ticks of the *Ixodes* genus. The presence of *Ixodes* ticks in the U.S. suggests the potential for this babesia species to become established in the country. *Babesia jakimovi* is the cause of Siberian piroplasmosis in cattle. *Babesia major* affects cattle in the United Kingdom and northern Europe. It is essentially nonpathogenic but can be induced to produce clinical effects and even death by serial passage in splenectomized calves. *Babesia ovata* and *B. occultans* have been described in cattle from Japan and southern Africa, respectively. The latter babesia is transmitted by the *Hyalomma marginatum rufipes* tick.

Babesia caballi, *B. equi* or both are the cause of *equine babesiosis*, also known as equine piroplasmosis and biliary fever. Currently some authorities have postulated that *B. equi* should be reclassified as *Theileria equi*. Equine babesiosis is widely distributed throughout the tropics and subtropics and to a lesser extent is known to occur in temperate regions.

Babesia motas and *B. ovis* are reported causes of babesiosis in sheep and, occasionally, in goats. Information on these species is limited, and few serological and cross-immunological studies have been performed to clarify the identity of these intraerythrocytic parasites. They don't seem to have the same epidemiological importance as the other babesias of domestic animals. More recently *B. sergenti*, *B. foliata* and *B. crassa* were reported as causing disease in sheep, but these organisms are limited to Algeria, India, and Iran, respectively. *Babesia trautmanni* and *B. perroncitoi* infect swine, and, on occasion, are responsible for serious losses following infections. Swine babesiosis has been described in the former Soviet Union, southern Europe, and Africa.

Babesia jakimovi can infect the Tartarean roe deer (*Capreolus capreolus*), Asian elk (*Alces alces*), and reindeer (*Rangifer tarandus*). In Africa, the wild pigs (*Potamochoerus porcus*) are thought to be reservoirs of *Babesia trautmanni* and *B. perroncitoi*.

Babesiosis is a zoonosis, and cases of disease induced by *B. microti*, *B. divergens* and *B. bovis* have been described in humans. At one time these human infections were thought to occur only in splenectomized individuals, or in those who were otherwise immunocompromised. This is not, however, the case regarding the rodent parasite *B. microti*, which has been reported as causing disease in immunocompetent persons.

5. EPIDEMIOLOGY

a. Transmission

The general mode of transmission of *Babesia* spp. is similar regardless of species, although minor differences exist. The disease is virtually always transmitted by ticks, but, as with most blood diseases, surgical procedures such as dehorning, castration and needle vaccination procedures are sometimes implicated in accidental transfer of blood from one animal to another, thereby transmitting infection.

Table 1 contains information on vector tick species implicated in the transmission of the various *Babesia* spp. Summarized here is the mode of transmission of *B. bigemina* and *B. bovis*. The infection occurs when nymphal and adult (for *B. bigemina*), and larval (for *B. bovis*) stages of the vector ticks feed on the host. After the inoculation as sporozoites, the parasite penetrates the RBCs in the definite host blood stream where they form a parasitophorous vacuole and evolve to trophozoites that later undergo binary division, usually forming a pair of merozoites. As merozoites leave the RBCs they cause rupture of their membrane and leakage of hemoglobin into the plasma (hemoglobinemia). Ticks acquire babesia infection during their feeding on infected animals. The infection is then passed on to the ovaries of the ticks, and thus the emerging larvae carry the infection.

Babesia caballi is transmitted by ticks of the genera *Dermacentor*, *Hyalomma*, and *Rhipicephalus* and is passed transovarially from one tick generation to the next. Experimental transmission of *B. caballi* under laboratory conditions has been reported using *Dermacentor nitens* and *D. variabilis*. The widespread prevalence of these ticks (*D. albipictus* and *D. variabilis*), plus the presence or past presence of *B. caballi* in the U.S., creates an unanswered question of why *B. caballi* has not become more widespread there. Transmission of *B. equi* appears only to occur transstadially. The vector or vectors of *B. equi* have not been identified in the Western Hemisphere.

b. Incubation period

The incubation period is 2-3 weeks for natural disease induced by both *B. bigemina* and *B. bovis* but can be as short as 4-5 (*B. bigemina*) and 10-14 (*B. bovis*) days in experimental inoculation depending on the size of the inoculum. The natural infection caused by *B. bovis* tends to present a longer incubation time than that caused by *B. bigemina.*

c. Morbidity

Several factors influence the infection and geographical distribution of bovine babesiosis. In general, the disease follows that of the vector ticks; this fact produces 3 epidemiological situations. In tick-vector *free areas* the disease does not occur and consequently cattle do not develop natural immunity. However, if tick-bearing and *Babesia* spp. carrier cattle are brought to these free areas, they can introduce the disease when the timing coincides with favorable weather conditions – thus allowing for the occurrence of a second generation of ticks infected by transovarian transmission. The larvae, nymphs or adults of this second generation will transmit the disease. Conversely, if cattle from free areas are introduced into areas where ticks are enzootic, the disease will occur unless the animals are vaccinated.

In areas of *enzootic instability* there is an alternation of warm and cold seasons. The cold periods prolong the free-living stages of the ticks, allowing cattle prolonged periods without tick contact. This results in a significant drop in antibodies due to the absence of babesia infection. When the warm periods return, the tick parasite load increases and outbreaks of the disease occur. In areas of *enzootic stability* (between parallels 32° S and 32° N) weather conditions allow the presence of ticks on cattle all year round. This fact confers high levels of lasting immunological protection to cattle.

Factors influencing the occurrence of babesiosis outbreaks include: (1) overinfestation by vector ticks resulting in a high inoculum of babesia; (2) long periods without ticks with resultant loss of immunity and vulnerability to infection (this situation might occur due to the prolonged use of acaricides and maintenance of cattle in pastures free of ticks); (3) stress factors and nutritional deficiencies which can induce a drop in immunity and vulnerability to the disease.

Excluding situations related to a drop in acquired immunity, calves are more resistant to infection by *Babesia* spp. than adult cattle, and zebu breeds are more resistant to tick infestation and babesiosis than European breeds.

d. Mortality

Among fully susceptible older cattle, mortality rates of 5-10% are common, even with treatment, and lethality rates in nontreated cattle can be up 50-100% in the case of *B. bovis* infection. Among cattle raised in an area of enzootic stability, few, if any, losses occur even though infection takes place. This phenomenon usually reflects early exposure of the neonates when they are more resistant, thus resulting in varying levels of protection.

6. CLINICAL SIGNS

a. Cattle

Bovine babesiosis caused by *B. bigemina* can occur as acute, subacute or chronic disease. In all cases affected cattle develop fever, depression, anorexia, paleness of mucous membranes, and hemoglobinuria. Icterus is also a clinical sign, mainly in the subacute form of the disease, but it can be minimal or absent in cases of peracute or acute disease. Cattle with the chronic form have marked emaciation, drop in milk yield, and abortions. Hematological findings in cattle with acute and subacute babesiosis are typical of hemolytic anemia and include hypochromic macrocytic anemia with excessive regeneration and high numbers of reticulocytes. Plasma of affected cattle has a tan-brown discoloration due to hemoglobinemia.

In many respects, infections of *B. bovis* resemble those seen with *B. bigemina*, but there are some characteristic differences. *B. bovis* tends to induce a more acute and severe disease in which peripheral circulatory disturbances with sequestration of parasitized RBCs in the peripheral circulation are unique features. Another unique feature of the disease caused by *B. bovis* infection in cattle are neurological disturbances characterized by incoordination, seizures, muscle tremors, opisthotonus, hyperexcitability, agressivity, blindness, head pressing, nystagmus, lateral recumbency with paddling limb movements, and coma. This form of disease is usually fatal.

The hematological findings observed in cattle with acute babesiosis by *B. bovis* are characterized by an intravascular hemolytic anemia with similar signs of erythroid regeneration as described previously for *B. bigemina* infection.

b. Horses

The severity of clinical disease in equine babesiois is variable, and in many cases spontaneous recovery may occur following a febrile response without marked hemoglobinuria or anemia. Clinical manifestations are characterized by depression, anorexia, fever, pale mucous membranes, icterus, hepatosplenomegaly and petechial hemorrhage in the mucous membranes. The prevalence of hemoglobinuria is controversial and described as commonly found by some and infrequent by others. Abortions have been also associated with equine babesiosis.

Chronic cases of equine babesiosis are characterized by weight loss, selective appetite and reduced performance. Horses with babesiosis develop normochromic normocytic anemia which becomes progressively more marked as the disease evolves, so the horses which do not die in the acute phase of the disease may become markedly anemic. Thrombocytopenia is another prevalent finding in equine babesiosis; it occurs in recurrent episodes that seem to be related to parasitemic peaks. Findings in blood smears are unremarkable because horses do not release reticulocytes in the blood stream. The plasma of affected horses can be markedly yellow (icterus) or have a tan-brown discoloration due to hemoglobinemia.

7. POST- MORTEM LESIONS

a. Cattle

Necropsy findings in cattle that die from *B. bigemina* infection include lesions related to intravascular hemolysis. Hemoglobin imparts a dark-red discoloration to the urine (red water) and this is a consistent finding in acute cases. Mucous membranes are pale or icteric and the blood has a watery aspect. The subcutaneous tissue, abdominal adipose tissue, and omentum may be yellowish. The spleen is markedly swollen, has a fleshy consistency, and the red pulp protrudes over the capsule on cut surface. The liver is swollen and orange-red. The kidneys are dark red or black and the intestinal serosa has a dark pink hue caused by ante-mortem hemoglobin imbibition.

Histologically, there is acute renal tubular necrosis associated with tubular proteinosis (casts of hemoglobin within tubular lumina). Centrilobular hepatocellular fatty changes and/or necrosis due to hypoxia can be observed in the liver associated with large amounts of bile pigment within bile ducts, canaliculi and hepatocyte cytoplasm. Erythrophagocytosis can be observed in the liver, spleen, lymph nodes, and bone marrow. In subacute and chronic cases of *B. bigemina* infection, erythrophagocytosis is accompanied by hemosiderosis in macrophages. In more marked cases, hemosiderin can be found within hepatocytes.

Necropsy and histopathological findings in infections of *B. bovis* are similar to those observed with *B. bigemina.* However, a unique aspect of the *B. bovis* infection in cattle is a cherry-pink discoloration of the grey matter of the brain. This aspect, which is very characteristic and even pathognomonic of the infection by *B. bovis*, results from the stasis of parasitized RBCs in the small capillaries of the cerebral gray matter due the cytoadherence between parasitized RBCs and capillary endothelial cells. The stagnation of blood that occurs in the encephalic capillaries results in the neurological signs so often observed in *B. bovis* infection, a condition that is recognized as *cerebral babesiosis*.

b. Horses

Similar to that described for bovine babesiosis, necropsy findings in horses dying from babesiosis are those of a hemolytic disease. The spleen is enlarged. The liver is frequently enlarged and pale due to anemia. In cases in which hemoglobinuria was a clinical sign, kidneys are dark red or black at necropsy. Histopathology includes erythrophagocytosis in the lymph nodes, spleen, liver and lung; centrilobular hepatocellular coagulation necrosis caused by hypoxia secondary to severe anemia; accumulation of bile pigment in bile ducts; and, canaliculi and hepatocytes and hyperplasia of Kupffer cells.

8. IMMUNE RESPONSE

a. Natural infection

The immune response in babesiosis apparently involves both the humoral and cell-mediated components of the immune system. Not all antibodies developed after infection are protective, and there is no complete correlation between antibody levels and protection. However, offspring of immune cows are resistant to *B. bigemina* and *B. bovis* infection while young calves (less than 2 months old) born to previously unexposed cows are susceptible, indicating a protective role for colostrally acquired antibodies. After the age of 2 months, calves are protected against babesiosis by a non-specific resistance independent of the immune status of the cow. After having recovered from natural infection, cattle will have durable, although not absolute, immunity that may last for life. Stress factors, old age, and calving (perinatal immunosuppression in cows) are situations that can precipitate the disease.

b. Immunization

Premunition is the oldest form of immunization of cattle against babesiosis and entails the inoculation of blood from carrier cattle containing live virulent organisms into susceptible young cattle. This is followed by chemotherapy, as needed, to modify the clinical effects inducing the so called premunition state. Although used in some parts of the world as a means to prevent disease when introducing cattle from free areas to enzootic areas, there are many limitations and this method has been abandoned in most countries.

There are now many good-quality attenuated vaccines available, usually from government supported laboratories. These vaccines are usually produced in monovalent form but an efficient multivalent vaccine containing *B. bigemina*, *B. bovis* and *Anaplasma marginale* is produced in Brazil.

9. DIAGNOSIS

a. Field diagnosis

i. Cattle

Fever, anemia, jaundice, and hemoglobinuria are suggestive clinical signs of babesiosis in cattle from enzootic areas. The clinical signs of hemolytic disease associated with the findings of the intraerythrocytic babesia organisms in blood smears, along with typical necropsy findings, permit a strongly presumptive field diagnosis. A practical field test to differentiate hemoglobinuria/myoglobinuria from hematuria is to allow a urine sample to settle in a transparent test tube for a few hours – in the case of hematuria, an erythrocyte sediment will be observed in the bottom of the tube.

To differentiate hemoglobinuria from myoglobinuria, tests using ammonium sulfate are possible. In the field the typical cherry pink discoloration observed in the grey matter of the brain is strongly indicative of *B. bovis* infection. Should this lesion be observed at necropsy, a brain squash smear from a small sample of cortical gray matter should be done and stained by Giemsa or any other Romanowski dye. Microscopically, capillaries distended by parasitized RBCs can be observed. Sequestration of parasitized RBCs in *B. bovis* infection, although striking in the encephalon, also occurs in various other organs of the body. Imprints from spleen, liver or kidney can also be adequate for cytologic diagnosis. It is important to bear in mind that treatment of babesiosis, mainly in cases of cerebral babesiosis by *B. bovis*, with adequate chemotherapy (e.g., imidocarb) can clean babesia organisms from the body within 12-24 hours, but the animal remains clinically ill due to circulatory disturbances. So it is not infrequent that babesia organisms may be absent in the blood and tissues from necropsied cattle that have been previously treated for babesiosis.

ii. Horses

The clinical diagnosis of equine babesiois is established by the association between clinical signs, hematological findings of anemia, and the detection of the etiological agent within RBCs. As described for bovine babesiosis, blood for the diagnosis of equine babesiosis should be sampled and sent to a reference lab for detection of babesia-bearing RBCs in the blood.

b. Laboratory diagnosis

i. Samples

Whole blood in anticoagulant and serum should be collected. Ideally, blood drawn from small caliber vessels is best for smears. Liver, spleen and brain are also optimal samples for collection.

ii. Laboratory procedures

Blood or impression smears should be taken and stained with Giemsa or any other Romanowski dye. ELISA test is used for serology. PCR assays are very sensitive for detecting the presence of babesia.

10. PREVENTION AND CONTROL

The control of bovine babesiosis is based on strategic tick control, vaccination, chemoprophylaxis and chemotherapy. For those species for which vaccination is not available, preventive management control, tick control and chemotherapy are advised.

The oldest and probably the most effective procedure for the control of babesiosis is to control and eradicate the vector tick. The eradication campaign in the U.S. conducted in the 1920's and 1930's relied largely on dipping all cattle every 2-3 weeks in vats charged with arsenical acaricides. These acaricides have been replaced by a wide variety of improved compounds, including the chlorinated hydrocarbons, carbamates, organophosphates, and natural and synthetic pyrethrins. Although eradication of tick vectors is the most desirable method in the control of babesiosis, it is not considered practical or economical in most countries. In some tropical countries, tick control rather than eradication is the goal. This approach attempts to establish an equilibrium in which the tick numbers are sufficient to maintain low-level infection in cattle and hence immunity to acute babesiosis, while trying to make sure the ticks are not in such abundance as to induce overt infection and clinical illness. Tick control in some areas has been complicated by the development of tick resistance to many of the common acaricides.

Successful treatment of *B. bigemina* depends on early diagnosis and the prompt administration of effective drugs. If medication is administered early, however, success is the rule since there are several effective compounds. The most commonly used compounds for the treatment of babesiosis are diminazene diaceturate (3-5 mg/kg), imidocarb (1-3 mg/kg), and amicarbalide (5-10 mg/kg). Imidocarb has been successfully used as a chemoprophylactic that will prevent clinical infection for as long as 2 months, while allowing mild subclinical infection to occur as the drug level wanes - thus resulting in premunition and immunity.

B. bovis is usually somewhat more difficult to treat because it induces a more severe disease. However chemotherapy for *B. bovis* is generally effective, with essentially the same drugs as used for *B. bigemina*.

Both *B. caballi* and *B. equi* respond to the imidocarb, which appears to be the drug of choice for eliminating carrier status of infected horses. In the case of *B. caballi*, 2 mg/kg given 2 times at a 24-hour interval appears effective. For the same effect in *B. equi*-infected horses, 4 mg/kg is given 4 times at 72-hour intervals. Side effects characterized by restlessness, abdominal pain, sweating, rolling and heavy breathing are not uncommon following imidocarb treatment at these higher levels.

Chemotherapy for the control of porcine and ovine babesiosis can be carried out using the same drugs as for bovine and equine babesiosis. Since no vaccines are commercially available for pigs, control of disease should rely on tick control and management measures such as avoiding contact between domestic and wild pigs.

Although safer vaccines are constantly being produced, vaccination against babesiosis is not a completely safe procedure and reactions may occur within 7 days (*B. bovis*) and 10-14 (*B. bigemina*) days after vaccination. Hence there is the recommendation to limit the vaccination to calves; in this animal category the risk of reactions is minimized by the non-specific immunity. When older cattle are to be vaccinated, closer attention should be given during the period of vaccine reactions.

Vaccinated animals in the stable or in the pasture should be observed clinically and kept free from ticks within 60 days post-vaccination, after which tick contact can be established but excessive tick loads must be avoided. In the absence of tick contact (and hence contact with the etiologic agent) vaccination immunity may break down, especially in case of *B. bigemina*.

The control of ovine babesiosis is similar to the control of bovine babesiosis and vaccination is performed by whole blood vaccines containing the attenuated organisms. No vaccine is yet available either for equine or porcine babesiosis.

Aside from efforts involved in the control and elimination of the tick vector, sanitation and disinfection do not contribute to an abatement of the disease incidence in enzootic areas. As with most blood diseases, however, care is recommended in routine surgery (dehorning, castration) and in needle vaccination procedures to prevent the accidental transfer of blood from one animal to another, thereby transmitting infection.

11. GUIDE TO THE LITERATURE

1. DE VOS, A.J., DE WAAL, D.T. and JACKSON, L.A. 2004. Bovine babesiosis. In: *Infectious Diseases of Livestock* 2nd ed., JAW Coetzer, RC Tustin, eds., Oxford University, Cape Town. pp. 406-424.
2. DE WAAL, D.T. and VAN HEERDEN, J. 2004. Equine piroplasmosis. In: *Infectious Diseases of Livestock* 2nd ed., JAW Coetzer, RC Tustin, eds., Oxford University, Cape Town. pp. 425-434.
3. DE WAAL, D.T. 2004. Porcine babesiosis. In: *Infectious Diseases of Livestock* 2nd ed., JAW Coetzer, RC Tustin, eds., Oxford University, Cape Town. pp. 435-437.
4. KESSLER, R.H., SOARES, C.O., MADRUGA, C.R. and ARAÚJO, F.R. 2002. Tristeza parasitária dos bovinos: quando vacinar é preciso. Campo Grande: Embrapa Gado de Corte. Série Documentos, n. 131, 27p.
5. RODRIGUES, A., RECH, R.R., BARROS, R.R., FIGHERA, R.A. and BARROS, C.S.L. 2005. Babesiose cerebral em bovinos: 20 casos. *Ciência Rural*, 35:121-125.
6. ROBERTSON, J.R. and ROONEY, J.L. 1996. Hemolymphatic system. In: *Equine Pathology*, Chap. 18, Iowa University Press, Ames, Iowa. pp. 348-366.
7. YERUHAM, I. and HADANI, A. 2004. Ovine babesiosis. In: *Infectious Diseases of Livestock* 2nd ed., JAW Coetzer, RC Tustin, eds., Oxford University, Cape Town. pp. 438-445.

See Part IV for photos

Claudio S.L. Barros, DVM, PhD and Rafael Fighera, DVM, PhD, Universidade Federal de Santa Maria, Santa Maria, Brazil, claudioslbarros@uol.com.br, anemiaveterinaria@yahoo.com.br

7

BLUETONGUE

1. NAME
Bluetongue. Sore muzzle, Ovine catarrhal fever

2. DEFINITION
Bluetongue (BT) is an acute, insect-transmitted, non-contagious disease of wild and domestic ruminants caused by BT virus (BTV). The outcome of BTV infection of ruminants is highly variable and is usually asymptomatic or subclinical in cattle, goats, and many sheep breeds. BT most commonly is encountered in specific sheep breeds and white-tailed deer, and is characterized by vascular injury that causes hemorrhage and edema in a variety of tissues. Outbreaks of BT continue to occur throughout the world, often as a result of the incursion of the virus into previously free regions.

3. ETIOLOGY
The BTV is the prototype member of the genus *Orbivirus*, family *Reoviridae.* The genus *Orbivirus* encompasses some 14 different serogroups, including those of African horse sickness, epizootic hemorrhagic disease, equine encephalosis, Eubanangee, Ibaraki, and Palyam viruses. At least 24 serotypes (1-24) of BTV occur worldwide, of which 4 (serotypes 10, 11, 13 and 17) are widely distributed throughout North America. BTV serotype 2 has been sporadically detected in Florida and adjacent regions of the southeastern United States since the early 1980s, and BTV serotype 1 very recently was detected in coastal Louisiana.

4. HOST RANGE
a. Domestic and wild animals
The BTV replicates in several species of *Culicoides* insects, ruminants, and certain carnivores. Bluetongue is almost exclusively a disease of sheep and white-tailed deer in North America, although the disease sporadically is reported in other ruminant species. The virus also is infectious to wild African carnivores and domestic dogs, and causes fatal infection of pregnant bitches.

b. Humans
There are no recorded infections of humans with BTV.

5. EPIDEMIOLOGY

a. Transmission

The global distribution of BTV coincides with that of competent vectors, which are specific species of *Culicoides* insects. Very few (< 20) of the approximately 1400 species of *Culicoides* insects that occur worldwide have been confirmed to be competent vectors of BTV. Infection with BTV occurs on all continents except Antarctica, although the serotypes and strains of BTV that occur in different continents vary markedly. The global distribution of BTV has traditionally been considered to include tropical and temperate regions located between latitudes 40°N and 35°S, but the virus clearly extends as far as 50°N in portions of North America, Asia and Europe.

The species of *Culicoides* that serve as principal vectors of the virus differ between regions. *Culicoides sonorensis* is the principal vector of BTV serotypes 10, 11, 13 and 17 in most of the U.S., whereas BTV serotypes 1 and 2 appear to be restricted to areas of the southeastern U.S. where *C. insignis* is present.

Culicoides insignis is the vector of numerous BTV serotypes (1, 2, 3, 4, 6, 8, 11, 12, 13, 14 and 17) in the Caribbean Basin and Central America. *C. imicola* is a major vector of BTV in Africa, Europe, the Middle East and portions of Asia, and *C. brevitarsis* (a relative of *C. imicola*) is considered the major vector of BTV in Australia and portions of Asia. Other vector species are also potentially important, including *C. bolitinos* in Africa, *C. obsoletus* and *C.pulicularis* in the Mediterranean Basin, *C. pusillus* in the Caribbean Basin and South America, and *C. actoni, C. fulvus, C. brevipalpis* and *C. wadai* in Australia.

b. Incubation period

The incubation period in BTV-infected ruminants is approximately 1 week, ranging from perhaps 2 to 10 days.

c. Morbidity

Morbidity can be very high, up to 100% of susceptible ruminants.

d. Mortality

Mortality is highly variable, even in susceptible breeds of sheep, ranging from 1-10% in severe epidemics.

6. CLINICAL SIGNS

The clinical manifestations of BT vary markedly in ruminants, reflecting the species of animal, virulence of the infecting strain of virus, and environmental stress factors such as intense sunlight. Many infections are subclinical or asymptomatic, even in susceptible breeds of sheep. Cattle, goats and many other species of ruminants rarely manifest any clinical signs of BTV infection. The course of clinical infections ranges from peracute to chronic.

BT in sheep is initially characterized by fever, hyperemia of the oral and nasal mucous membranes, salivation, and edema of the head. Petechial and ecchymotic hemorrhages then develop on the muzzle and lips, and throughout the oral cavity with subsequent erosion and/or ulceration of the overlying epithelium. The cyanotic tongue for which the disease is named is an extremely uncommon manifestation. Affected animals are anorectic and develop a nasal discharge that is initially serous but rapidly becomes mucopurulent, leading to the accumulation of crusty exudates around the nares. Foot lesions appear toward the end of the febrile period, and hyperemia and hemorrhage are often prominent along the coronary band. The feet of affected animals become hot and painful, thus affected sheep are reluctant to move and their gait is stiff and difficult. Sheep that survive severe BT often exhibit marked loss of condition, with extensive muscle loss (leading to torticollis), hoof deformities, loss of wool (wool fiber break), and generalized weakness and ill-thrift.

Serological studies indicate that large African carnivores commonly are infected with BTV whereas smaller predators that co-habit with them are not, suggesting that large carnivores are infected through feeding on BTV-infected ruminants. Furthermore, inadvertent contamination of a canine vaccine with BTV confirmed that dogs are susceptible to BTV infection; indeed, pregnant bitches that received this contaminated vaccine typically aborted and died. There is no evidence, however, that dogs or other carnivores are important to the natural cycle of BTV infection.

7. POST- MORTEM LESIONS

a. Gross

The clinical signs and lesions of BT reflect vascular injury as BTV replicates in endothelial cells causing cell injury and necrosis, sometimes leading to consumptive coagulopathy (disseminated intravascular coagulation). Endothelial injury is likely responsible for increased vascular permeability (leading to edema in tissues such as the lung) and thrombosis, which in turn leads to tissue infarction. Lesions present at postmortem of affected sheep include hyperemia, hemorrhages, erosion and ulceration of the mucosa of the upper gastrointestinal tract (oral cavity, esophagus, forestomachs); subintimal hemorrhages in the pulmonary artery; pulmonary edema; pleural and/or pericardial effusion; edema within the fascial planes of the muscles of the abdominal wall; and, necrosis of skeletal and cardiac muscle, with the papillary muscle of the left ventricle being an especially characteristic site.

Strains of BTV that are modified by cell culture passage, which typically are live vaccine strains of the virus, can cross the placenta of pregnant ruminants whereas field strains of the virus that have not experienced cell culture passage are apparently unable to do so.

Infection of the ruminant conceptus with BTV in early gestation leads to either embryonic/fetal death or to teratogenic defects in progeny that survive. The virus replicates within cells of the developing central nervous system of infected ruminant fetuses to produce a cavitating encephalopathy, which ranges at birth from severe hydranencephaly in which the cerebral hemispheres persist only as fluid filled sacs to so-called false porencephaly that is characterized by the presence of dilated lateral ventricles with bilaterally symmetrical cerebral cysts in the persisting cortical tissue.

b. Key microscopic

The expected histologic lesions of BT include: marked pulmonary edema; hemorrhage; necrosis of skeletal and cardiac muscle; and, necrosis, hemorrhage and ulceration of the mucous membranes of the upper gastrointestinal tract.

8. IMMUNE RESPONSE

a. Natural infection

Natural BTV infection of ruminants results in a prompt and high-titered humoral immune response to a variety of viral proteins. Neutralizing antibodies are directed against serotype-specific epitopes on the outer capsid protein VP2, and these antibodies confer long-term resistance to reinfection with the homologous BTV serotype but not other serotypes. The other outer capsid protein (VP5) exerts a conformational pressure on the neutralizing epitopes on VP2 and so affects virus neutralization. Antibodies to other viral proteins are not neutralizing and do not confer resistance to re-infection. Antibodies to epitopes on viral proteins such as VP7 that are conserved amongst all members of the BTV serogroup form the basis of serogroup-specific serological diagnostic assays. Infected ruminants also develop cell-mediated immune responses to a variety of viral proteins, and these can confer transient, partial immunity against reinfection with homologous as well as to a variety of heterologous BTV serotypes.

Despite the prompt humoral and cellular immune response of BTV-infected ruminants, viremia often persists for several weeks, especially in cattle. This prolonged but not persistent infection of BTV ruminants appears to result from the intimate association of BTV particles with erythrocytes, an interaction that both protects the virus from immune clearance and also facilitates infection of hematophagous vector insects that feed on the viremic animal.

b. Immunization

Modified live BTV vaccines are useful in preventing losses attributable to BT, but they have some deficiencies (see Prevention and Control).

9. DIAGNOSIS

a. Field diagnosis

Field diagnosis of BT is based upon the characteristic clinical signs and lesions in affected animals, especially sheep, coupled with environmental and/or surveillance data documenting the presence or likely presence of competent *Culicoides* insect vectors. It is to be stressed that BT disease is most prevalent amongst ruminants located adjacent to the upper and lower global margins of the virus' range (latitudes of approximately 45°N and 35°S), and that outbreaks frequently are associated with extension of the virus into areas that were previously free from infection for substantial periods of time (years). In marked contrast, BT rarely is encountered amongst ruminants in areas where the virus is endemic and present throughout much of the year.

b. Laboratory diagnosis

i. Samples

Laboratory diagnosis of BTV infection of ruminants is either by virus detection (blood) or serological methods (serum).

ii. Laboratory procedures

Serological assays:

Ruminants infected with BTV develop a prompt and high-titered antibody response to a variety of viral proteins. Antibodies directed against core protein VP7, as well as other structural and nonstructural proteins, may be detected with serogroup-reactive assays such as the agar gel immunodiffusion and competitive enzyme-linked immunosorbent assay (cELISA). Serotype-specific neutralizing antibodies are directed against outer capsid protein VP2, and these readily can be detected by serum neutralization test. A positive serological result confirms only that an animal previously was infected with BTV. Furthermore, although BTV infection of cattle and sheep often is prolonged, there is no credible evidence of long-term persistent BTV infection of ruminants. Thus, the vast majority of seropositive cattle and sheep from BTV-endemic regions are not infected with the virus and pose no threat for movement.

Virus detection assays:
The presence of BTV in the blood of ruminants can be determined by isolation in embryonated chicken eggs, cell culture, or by inoculation of susceptible sheep. Nested PCR (nPCR) assay increasingly is used for screening of ruminants for the presence of BTV nucleic acids because it is highly sensitive and specific if performed properly; also, it is an extremely conservative assay in that BTV nucleic acid may be detected in the blood of sheep and cattle long after infectious virus has been cleared. Ruminants whose blood is negative by nPCR assay pose no threat for inadvertent movement of BTV by trade. nPCR assays have been developed for genes encoding several BTV proteins, but the OIE prescribed test amplifies a portion of gene encoding the conserved NS1 protein.

10. PREVENTION AND CONTROL

The prevention and control of BTV infection of ruminants is complicated by the fact that the virus is not contagious, rather, it strictly is transmitted by *Culicoides* insects. As control of insects is usually difficult or impossible, vaccination typically is used to prevent outbreaks of BT and to control incursions of BTV. Only modified live BTV vaccines are in widespread use, particularly in Africa, the U.S. and, most recently, southern Europe. These vaccines have proven useful in preventing losses attributable to BT, but they also suffer from a number of serious potential deficiencies including: the introduction of novel virus strains into the environment, perhaps leading to infection of vector insects; quasispecies evolution with possible reversion to virulence or creation of new strains of BTV; reassortment of gene segments with indigenous viruses to generate potentially novel recombinants; fetal infection; and, teratogenesis. In fact, it is increasingly clear that only strains of BTV that have been modified by growth in cell culture, such as MLV vaccine strains, have the capacity to cross the ruminant placenta. New generation vaccines, such as the virus-like particles produced by baculovirus expression of different viral proteins, have not yet been used extensively because of their cost, but recombinant vaccines are now being developed and evaluated.

11. GUIDE TO THE LITERATURE

1. BARRATT-BOYES, S.M. and MacLACHLAN, N.J. 1995. Pathogenesis of bluetongue virus infection of cattle. *J Am Vet Med Assoc*. 206:1322-1329.
2. BROWN, C.C., RHYAN, J.C., GRUBMAN, M.J. and WILBUR, L.A. 1996. Distribution of bluetongue virus in tissues of experimentally infected pregnant dogs as determined by in situ hybridization. *Vet Pathol.* 33:337-40.
3. DeMAULA, C.D., LEUTENNEGGER, C.M., BONNEAU K.R. and MacLACHLAN N.J. 2002. The role of endothelial cell-derived inflammatory

and vasoactive mediators in the pathogenesis of bluetongue. *Virology*. 296:330-337.
4. MacLACHLAN, N.J. and PEARSON, J.E., eds.. 2003. Bluetongue, proceedings of the Third International Symposium. *Veterinaria Italiana.* 40:1-730.
5. VERWOERD, D.W. amd ERASMUS, B.J. 2004. Bluetongue. In: *Infectious diseases of livestock.* 2nd ed., Coetzer JAW, Tustin RC, eds. Cape Town: Oxford Press, pp.1201-1230.

See Part IV for photos.

N. James MacLachlan, DVM, PhD, University of California at Davis, Davis, CA, 95616, njmaclachlan@ucdavis.edu

8

BORNA DISEASE

1. NAME

Borna disease, BD. Bornasche Krankheit, Hitzige Kopfkrankheit der Pferde

2. DEFINITION

Borna disease (BD) is a progressive meningopolioencephalitis, first reported more than 200 years ago, affecting mainly horses and sheep, more rarely other Equidae, cattle, goats, rabbits and exceptionally a variety of other animal species and possibly humans.

3. ETIOLOGY

Borna disease virus (BDV) is a highly neurotropic, enveloped virus with a non-segmented negative-sense, single-stranded (ss) RNA genome of 8.9 kilobases. Its unique replication and transcriptional strategy required classification into the new family *Bornaviridae* within the order *Mononegavirales.* The BDV represents the prototype of the genus *Bornavirus.* Viral particles are spherical, enveloped, and approximately 130 nm in diameter, with spikes of 7nm in length. The particles are released by budding from the cell surface. Infectivity can be destroyed by treatment with disinfectants used for enveloped viruses or with lipid solvents such as ether, chloroform or acetone. The virus is labile at pH 3.0 and inactivated by heating at 56°C for 30 minutes. UV treatment decreases virus infectivity at the same rate as for other conventional viruses.

The complete sequence of the genome from several BDV isolates has been determined. There is a high degree of genetic stability and homology among wild-type and experimentally host-adapted viruses. This might imply that BDV adapts easily to various animal species without significant genetic change. However, natural BDV mutations with changes in antigenic epitopes might occur; this should be kept in mind when diagnosing BD using monoclonal antibodies. Recent phylogenetic analysis of wild-type and laboratory strains indicate that distinct virus clusters corresponding to geographical endemic areas exist in Central Europe.

Further characteristics include the induction of a persistent noncytopathic infection *in vitro* and *in vivo*, and a low replication rate.

4. HOST RANGE

a. Domestic animals

Horses and sheep are predominantly affected by BD and have historically been known as the main natural hosts. However, the host spectrum is much wider. Reports describe the sporadic occurrence of natural BD in additional farm animals such as other Equidae besides horses, cattle, goats, and rabbits, or in companion animals, e.g., dogs and cats. To date, clinical BD has been recognized only in Germany, Switzerland, Liechtenstein and Austria, but BDV infections in animals seem to be more widely distributed; virus-specific antibodies have been demonstrated in horses from several continents.

b. Wild animals

A variety of zoo animals have been found to be susceptible to natural BDV infections. So far, BDV has been demonstrated in alpacas, sloths, monkeys and hippopotami. Recently, BD was detected in a free ranging lynx.

c. Humans

During the last decade, the matter whether BDV can infect humans and cause psychiatric disease has been discussed controversially; this debate is still ongoing. It was proven that antibodies recognizing BDV-specific antigens are present in sera of patients with various psychiatric diseases, but these antibodies can also occasionally be found in clinically healthy people. Reports describing detection of viral protein or RNA in human samples are still very controversial because re-investigations revealed the presence of laboratory contaminations.

5. EPIDEMIOLOGY

a. Transmission

The natural route of transmission of BDV still remains unknown, but investigations in experimentally infected rats demonstrated that BDV can be shed in urine, nasal, and lacrimal secretions, suggesting a nasopharyngeal transmission via open nerve endings. Animals may become infected by direct contact with infectious secretions or indirectly via fomites.

Wild rodents most likely represent the natural reservoir because in endemic areas a seasonal accumulation of cases is observed in spring and early summer with a significant decrease in late fall and winter. Recently, the bicolored white-toothed shrew (*Crocidura leucodon*) was identified as a BDV reservoir species in an area in Switzerland with endemic BD.

b. Incubation period

The incubation period for BD after natural infection is variable, ranging from 4 weeks to several months. In a large number of experimentally infected animals, the average incubation period was found to be 2-3 months. More recently, experimental intracerebral infection of 3 ponies with various doses of BDV resulted in an incubation period of 15-26 days.

c. Morbidity

The incidence of BD is relatively low, as less than 100 horses or sheep are recorded in the BD-endemic areas per year. Clinical BD mostly occurs in single horses while in sheep flocks sometimes several animals can be affected. This might indicate that onset of disease after natural infection most likely depends on host genetics, age and immune status of the host as well as the genetic characteristics and pathogenicity of the virus.

As shown by seroepidemiological studies, BDV infections mostly run an inapparent course. Seroprevalence of BDV-specific antibodies in clinically healthy German horses has been found to be 11.5%, increasing significantly to 22.5% in endemic regions and to 50% in stables with diseased horses.

d. Mortality

Typically, natural BDV infections in horses lead to death within 1-4 weeks after onset of initial signs in more than 80% of animals. In cattle and sheep, death occurs within 1-6 weeks or 1-3 weeks, respectively, in more than 50% of animals.

In less severe cases in horses, spontaneous recovery can occasionally be observed despite a persistent infection of the CNS.

6. CLINICAL SIGNS

a. Horses

Natural BDV infection can result in peracute, acute, or subacute disease. In general, the occurrence of clinical signs correlates with the appearance of inflammatory lesions in the brain. Clinical signs may vary but typically simultaneous or consecutive changes in psyche, sensorium, sensibility, motility and in the autonomous nervous system are noted. Early stages are characterized by alterations in behavior and consciousness and these signs progressively worsen. Slow motion eating, eating arrest with chewing movements (called "Pfeifenrauchen" = "pipe smoking"), recurrent fever, lethargy, somnolence, stupor, hyperexcitability, fearfulness and unusual aggressiveness may be present to varying degrees.

The neurological signs are variable and depend on the course of disease and the brain areas affected by inflammatory lesions. They range from hypokinesia and abnormal postures (postural unawareness) up to hyporeflexia of spinal reflexes, head tilt and hypoesthesia with disturbances in proprioceptive sensory functions in more advanced stages of BD. In advanced stages, horses may exhibit ataxia and imbalance. In final stages of BD, neurogenic torticollis, compulsive circular walking and slight head tremor followed by convulsions with head pressing and coma are noted. Blindness is also observed regularly in acute BD. In more than 50% of animals with acute BD, disturbances in chewing and swallowing of food and water develop, which could limit the duration of the disease.

b. Sheep

Clinical signs in sheep are similar to the ones described for horses; mainly disturbances in behavior and movement are observed. In addition, mild or inapparent courses might also occur.

c. Cattle

The few recorded cases of natural BD in cattle exhibited a classical clinical course as described for horses.

d. Rabbits

Rabbits, especially experimentally infected animals, develop an acute and fatal paralytic disease accompanied by blindness.

7. POST- MORTEM LESIONS

a. Gross

If present, the only gross findings in horses consist of leptomeningeal hyperemia, brain edema or hydrocephalus internus in later stages of the disease

b. Key microscopic

Histopathological changes associated with BDV-infection are similar in all naturally infected animals. They are restricted to the brain (mainly in the gray matter), spinal cord and retina. Typically, a severe nonsuppurative poliomeningoencephalomyelitis with massive perivascular and parenchymal infiltration can be found. Neuronal degeneration is not prominent, however, a reactive astrogliosis is regularly present. The perivascular cuffs consist predominantly of macrophages, T-lymphocytes (CD4+ and CD8+) and, late after infection, some plasma cells.

In horses, brain areas mainly affected are the olfactory bulb, basal cortex, caudate nucleus, thalamus, hippocampus and periventricular areas, mainly in the medulla oblongata. If present, the occurrence of intranuclear eosinophilic, "Joest-Degen" inclusion bodies is regarded as a pathognomonic finding. A peculiarity of BDV-infected rats and rabbits is the nonsuppurative chorioretinitis with degeneration of rods and cones. Interestingly, horses do not develop inflammatory lesions in the retina despite clinical signs of blindness.

8. IMMUNE RESPONSE

a. *Natural infection*

Borna disease is caused by a T cell-mediated virus-induced immunopathological reaction in the CNS. Most of the currently available information regarding the immunopathogenesis of BDV-infection is derived from studies of experimentally infected rodents. Taken together, there is convincing evidence that T cells play a crucial role in the immunopathogenesis of BD, and that BD encephalitis most likely represents a delayed type hypersensitivity reaction.

In natural BD of horses and sheep, the meningoencephalitis with mononuclear infiltrates as well as the composition of the immune cell infiltrates is quite similar to that in experimentally infected rats, indicating a similar immunopathogenesis.

b. *Immunization*

Since BD represents an immune-mediated disease, vaccination of animals with inactivated virus preparations or viral antigens does not protect from subsequent infection and disease. However, it was shown that protection against BD could successfully be achieved by using attenuated virus preparations. Immunoprophylaxis of horses with a lapinized live BDV vaccine was practiced for many years in Germany but its efficacy was questionable so that it was abandoned in 1992. New data from experimental studies applying recombinant parapoxvirus constructs are promising, but its use in naturally infected animals needs further investigations. Nevertheless, in view of the T cell-mediated immunopathogenesis of BD, artificial stimulation of immune reactions should be handled with caution.

9. DIAGNOSIS

a. *Field diagnosis*

A clinical diagnosis of BD must remain tentative until it is confirmed with laboratory tests. Neurological signs of BD might exhibit a complex pattern due to multifocal or disseminated brain lesions. Central nervous system infection with a variety of pathogens may result in similar signs.

b. Laboratory diagnosis

Most data regarding the reliability of methods for diagnosing BD result from studies in horses.

i. Samples

For the *in vivo* diagnosis of BD, serum and CSF is needed for the detection of BDV-specific antibodies. If possible, seroconversion should be demonstrated in a second investigation.

For the post mortem diagnosis of BD, serum and CSF should be collected along with fresh frozen samples of brain (N. caudatus, hippocampus, cerebral cortex), eye (retina), lacrimal and parotid gland, trigeminal ganglion, hypophysis and spinal cord; the latter samples are used for the detection of virus antigen, virus specific RNA and infectious virus. Similar tissues should be submitted fixed in 10% formalin for subsequent histological evaluation.

ii. Laboratory procedures

Ante-mortem diagnosis

Hematological and biochemical parameters are normal in horses with clinical BD. Nonspecific hyperbilirubinemia is frequently observed. In acute stages of the disease, the protein concentration in the CSF is increased, accompanied by lympho-monocytic pleocytosis. Using nested RT-PCR, it is possible to detect BDV RNA in cells isolated from the CSF.

Demonstration of BDV-specific antibodies can be carried out in the serum and/or CSF using Western blot analysis, enzyme-linked immunosorbent assay, or an indirect immunofluorescence assay (IFA). The IFA is acknowledged to be the most reliable method for the detection of BDV-specific antibodies.

Post-mortem diagnosis

The post-mortem diagnosis of BD requires histopathology, immunohistochemistry (IHC) and/or virological methods. IHC can be performed using monoclonal or polyclonal antibodies specific for various BDV-proteins. The same monoclonal and polyclonal antibodies can be used in Western blot analysis. Comparative studies indicate that all 3 methods as well as nested RT-PCR give identical diagnostic results in acute cases of BD. The isolation of infectious BDV and the demonstration of virus-specific RNA by *in situ* hybridization can complete diagnosis of BD, however, both are less reliable when the material is autolyzed. In this case, IHC and Western blotting are preferred.

10. PREVENTION AND CONTROL

To date, there is no specific therapy for treatment of BD. Recently, the use of amantadine sulfate, a drug with antiviral activity against influenza A, was recommended for treatment of BD in horses, however, its efficacy for the treatment of persistently BDV-infected cell cultures and animals revealed inconsistent results. Filtration of CSA was proposed as another new therapeutic strategy, but its usefulness for anti-BD therapy needs evaluation of a larger number of BD cases.

11. GUIDE TO THE LITERATURE

1. DANNER, K. 1982. *Borna-Virus und Borna-Infekionen,* Stuttgart. Enke Copythek.
2. De la TORRE, J.C. 1994. Molecular biology of Borna disease virus: prototype of a new group of animal viruses, *J Virol.* 68:76697675.
3. GOSZTONYI, G., and LUDWIG, H. 1984. Borna disease of horses: an immunohistological and virological study of naturally infected animals, *Acta Neuropathol (Berl).* 64:213-221.
4. HEINIG, A. 1969. Die Bornasche Krankheit der Pferde und Schafe. In Röhrer H, ed. *Handbuch der Virusinfektionen bei Tieren,* Jena, Fischer, p.83.
5. HILBE, M., HERRSCHE, R., KOLODZIEJEK, J., NOWOTNY, N., ZLINSKY, K., and EHRENSPERGER, F. 2006. Shrews as reservoir hosts of borna disease virus. *Emerg. Infect. Dis*. 12:675-677.
6. KOPROWSKI, H., and LIPKIN, W.I. 1995. Borna disease, *Curr Top Immunol Microbiol.* 190:1.
7. RICHT, J.A., GRABNER, A., HERZOG, S., GATERN, W., and HERDEN, C. 2007. Borna Disease in Equines. In: *Equine Infectious Diseases*, D. Sellon, M. Long, eds., Saunders, Missouri, pp. 207-213.
8. RICHT, J.A., and ROTT, R. 2001. Borna disease virus: a mystery as emerging zoonotic pathogen, *Vet J.* 161:24-40.

See Part IV for photos.

Jürgen A. Richt, DVM, PhD, National Animal Disease Center, USDA, ARS, Ames, IA, 50010, jricht@nadc.ars.usda.gov

- and -

Christiane Herden, DVM, PhD, Institut fur Pathologie, Tierarztliche Hochschule Hannover, Bunteweg 17, D-30559 Hannover, Germany, christiane.herden@tiho-hannover.de

9

BOVINE EPHEMERAL FEVER

1. NAME

Bovine ephemeral fever, Ephemeral fever, Three-day stiffsickness, Bovine epizootic fever, dragon boat disease, Lazy man's disease, and dengue of cattle.

2. DEFINITION

Bovine ephemeral fever (BEF) is a non-contagious, arthropod-borne, viral disease of cattle and water buffalo characterized by sudden onset of fever, depression, stiffness and lameness.

3. ETIOLOGY

The causal agent, bovine ephemeral fever virus (BEFV), is an ether-sensitive, single-stranded, negative sense RNA virus belonging to the family *Rhabdoviridae*. The virions consist of 5 structural and 1 non-structural protein, and can range from bullet to blunt cone shaped. They have a diameter of approximately 73nm but the length can vary from 70-183nm. The shorter bullet and conical forms are considered to be defective particles that probably interfere with virus growth in tissue culture. The virus is readily inactivated at pH levels below 5 and above 10. There is no evidence of immunogenic diversity within the BEFV population, but antigenic variation has been demonstrated using panels of monoclonal antibodies and by epitope mapping.

The BEFV is antigenically related to 5 other viruses: Puchong virus, which may cause an ephemeral fever-like disease in cattle; Kimberley, Berrimah and Adelaide River viruses, which are nonpathogenic for cattle; and, Malakal virus, for which the cattle pathogenicity is unclear. Together these 6 viruses constitute the genus *Ephemerovirus*. The antigenic relationships between the BEF-related rhabdoviruses and BEFV itself are more than academic because, although they do not provide cross-protection against BEFV, they can induce an anamnestic antibody response in cattle following a subsequent BEFV infection.

4. HOST RANGE

a. Domestic animals

Clinical disease has only been reported in cattle and water buffalo, although neutralizing antibodies to BEFV have been found in domesticated deer. There is also one report of BEFV neutralizing antibodies in goats. Antibodies can be produced in various small laboratory animals following the intravenous or subcutaneous injection of BEFV.

b. *Wild animals*

Few serological surveys of wildlife have been undertaken. Antibodies have been detected by virus neutralization (VN) in several species of deer and by VN and liquid-phase ELISA in a number of different sub-families within the family *Bovidae*. The most consistently detected sero-positive species are Cape buffalo, wildebeest, hartebeest and waterbuck, although low levels of antibody have also been detected in kudu, sable antelope, eland, impala, topi, tsessebe, and giraffe. The specificity of the antibodies detected in African wildlife has not been confirmed.

c. *Humans*

There is no evidence that humans can be infected, despite the fact that thousands of people have been in contact with infected cattle and insect vectors. Limited studies with sera from farmers in contact with infected cattle and from laboratory workers handling the BEFV have also yielded negative results.

5. EPIDEMIOLOGY

a. *Transmission*

In the sub-tropical and temperate climates of Africa, Asia and Australia, epizootics of ephemeral fever occur during the summer and autumn months and disappear in winter when the temperature drops significantly. Disease often spreads rapidly across considerable distances in a northerly or southerly direction away from the equator, but the extension of the virus appears to be limited by latitude rather than a lack of susceptible hosts or the topography. Epizootics occurring in tropical regions seem to have a strong association with the rainy season as evidenced by the sporadic outbreaks that occur after heavy rainfall. These patterns of disease transmission are analogous with other vector-borne diseases.

Although transmission of BEFV by arthropods has not been demonstrated, the circumstantial evidence for spread by biting insects is overwhelming, and in addition, therc are no reports of either vertical or horizontal transmission. So far, BEFV has been isolated from field collections of *Culicine* and *Anopheline* mosquitoes in Australia and from several species of biting midge belonging to the genus *Culicoides* in Africa and Australia. These virus isolations were made from insects without any obvious trace of blood in the gut and therefore it is likely they were true infections and were not due to coincidental contamination.

Mosquitoes, often referred to as vessel feeders, use their proboscis to probe for micro-capillaries within the dermis and therefore are in direct contact with the bloodstream. *Culicoides* biting midges, on the other hand, lacerate the skin and imbibe from the collected blood and are therefore referred to as pool feeders. Because disease can only be reproduced in cattle by intravenous inoculation of the virus, and since virus is absent in the lymph during early viremia, mosquitoes are currently favored as the major vectors. Given that the range of BEFV is beyond the distribution of the arthropod species identified to date, other species must be involved.

b. Incubation period

The incubation period following experimental infection usually varies between 3-5 days but in extreme cases can be as short as 29 hours and as long as 10 days. The variations in incubation time are probably a reflection of the strain and/or dose of virus used. Viremia can be detected approximately 24 hours before the onset of fever and usually lasts for 4-5 days. However, in extreme cases it can persist for up to 13 days. There is no evidence of a virus carrier status in cattle. The natural incubation period remains unknown, but is probably similar to that of experimental infection. In an epizootic, the index case or cases usually occur approximately 1 week before the main wave of infection within a herd. The course of the disease in the herd may range from 3-6 weeks. Single, double or even triple bouts of disease have been reported within a single epidemic. A number of explanations for this have been proposed but the true cause is not entirely clear.

c. Morbidity

Morbidity may be as high as 80% and is influenced partly by the number of susceptible cattle in the herd and partly by the intensity of the epidemic.

d. Mortality

Mortality is generally low and seldom exceeds 2-3% in uncomplicated cases but it can be as high as 30% in very fat cattle. Dairy cows in early and mid-lactation are more likely to die than are dry cows. Death can occur suddenly either during the acute stage of disease or when most animals appear to be recovering.

6. CLINICAL SIGNS

The appearance and severity of the clinical signs of BEF vary considerably from outbreak to outbreak and even within a herd. Individual animals may not exhibit the entire range of clinical signs characteristic of the disease. Cattle and water buffalo exhibit similar signs but the clinical course in buffaloes is milder. In cattle, the disease is often more severe in adults than in calves, in heavy bulls than in light steers, in fat than in lean animals, and in lactating cows than in dry cows.

The course of disease can be broadly described under 4 main headings: Onset of fever, period of disability, recovery and sequelae.

i. Onset of fever

Fever is usually biphasic but sometimes triphasic and occasionally multiphasic with peaks spaced 12-24 hours apart. The onset of fever is sudden, reaching a peak of 40°-42°C within a few hours and lasting for 12-24 hours. Clinical signs are generally mild at this stage, except for a dramatic fall in milk production in lactating cows, and so the first febrile response may be missed. Some cattle may show depression, reluctance to move and some stiffness, although appctite is maintained. This stage of the infection may last for up to 12 hours.

ii. Period of disability

The more characteristic signs of BEF occur during the second febrile phase and may persist for 1-2 days.

Although the first peak of fever has passed, the rectal temperature may still be elevated. Animals become anorexic, severely depressed and show signs of general muscle stiffness. The accompanying lameness, often referred to as a shifting lameness, may or may not include joint swelling. Serous or mucoid nasal discharge is common but ocular, watery discharge is less so. There may be periorbital and/or submandibular edema, which may be patchy elsewhere on the head. The heart and respiration rates are accelerated and the animal may exhibit waves of shivering or muscle twitching. Dry rales progressing to moist rales may be detected in the lungs. By this stage, milk production may have virtually ceased and the quality of the remaining milk will be substandard.

Sternal recumbency is an obvious feature of BEF and may last for 12-24 hours or even longer in some cases. Initially, animals are able to rise if sufficiently stimulated but as the disease progresses they may be unable to do so, remaining sternally recumbent for hours or even days with the head turned to the flank. Loss of the swallowing reflex, bloat, ruminal stasis, constipation, and excessive salivation may be evident. In severe cases, animals may lapse into lateral recumbency with further reflex loss, progressing to coma and death.

iii. Recovery

In the majority of cases the disease is mild to moderately severe. Recovery can be either gradual or dramatic but is usually well advanced by the third day after the appearance of overt clinical signs. The early signs of improvement are evident in most cattle a few hours after fever disappears. Lactating cows, bulls in good condition, and fat steers are the worst affected and their recovery may take up to a week or longer, particularly if complications set in.

More specifically, bulls and fat cows that have lost condition are slow to regain weight. In lactating cows, milk yields steadily increase as animals recover but unless cows are in the early stages of lactation, they do not return to pre-illness levels. With the exception of pregnant cows that abort in late term, milk production usually returns to 85-90% of the pre-illness levels within 10 days of disease. The remaining 10-15% loss of production may persist in affected animals for the rest of the lactation period. There is no effect on the long-term fertility of the female and subsequent lactations are normal unless the cows develop a secondary bacterial mastitis.

iv. Sequelae

Complications occur in a small number of animals and include aspiration pneumonia, mastitis, abortion during late pregnancy, and temporary infertility in bulls for up to 6 months. Hindquarter paralysis may also occur and persist for days, weeks, or even permanently. Recovery from longer-term paralysis can be complete, but some cattle may retain an abnormal gait.

In general, if animals with clinical signs of BEF are forced to work or are subjected to other forms of physical stress or severe climatic conditions they are more likely to suffer severely or die. Infections with BEFV also have significant and serious socio-economic consequences in terms of animal protein production, reproduction, and national and international trade.

7. POST- MORTEM LESIONS

Few descriptions of the pathological changes that occur following a natural BEFV infection are available but they appear to be consistent with those described for experimental infection. The sporadic mortality associated with field outbreaks is the probable reason for the paucity of information.

The most obvious pathological changes include serofibrinous polysynovitis, polyarthritis, polytenovaginitis, cellulitis and focal necrosis of skeletal muscles. These lesions may vary in severity. They tend to be more severe in the joint capsules and muscles in the limbs, particularly those on which the animal was limping, but even the synovial surfaces of the spine may have fibrin plaques. Fluid flecked with fibrin clots may be present in the pericardial, pleural, and peritoneal cavities. A generalized edema of the lymph nodes is invariably evident but petechial hemorrhages are less frequent. The lungs may have patchy edema, lobular congestion, and atelectasis, and in a small number of cases severe alveolar and interstitial emphysema.

8. IMMUNE RESPONSE

Natural immunity

Following a natural BEFV infection most animals remain immune for many years, if not for life. Almost all animals that have experienced an infection are immune to any subsequent natural or experimental challenge. Challenge with BEFV strains of a different origin does not cause disease in immune animals. However, in some cases, particularly in older animals, immunity may be lost after a few years.

Immunization

Live vaccines with adjuvant confer at least 12 months protection after 2 doses given 4 weeks apart. In contrast, the killed vaccines provide approximately 6 months protection when given in 2 doses, 4 weeks apart. Animals can be vaccinated from 6 months of age and should be revaccinated annually to ensure continued immunity.

Maternal Antibody

Maternal antibodies, derived from the dam after either natural infection or vaccination, appear to be protective during an epidemic. However, these calves become fully susceptible after the normal decline of maternal antibody.

9. DIAGNOSIS

a. Field Diagnosis

A presumptive diagnosis can and often is made from clinical observations and the history of the outbreak. Single cases are difficult to diagnose In endemic areas where immunity is high and clinical cases are less common, laboratory confirmation is required.

In animals that die, the serofibrinous inflammation observed in the tendon sheaths, fasciae and joints, together with the pulmonary lesions are, as a rule, sufficiently specific to substantiate a presumptive diagnosis of BEF.

b. Laboratory diagnosis

i. Samples

A minimum of 2 blood samples are required, one taken during the period of fever and a second 1-3 weeks later. Each blood sample should be divided to provide a serum and a whole blood sample in a suitable anticoagulant. Samples should be taken from different animals in various stages of the disease to facilitate a rapid laboratory confirmation. Blood taken during illness often fails to coagulate, even after standing for several days, and may be streaked with fibrin.

ii. Laboratory procedures

A differential leukocyte count on a blood smear made from the whole blood containing anticoagulant is the most rapid test for BEF and can provide strong supporting evidence for the field diagnosis. Although a high percentage of neutrophils with at least 30% immature or banded forms is not itself pathognomonic for ephemeral fever, if not present, the field diagnosis is not correct.

BEFV can be isolated from whole blood in *Aedes albopictus* cells and by intracerebral inoculation of 1-2 day old mice. Baby hamster kidney and VERO cells are best suited for amplification of the already isolated virus. Confirmation of the virus identity is achieved using assays that utilize specific BEFV antisera or monoclonal antibodies. Immunofluorescence and complement fixation are ephemerovirus group-specific assays and have limited value.

Laboratory confirmation of a BEFV infection may be achieved serologically by demonstrating a 4-fold rise in the neutralizing antibody titer between 2 sequentially collected serum samples. However, this 4-fold rise in titre can be subjective if animals have been exposed previously to antigenically related ephemeroviruses. In such cases, animals may exhibit an anamnestic rather than primary response following the first BEFV infection. The serum neutralization test (SNT) is probably the most widely available and favored test. Also, an antibody blocking ELISA that gives similar results to the SNT and distinguishes between the antibodies induced by BEFV and antigenically related viruses has been reported.

10. PREVENTION AND CONTROL

a. Treatment

Ephemeral fever is one of the rare viral diseases for which interceptive treatment is effective. The inflammatory nature of the disease process means that it is responsive to anti-inflammatory drugs. These drugs must be given in repeated doses for the expected course of the clinical disease. Oral administration should be avoided if the swallowing reflex is not functional. If signs of hypocalcemia are observed, the paresis or paralysis responds to injected calcium borogluconate, which should be administered every 6 hours for up to 1 day. Early treatment is more effective than late, and relapses may occur if anti-inflammatory treatment is discontinued too early or if animals are stressed during the recovery period. Viremia and subsequent immunity are not significantly affected by treatment.

Rehydration with isotonic fluids may also be warranted, particularly for valuable stock and antibiotic treatment may be required to combat secondary infection.

b. Prevention

Not all of the arthropod vectors of BEFV have as yet been identified. Furthermore, the relative importance of mosquitoes and/or *Culicoides* biting midges in the spread of BEFV has not been properly assessed. Hence, the value of any vector control program at this time is uncertain, although housing cattle in insect-proof facilities at times of vector activity may be worthwhile for valuable animals.

c. Vaccination

Vaccination is currently the only useful preventive measure. Live attenuated vaccines that incorporate an adjuvant have been produced in South Africa, Japan and Australia. These vaccines appear to protect cattle against severe laboratory challenge, but evidence of their effectiveness in the field in the face of an outbreak may be variable. In Japan, killed vaccines that are cheaper to produce have also been developed and used to boost the initial immunity produced by live virus vaccines. A subunit vaccine that protects against laboratory and field challenge has been described but is not commercially available.

In epizootic areas, it is important to vaccinate cattle annually, particularly dairy, feedlot and valuable breeding stock, to maintain immunity and to help prevent the production losses caused by BEFV. Vaccine should be given in the spring to ensure a high level of immunity during the summer and autumn when the vectors are likely to be most abundant and active. Maternally derived antibody in calves younger than 6 months of age may interfere with the vaccinal response.

11. GUIDE TO THE LITERATURE

1. ANDERSON, E.C. and ROWE, L.W. 1998. Thc prevalence of antibodies to thc viruses of bovine virus diarrhea, bovine herpes virus 1, Rift Valley fever, ephemeral fever and bluetongue and to *Leptospira* sp. in free-ranging wildlife in Zimbabwe. *Epidemiol. Infect*. 121:441-449.
2. BASSON, P. A., PIENAAR, J. G. and VAN DER WESTHUIZEN, B. 1970. The pathology of ephemeral fever: A study of the experimental disease in cattle. *J. S. Afr. Vet. Med. Assoc*. 40:385-397.

3. KIRKLAND, P.D. 1993. The epidemiology of bovine ephemeral fever in south-eastern Australia: Evidence for a mosquito vector. In: *Bovine Ephemeral Fever and Related Arboviruses*. T.D. St.George, M.F. Uren, P.L.Young, D. Hoffman, eds. *ACIAR Proc*. No. 44 Canberra, Australia. pp. 33-37.
4. MELLOR, P. S. 2001. Bovine ephemeral fever. In *Encyclopedia of Arthropod-Transmitted Infections of Man and Domesticated Animals*. M.W. Service, ed., CABI Publishing, pp. 87-91.
5. NANDI, S. and NEGI, B.S. 1999. Bovine ephemeral fever: a review. *Comp. Immunol. Microbiol. Inf. Dis*. 22:81-91.
6. ST. GEORGE, T. D. and STANDFAST, H. A. 1988. Bovine ephemeral fever. In: *The Arboviruses: Epidemiology and Ecology.* T. P. Monath, ed., Boca Raton, Florida: CRC Press, Vol. 2, pp. 71-86.
7. St GEORGE, T.D. 1990. Bovine ephemeral fever virus. In *Virus Infections of Ruminants.* Z. Dinter, and B. Morein, eds., Amsterdam: Elsevier, Vol. 3, Ch. 38, pp. 405-415.
8. ST. GEORGE, T. D. 2005. Bovine ephemeral fever. In: *Infectious Diseases of Livestock*, 2nd ed., J.A.W. Coetzer, R.C. Tustin, eds. Capetown: Oxford University Press, pp. 1183-1193.
9. THEODORIDIS, A., GIESECKE, W. H. and DU TOIT, I. J. 1973. Effect of ephemeral fever on milk production and reproduction of dairy cattle. *Onderstepoort J. Vet. Res.,* 40:83-91.
10. UREN, M.F., ST. GEORGE, T.D. and MURPHY, G.M. 1992. Studies on the pathogenesis of bovine ephemeral fever III: Virological and biochemical data. *Vet. Microbiol*. 30:297-307.

Christopher Hamblin, DVM, PhD, 94 South Lane, Ash, Near Aldershot, Hampshire, GU12 6NJ, England, chris.hamblin@ntlworld.com

10

BOVINE SPONGIFORM ENCEPHALOPATHY

1. NAME

Bovine spongiform encephalopathy, “mad cow disease”, vaca loca

2. DEFINITION

Bovine spongiform encephalopathy (BSE), also referred to as “mad cow disease” is a chronic, non-febrile, neuro-degenerative disease affecting the central nervous system (CNS).

The transmissible spongiform encephalopathies (TSEs) of domestic animals, of which BSE is a member include: scrapie of sheep and goats; chronic wasting disease (CWD) of elk, moose, mule, and white tailed deer; feline spongiform encephalopathy; and, transmissible mink encephalopathy.

The central factor in BSE pathogenesis is the accumulation in the CNS of a misfolded version of the normal prion protein. The key component of BSE pathogenesis which helps explain the worldwide decline in BSE cases is the absence of horizontal (contact) transmission. This characteristic of BSE pathogenesis is in contrast to the efficient horizontal transmission of CWD and the less efficient but clearly horizontally transmitted scrapie of sheep.

3. ETIOLOGY

The prion protein, a glycosylphosphatidylinositol-linked surface glycoprotein is at the center of the etiology of the TSEs. The normal form of this protein is expressed by numerous cells, including those of the central and peripheral nervous symptoms as well as those of the lymphoreticular system. Without doubt research will continue to uncover new components in the conversion of the normal cellular prion molecule into the misfolded disease form. The misfolded prion, which partially comprises protease resistant aggregates, accumulates in the course of BSE and is a major component, if not the sole component, of the transmissible agent. Research will continue to explore the potential role of viruses, bacteria, and other infectious agents as well as components of cellular metabolism in the pathogenesis of BSE. However, critical to the current successful control strategies is the detection of the misfolded prion protein within diagnostic samples.

4. HOST RANGE

Consideration of host range of the TSEs and BSE in particular must take into consideration the route used during experimental transmission. This distinction is critical for accurate determination of cross-species transmission risk and implementation of control strategies. The natural route of BSE transmission is ingestion. Unique among the TSEs, BSE has extensively crossed species barriers. In addition to the strong epidemiological ties between BSE and variant Creutzfeldt-Jakob disease (vCJD) in humans, BSE has occurred naturally in 7 species of the family Bovidae, 4 Felidae and 4 non-human primates. BSE has been experimentally transmitted orally to cattle, sheep, goats, mice and mink.

5. EPIDEMIOLOGY

a. Transmission

Transmission of BSE is by ingestion of feedstuffs contaminated with recycled protein from the central nervous system of infected cattle. Thus far there is no evidence for horizontal or vertical transmission, however, increased risk for BSE in the offspring of clinical cases has been reported. Whether "spontaneous" cases of BSE occur, as is the case with Creutzfeldt-Jakob disease in humans, is unknown. Current knowledge does not offer a method to differentiate cases of spontaneous BSE with those acquired by ingestion of contaminated feed. With the exception of limited involvement of the Peyer's patches in the distal portion of the ileum, eyes and spleen, BSE is confined to the CNS. Therefore current regulations which restrict ruminant protein from ruminant feed preclude the possibility of transmission from spontaneous cases.

b. Incubation period

The incubation period of BSE varies from 2-8 years with the majority of cases occurring in dairy or dairy crossbreeds between 3-5 years old.

c. Morbidity

Usually only a single animal in a herd will be noted to be infected. However, because of the serious public health nature of the disease, the entire herd is depopulated when the disease is discovered, so it is difficult to know if there would be more extensive morbidity.

d. Mortality

Once clinical signs appear death ensues within 1-6 months. All evidence to date indicates 100% mortality.

6. CLINICAL SIGNS

Clinical signs of BSE are referable to the central nervous system (CNS) and present as behavioral changes such as hyper-reactivity, apprehension, gait abnormalities, nervousness or aggression, and abnormal posture. Early BSE clinical signs are suggestive of metabolic or deficiency diseases such as nervous ketosis and hypomagnesemia. Experience to date indicates that BSE is ultimately fatal.

7. POST- MORTEM LESIONS

There are no gross post-mortem lesions associated with BSE. Histological changes most characteristic of BSE are bilaterally symmetrical vacuolation of the gray matter neuropil. In cattle the areas most consistently affected are the solitary tract nucleus, the spinal tract nucleus of the trigeminal nerve, and the central gray matter of the midbrain. Vacuolation of neuropil of these targeted nuclei are considered pathognomonic for BSE.

8. IMMUNE RESPONSE

There is no known measurable response to the misfolded prion protein of BSE or other TSEs. The disease form of the prion is derived from a normal host protein which is recognized as self. However tolerance can be broken through the use of certain immunization strategies and current research is exploring the potential role of anti-prion immunity in blocking transmission and/or disease progression.

9. DIAGNOSIS

Currently there is not an economical method to definitively diagnose BSE infection in live cattle. Therefore, diagnosis of BSE is dependent on laboratory evaluation of post-mortem samples from the central nervous system, in particular the midbrain. The laboratory methods utilized to diagnose BSE are microscopic examination to identify the pathological changes described above, and methods to detect partially-proteinase resistant misfolded prion protein. These methods are ELISA, immunohistochemistry and immunoblotting.

10. PREVENTION AND CONTROL

A BSE risk assessment published in 2001 and revised in 2003 by Harvard University concluded that the “multiple firewall” system put in place by government would keep BSE from spreading and that it would eventually die out. The firewalls consist of a feed ban, import controls and a surveillance program. The Harvard study identified the feed ban as the most important component of the control firewalls. Based on the fact that data to date indicate BSE is not horizontally transmitted, and that the primary, if not exclusive route of transmission is ingestion of contaminated feed, a ban on feeding cattle and

other ruminants mammalian protein was instituted. Also since 1989, the USDA has banned the import of live ruminants and most ruminant products from countries where BSE has been reported or thought to be of high risk. The USDA has also initiated a surveillance program. The current details of BSE testing in the U.S. can be found at the following website:
http://www.aphis.usda.gov/newsroom/hot_issues/bse/surveillance/ongoing_surv_results.shtml.

11. GUIDE TO THE LITERATURE

1. BROWN, P. 2003. Transmissible spongiform encephalopathy as a zoonotic disease. International Life Sciences Institute, ILSI Press, One Thomas Circle, NW, Ninth Floor, Washington DC 20005-5802, U.S.
   e-mail: publications@ilsieurope.be
2. COHEN, J. T., DUGGAR, K., GRAY, G. M. and KREINDEL, S. 2001, revised 2003. Evaluation of the potential for bovine spongiform encephalopathy in the U.S., Harvard Center for Risk Analysis,
   Harvard School of Public Health and the Center for Computational Epidemiology, College of Veterinary Medicine, Tuskegee University http://www.aphis.usda.gov/lpa/issues/bse/madcow.pdf
3. FRANCO, D. A. 2005. An introduction to the prion diseases of animals, assessing the history, risk inferences, and public health implications in the U.S. The National Renders Association, Prion Diseases Booklet, National Renderers Association, 801 N. Fairfax St., Suite 205, Alexandria, VA 22314 (703-686-0155).
4. GAVIER-WIDEN, D., STACK, M. J., BARON, T., BALACHANDRAN, A., and SIMMONS, M. 2005. Diagnosis of transmissible spongiform encephalopathies in animals: a review. *J. Vet. Diagn. Invest*. 17: 509-527.
5. WATTS, J. C. BALACHANDRAN, A., and WESTAWAY, D. 2006. The expanding universe of prion diseases.
   PLoS Pathogens/www.plospathogens.org, Volume 2. Issue 3:e26

See Part IV for photos.

Donald Knowles, DVM, PhD, USDA-ARS, Pullman, Washington 9914-6630,
dknowles@vetmed.wsu.edu

11

CAPRIPOXVIRUSES:

SHEEP AND GOAT POX (SGP) *and* LUMPY SKIN DISEASE (LSD)

1. NAME

Capripoxviruses; Knopvelsiekte (LSD)

2. DEFINITION

Sheep pox, goat pox and lumpy skin disease are all systemic infections caused by capripoxviruses.

Sheep and goat pox (SGP) are systemic pox diseases of sheep and goats characterized by fever, macules developing into papules and necrotic lesions in the skin and nodular lesions in internal organs, secondary infections and death in susceptible stock. The diseases in sheep and goats are caused by strains of capripoxvirus which are indistinguishable serologically, and although strains derived from sheep may show different DNA restriction enzyme digest patterns from those derived from goats, there are strains with characteristics of both sheep and goat strains. Similarly, while most strains cause disease in either sheep or goats, there are strains equally pathogenic in both species.

Lumpy skin disease (LSD) is a pox disease of cattle characterized by fever, nodules on the skin, mucous membranes, and internal organs, emaciation, enlarged lymph nodes, edema of the skin, and sometimes death. The disease is of economic importance because it causes reduced production, particularly in dairy herds, and also causes damage to the hide.

Capripox is a major constraint to the introduction of exotic breeds of sheep and goats, and to the development of intensive livestock production. Strains of capripoxvirus that cause lumpy skin disease (such as Neethling), are also found in cattle, but there is no evidence that these strains will naturally cause disease in sheep and goats. The geographical distribution of lumpy skin disease differs from that of sheep pox and goat pox.

3. ETIOLOGY

Capripoxviruses are large brick-shaped, double stranded DNA viruses, morphologically indistinguishable from orthopoxviruses, measuring 295 by 265 nm. The virion is covered in short tubular elements which give it a different appearance from orf (contagious pustular dermatitis) virus (a parapoxvirus), which is more oval in shape and covered in a continuous filament. Replication of poxviruses is very faithful, and DNA restriction patterns of isolates collected in 1959 appear the same as 1986 isolates, indicating very little change in the genome over time. However, there is evidence for recombination events occurring between strains of capripoxvirus in the field, as has also been seen *in vitro*, and this could result in changes in host range or virulence.

The virus is susceptible to sunlight and detergents containing lipid solvents, but in dark environmental conditions, such as contaminated animal sheds, it can persist for many months. The scabs shed from recovered animals contain large amounts of virus in association with antibody, but it is not known whether these scabs could remain a source of infection. Certainly it is difficult to recover live virus on tissue culture from scab material, but the presence of the virus within type A inclusion protein could protect the virus in the environment.

Strains of capripoxvirus do pass between sheep and goats, although most cause more severe clinical disease in only one species. Recombination also occurs between these strains, producing a spectrum showing intermediate host preferences and a range of virulence. Some strains are equally pathogenic in both sheep and goats.

4. HOST RANGE

a. *Domestic animals and wild animals*

Sheep and goat pox affects all breeds of domestic and wild sheep and goats. In endemic areas, native breeds are far less susceptible to severe disease than introduced breeds. When European or Australian breeds are imported into endemic regions without protection from contact with local animals, morbidity and mortality can approach 100%.

LSD affects all breeds of cattle. *Bos taurus* is more susceptible than *Bos indicus*; the Asian buffalo (*Bubalus bubalis*) has also been reported to be susceptible. Within *Bos taurus*, the fine-skinned Channel Island breeds develop more severe disease, with lactating cows appearing to be thc most at risk. While wildlife species such as kudu and giraffe have been experimentally infected with capripoxvirus, their involvement in the epidemiology of LSD or sheep and goat pox has not been shown.

b. *Humans*

Although there have been 2 reports in the literature that Capripoxvirus can transmit to man, these can be discounted. Capripoxvirus does not infect people.

5. EPIDEMIOLOGY

a. Transmission

Transmission of sheep pox is usually by aerosol following close contact between a susceptible and clinically affected animal. Animals that die with acute disease before showing typical clinical signs and animals with only very mild, localized infections rarely transmit disease. The epidemiology is very similar to that of human smallpox and, like smallpox, most transmission occurs from severely affected individuals during the stage when ulcerated papules are present on the mucous membranes to the start of papule necrosis, when the antibody response can be detected. There is no transmission during the 'pre-papular' stage, and the infectivity of the virions present in the scab material is not known. Although biting insects have been shown to transmit Capripoxvirus under experimental conditions, there is no evidence that they are significant in the epidemiology of sheep pox.

The involvement of goats in the epidemiology of sheep pox and of sheep in the epidemiology of goat pox has been controversial. Certainly, there are strains of Capripoxvirus that can transmit between the 2 species and cause disease in both. However, the majority of strains show a host preference.

Transmission of LSD virus is thought to be predominantly by insects, as natural contact transmission in the absence of insect vectors is inefficient. Transmission by insects is mechanical, with no replication of virus occurring in the vector, and it is probable that a large number of biting and blood feeding insects are capable of spreading the virus. This is sometimes reflected in the distribution of the lesions, with more papules being seen on the legs if the insect vector is predominantly feeding on those sites. It is probable that mosquitoes are responsible for transmission in the more severe outbreaks because they are likely to inoculate virus directly into the bloodstream, and this method of infection is more likely to produce generalized clinical signs.

b. Incubation period

The incubation period of sheep- and goat pox following contact between an infected and susceptible animal is 8-13 days. Infection is usually by the aerosol route, but virus also can be spread mechanically by insect bite or experimentally by intradermal or subcutaneous injection, in which case a reaction can be seen at the site of infection within 4 days.
For LSD, the incubation period under field conditions has not been reported, but following inoculation is 6-9 days until the onset of fever.

c. Morbidity

With SGP, morbidity depends on many factors including viral load in the environment, level of vaccination, and breed predisposition. Morbidity can vary from 10-100%. Morbidity with LSD is dependent upon breed and presence of insect vectors, therefore there is some seasonal variation.

d. Mortality

Mortality with SGP is also dependent on breed, but can be up to 100% in non-native breeds. The mortality rate due to LSD may be as high as 10%, and is more likely to occur in non-native cattle.

6. CLINICAL SIGNS

a. Sheep and goats

The rectal temperature rises to 40°C or higher and, in the following 2-5 days, macules (1-3 cm-diameter areas of hyperemia) appear on the skin, particularly in the groin, axilla and perineum. Macules develop into hard swellings or papules, which rarely have a fluid-filled cap. Some particularly susceptible breeds of sheep, such as the Soay, die before the appearance of the characteristic clinical signs. The papules on the mucous membranes of the nose, mouth, mammary glands, vulva, and prepuce quickly ulcerate, leading to rhinitis, conjunctivitis, and blepharitis and a secondary mucopurulent discharge and mastitis. The superficial lymph nodes, in particular the prescapular nodes, become enlarged and pressure on the trachea from the enlarged retropharyngeal nodes can interfere with breathing. In addition, animals may be dyspneic because of extensive viral damage in the lungs.

If the affected animal survives the initial acute stage of the disease, between 5-10 days after their initial appearance, the papules form scabs, which persist on the skin for a further 2-3 weeks and then are shed, leaving scars, most obvious on the face. In tropical regions, secondary fly strike can result in permanent blindness or death. Recovery from severe disease is slow and is characterized by bouts of fever, pneumonia and anorexia. Abortion is rare unless there are secondary or additional infections.

b. Cattle

The severity of clinical signs of LSD depends on the strain of capripoxvirus and the breed of host. In the acutely infected animal, there is an initial pyrexia, which may exceed 41°C and persist for 1 week. Rhinitis and conjunctivitis develop, and in lactating cattle there is a marked reduction in milk yield. Nodules of 2-5 cm in diameter develop over the body, particularly on the head, neck, udder, and perineum between 7-19 days after virus inoculation. These nodules involve the dermis and epidermis and may initially exude serum, but over the following 2 weeks may become necrotic plugs that penetrate the full thickness of the hide. All the superficial lymph nodes are enlarged, the limbs may be edematous and the animal is reluctant to move. The nodules on the mucous membranes of the eyes, nose, mouth, rectum, udder, and genitalia quickly ulcerate, and by then all secretions contain LSD virus. On the appearance of clinical signs, the discharge from the eyes and nose becomes mucopurulent, and keratitis may develop. Pregnant cattle may abort, and there are reports of aborted fetuses being covered in nodules. Bulls may become permanently or temporarily infertile and the virus can be excreted in the semen for prolonged periods. Recovery from severe infection is slow; the animal is emaciated, may have pneumonia and mastitis, and the necrotic plugs of skin, which may have been subject to fly strike, are shed leaving deep holes in the hide.

7. POST- MORTEM LESIONS

a. Gross

Sheep and goats: On post-mortem examination of the acutely infected sheep or goat, the skin lesions are often less obvious than on the live animal. The mucous membranes have multiple necrotic foci and all the body lymph nodes are enlarged and edematous. Papules, which may be ulcerated, can usually be found on the abomasal mucosa, and sometimes on the wall of the rumen and large intestine, on the tongue, hard and soft palate, trachea and esophagus. Pale areas of approximately 2 cm in diameter may occasionally be seen on the surface of the kidney and liver, and have been reported to be present in the testicles. Throughout the lungs, but particularly in the diaphragmatic lobes, there are numerous hard lesions of up to 5 cm in diameter.

Cattle: In cattle dying of LSD, nodules penetrate the thickness of the skin and involve the subcutaneous tissue which appears edematous, and is infiltrated with blood tinged fluid. Nodules may be present in the mouth, subcutis and muscle, in the trachea, reproductive tract, and alimentary tract (particularly the abomasum), and in the lungs, resulting in primary and secondary pneumonia. The lymph nodes may be grossly enlarged due to lymphoid hyperplasia and edema. Secondary pneumonia and mastitis may be seen.

b. Key microscopic

In both SGP and LSD, histopathology is characterized by the presence of intracytoplasmic inclusion bodies in infected cells, which are sites of virus replication (type B inclusions). Infected cells appear as large polygonal cells with a large poorly defined eosinophilic inclusion and a vasicular nucleus. These cells are particularly apparent in histological sections of the skin (dermis and subcutis) and mucous membranes in glandular cells. The virus also replicates in macrophages, and can be isolated from the blood buffy coat during the viremic stage of the disease. During the later stage, as the affected animal develops an immune response, the blood vessels supplying the papules thrombose, the papules become necrotic, and the virions in the affected cells concentrate in a matrix of virus-induced protein within the cytoplasm (type A inclusions). This material may protect the virions as the cells are shed into the environment and could provide a future source of infection if inhaled by a susceptible animal.

8. IMMUNE RESPONSE

a. Natural infection

Following recovery from capripoxvirus infection in sheep, goats or cattle, immunity is life-long and prevents infection with all strains of virus.

b. Immunization

A variety of attenuated live and inactivated capripoxvirus vaccines has been used to provide protection to sheep and goats against capripox. All strains of capripoxvirus of ovine, caprine, or bovine origin examined so far share a major neutralizing site, so that animals recovered from infection with one strain are resistant to infection with any other strain. Consequently, it is possible to use a single strain of capripoxvirus to protect sheep, goats, and cattle against all field strains of virus, regardless of whether their origin was in Asia or Africa.

There are 2 antigenic forms of capripoxvirus, the intact virion covered in short tubular elements, and the intact virion additionally covered in a host-cell-derived membrane. The latter is the form usually produced by the infected animal, whereas the former is that seen when virus is produced by freeze–thawing infected tissue culture. Dead vaccines produced from tissue culture are almost entirely naked virions, and, when used as a vaccine, do not stimulate immunity to the membrane-bound virion. This in part explains the poor success of inactivated vaccines. An additional factor is that inactivated vaccines are less effective than live, replicating vaccine virus in stimulating the cell-mediated immune response, which is the predominant protective response to poxvirus infection. Dead capripox vaccines provide, at best, only temporary protection. A number of strains of capripoxvirus have had widespread use as live vaccines, for example the 0240 Kenya sheep and goat pox strain used in sheep and goats, the Romanian and RM-65 strains used mainly in sheep, and the Mysore and Gorgan strains used in goats. Immunity in sheep and goats against capripox following vaccination with the 0240 strain lasts over a year, and will probably provide lifelong protection against lethal challenge. The 0240 strain should not be used in *Bos taurus* breeds of cattle.

The attenuated Neethling strain is used in South Africa to protect cattle, but sheep strains are also effective. The Romanian strain was used in Egypt when LSD first was introduced. Both the Neethling and 0240 strains can cause reactions in cattle.

9. DIAGNOSIS

a. Field diagnosis

The clinical signs of severe sheep and goat pox and LSD are highly characteristic. However both diseases in their mild form can be confused with parapoxvirus infections causing orf in sheep and goats or pseudocowpox and bovine papular stomatitis in cattle. Multiple insect bites or urticaria may also appear similar to capripoxvirus infections.

b. Laboratory diagnosis

i. Samples

Material for virus isolation and antigen detection should be collected by biopsy or at post-mortem from skin papules, lung lesions or lymph nodes, preferably during the first week of clinical signs. Samples for genome detection by polymerase chain reaction (PCR) may be collected later, even when neutralizing antibody is present.

ii. Laboratory procedures

Laboratory confirmation of capripox is most rapid by the demonstration of typical capripox virions using the transmission electron microscope. A precipitating antigen can be identified by an agar gel immunodiffusion test (AGID) using lymph node biopsy material taken from an early case of capripox and specific immune sera, however, there is a cross-reaction with parapoxvirus. Capripoxvirus will grow on tissue culture of ovine, caprine, or bovine origin, although field isolates may require up to 14 days to grow, or require one or more additional tissue culture passage(s). An antigen-detection enzyme-linked immunosorbent assay (ELISA) using a polyclonal detection serum raised against a recombinant immunodominant antigen of capripoxvirus has been developed. Genome detection using capripoxvirus-specific primers for the fusion protein gene and attachment protein gene has also been reported.

The virus neutralization test is the most specific serological test, but because immunity to capripox infection is predominantly cell-mediated, the test is not sufficiently sensitive to identify animals that have had contact with the virus and have developed only low levels of neutralizing antibody. The AGID and indirect immunofluorescence tests are less specific due to cross-reactions with antibody to other poxviruses. Western blotting using the reaction between the P32 antigen of capripoxvirus with test sera is both sensitive and specific, but is expensive and difficult to carry out. The use of this antigen, expressed by a suitable vector in an ELISA, offers the prospect of an acceptable and standardized serological test.

10. PREVENTION AND CONTROL

Prevention in countries free of sheep and goat pox and LSD is directed towards prohibiting the importation of live sheep, goats, or cattle, or their products from pox-endemic regions. Prevention in endemic areas relies heavily on vaccination.

Protection against sheep and goat pox is effectively lifelong following vaccination with the 0240 strain, although, experimentally, virulent challenge virus will replicate at the site of subcutaneous inoculation in an animal vaccinated 2 years previously. Immunity is sufficient to prevent any generalization. The vaccine is very cheap to produce, and in control programs should be used annually in all sheep and goats, regardless of age.

11. GUIDE TO THE LITERATURE

1. BLACK, D.N., HAMMOND, J.M. and KITCHING, R.P. 1986. Genomic relationship between Capripoxviruses. *Virus Res*. 5:277-292.
2. CARN, V.M. 1993. Control of capripoxvirus infections. *Vaccine*. 11:1275-1279.
3. CARN, V.M. and KITCHING, R.P. 1995. An investigation of possible routes of transmission of lumpy skin disease virus (Neethling). *Epidemiol. Infect*. 114:219-226.
4. CARN, V.M., and KITCHING, R.P. 1995. The clinical response of cattle following infection with lumpy skin disease (Neethling) virus. *Arch. Virol*. 140:503-513.
5. COETZER, J.A.W. 2004. Lumpy skin disease. In: *Infectious Diseases of Livestock*, JAW Coetzer, RC Tustin, eds., Cape Town: Oxford University Press 2nd Edition. pp. 1268-1276.
6. HEINE, H.G., STEVENS, M.P., FOORD, A.J., and BOYLE, D.B. 1999. A capripoxvirus detection PCR and antibody ELISA based on the major antigen P32, the homolog of the vaccinia virus H3L gene. *J. Immunol. Methods*. 227:187-196.
7. KITCHING, R.P., HAMMOND, J.M. and BLACK, D.N. 1986. A single vaccine for the control of capripoxvirus infection in sheep and goats. *Res. Vet. Sci*. 42:53-60.

See Part IV for photos.

Paul Kitching, DVM, PhD, Director, National Centre for Foreign Animal Disease, Winnipeg, Manitoba R3E 3M4, Canada,
kitchingp@inspection.gc.ca

12

CLASSICAL SWINE FEVER

1. NAME

Hog cholera, peste du porc, Virusschweinepest

2. DEFINITION

Classical swine fever (CSF) is a highly contagious and often fatal viral disease of pigs. In pigs that are infected postnatally, the disease can run an acute, subacute or chronic course. The acute form of disease is characterized clinically by high fever, severe depression, skin hemorrhages, and high morbidity and mortality. Pigs with subacute and chronic forms of the disease present with clinical signs that are less severe than those seen with animals exhibiting the acute course. Apparent recovery can occur in older animals that develop the subacute and chronic forms. Prenatal infection with virus strains of low to moderate virulence can lead to the birth of persistently infected piglets that may subsequently go on to develop late onset disease. Persistently infected pigs that result from congenital infections constitute a major risk for virus spread to non-infected farms.

3. ETIOLOGY

Classical swine fever virus (CSFV) belongs to the family Flaviviridae, genus *Pestivirus*. Although antigenic variants of CSFV occur there is only one serotype. The CSFV is antigenically related to bovine viral diarrhea virus (BVDV) and border disease virus (BDV), 2 other Pestiviruses of veterinary importance. Because CSFV is an enveloped virus, it has been regarded as moderately fragile, however, depending on physical conditions it can survive for variable periods of time in the environment. It is considered relatively stable at a pH range of 5-10, but is rapidly inactivated at pH values below 3 and above 10. It has been reported that virus inactivation occurs in less than a minute at 100°C, within 1 minute at 90°C, 2 minutes at 80°C and 5 minutes at 70°C. The virus is relatively stable at lower temperatures and is capable of surviving in infected pig meat for weeks when refrigerated and for years when frozen.

4. HOST RANGE

The pig is the only natural host for CSFV and this includes domestic pigs and wild boar. Humans are not susceptible to infection.

5. EPIDEMIOLOGY

a. Transmission

Infected pigs shed virus for variable but considerable periods of time in their saliva, lacrimal secretions, urine, feces, semen and blood. The length of time that virus is shed is dependent on the virulence of the strain involved and the development of neutralizing antibodies. It is likely that direct pig to pig contact is the major route of CSFV transmission. Oronasal exposure to contaminated secretions or excretions is probably the most common route of infection however, entry of virus via mucous membranes, conjunctiva, skin abrasions, insemination, or the use of contaminated needles are other potential routes by which infection can take place. Airborne spread of the virus may occur where there is a large concentration of infected animals. This is most likely to take place between mechanically ventilated buildings that are in close proximity to one another. Fomites (implements, vehicles, clothing, instruments, etc.) are another important means of spread which can be associated with activities such as livestock auctions, visits by feed dealers, rendering trucks and veterinary services. Mechanical spread by insects is also possible especially during the warm seasons. However, there is no evidence that CSFV replicates in invertebrates. The feeding of raw or insufficiently cooked garbage containing pork scraps is regarded as another important means of transmission and as a consequence is prohibited in many countries.

b. Incubation period

The incubation period can range from 2 to 14 days but is typically 3-4 days. The length of the incubation period, to a large degree, is dependent on virus strain with high virulence strains having shorter incubation periods than strains of lower virulence. Other factors that can affect the length of the incubation period include virus dose and route of exposure.

c. Morbidity

Morbidity can be highly variable and dependent on a number of viral and host factors. These include the virulence of the virus strain involved, ages of animals in the herd, general health status and concurrent infections within the herd, and nutritional factors. In general, high virulence strains result in more prolonged virus shedding than low virulence strains. As a result, infection with high virulence strains will spread faster through a herd causing higher morbidity than is seen with low virulence strains.

d. Mortality

Mortality also varies depending on the viral and host factors listed above. Higher levels of mortality are typically seen in younger animals regardless of viral strain virulence. In acute CSF death will occur within 10-20 days. In subacute and chronic CSF, morbidity and mortality are much lower and clinical signs can be intermittent and protracted lasting for weeks or even months.

6. CLINICAL SIGNS

Acute Classical Swine Fever

Pigs with acute CSF will initially present with nonspecific signs that include a reduction of appetite and activity, a hunched posture and drowsiness along with droopiness of the head. Fever is usually present at the first sign of inactivity and may reach as high as 42.2°C (108°F) but more typically is between 41°C and 42°C (105.8°F and 107.6°F). Reddening of the skin may accompany the fever. The affected pigs will huddle together and pile up on each other and will rise only when prompted. Discharge from the eyes with conjunctivitis is frequently observed and may progress to involve encrustation of the eyelids to the point where they are completely adhered. Gastrointestinal signs include vomiting of yellow, bile containing fluid as well as transient constipation that is followed by a severe watery, yellow-grey diarrhea. As the disease progresses pigs will have a gaunt tucked up appearance and a staggering gait that is due to hind-end weakness. This will often progress to posterior paresis. Skin hemorrhages involving the ears, abdomen, and inner thighs as well as cyanosis of the skin of the ears, tail and snout may occur near the terminal stage of the disease.

Chronic Classical Swine Fever

Three phases have been described for chronic CSF. The first or acute phase is characterized by anorexia, depression and fever. The second phase is characterized by a general improvement in the clinical condition of the animal. The third phase is characterized by clinical relapse and death. Gastrointestinal signs during the active phase of the disease include alternating diarrhea and constipation. Diarrhea may last for weeks or even months. Affected animals grow poorly and may have a disproportionately large head relative to their trunk. Runted pigs may stand with arched backs and have skin lesions. The skin lesions, which tend to be multifocal and necrotic, can be confused with those seen with porcine dermatitis and nephropathy syndrome (PDNS). Pigs with chronic CSF may survive for longer than 3 months before dying.

Congenital Classical Swine Fever

When CSFV infects pregnant sows it can cross the placenta to infect the developing fetus. The outcome depends on the virulence of the strain of virus infecting the sow and the stage of gestation. A transient fever and anorexia accompanied by poor reproductive performance may be all that is seen in the affected sows.

Infection with virulent strains of CSFV usually results in abortions or early postnatal deaths. Infection with low or moderately virulent strains can result in fetal mummification, stillbirths, or the birth of weak piglets with tremors. Malformations involving the visceral organs and central nervous system can occur. Infection with a low virulent strain during the first trimester of pregnancy can lead to the birth of healthy-looking, persistently infected pigs. These pigs do not produce neutralizing antibodies, have a persistent viremia, and can excrete large amounts of virus over their lifespan. Such persistently infected pigs may remain disease free for several months before going on to develop mild anorexia, depression, conjunctivitis, dermatitis, diarrhea, and posterior paresis. Pigs affected in this way can survive 2-11 months, but all will eventually die.

7. POST- MORTEM LESIONS

Acute Classical Swine Fever

The predominant lesion of acute CSF is hemorrhage. This can be seen as enlarged, marbled red lymph nodes and widespread petechial or ecchymotic hemorrhages involving the skin, subcutis, bladder, kidneys, and serous and mucous membranes. These lesions are due in large part to the direct effect the virus has on vascular endothelial cells. Necrotic foci may also be present on the palatine tonsils. The submandibular and pharyngeal lymph nodes are the first to become edematous and hemorrhagic owing to the fact that the primary site of CSFV replication is the epithelial cells that line the tonsillar crypts. From there, CSFV spreads via lymphatic vessels to lymph nodes that drain the tonsillar region. As the disease progresses, lesions will spread to involve other lymph nodes. Infarction of the spleen, expressed as raised, dark wedge-shaped areas along the edge and surface of the organ, occurs infrequently in pigs that are infected with the currently circulating strains of CSFV, but when it does, it is considered to be almost pathognomonic for the disease.

Infarctions and hemorrhages may also be present in the lung. Catarrhal to fibrinous bronchopneumonia and pleuritis may develop as a result of secondary bacterial infections. In addition to the accumulation of straw-colored fluid in the thoracic cavity, similar fluid accumulations may be seen in the peritoneal cavity and pericardial sac.

Encephalitis is a common finding in CSFV-infected pigs at necropsy. Gross lesions associated with encephalitis include hyperemia and congested blood vessels. The principal microscopic lesion is perivascular cuffing although mild degeneration of endothelial cells, microgliosis, and areas of focal necrosis can also be seen.

Chronic Classical Swine Fever

In chronic CSF the lesions are the same as observed with acute CSF but less severe. Hemorrhages and infarctions can be completely absent. Secondary bacterial infections often complicate the primary CSF lesion. Necrosis and ulcerations in the form of button ulcers can be found in the mucous membrane of the gastrointestinal tract, particularly the cecum and colon. In growing pigs that survive for longer than 30 days, transverse striations which consist of residual unmodelled growth cartilage can be seen in ribs that are split longitudinally along the costochondral junction.

Congenital Classical Swine Fever

The lesions associated with congenital CSFV infections include fetal mummification, malformations and stillbirths. The most commonly seen malformations are deformities of the head and limbs, hypoplasia of the cerebellum and lungs, microencephaly, and hypomyelinogenesis. Generalized subcutaneous edema, ascites and hydrothorax are the most pronounced lesions in stillborn pigs.

8. IMMUNE RESPONSE

a. Natural infection

CSFV is known for its ability to evade and compromise the host's immune system. Thymic and bone marrow atrophy and destruction of germinal centers in secondary lymphoid organs are common lesions in CSFV infected pigs. Cells of the monocyte-macrophage lineage are the early targets for virus infection and replication; however, other cells which comprise the reticuloendothelial system are also affected. Leukopenia and in particular lymphopenia is a characteristic early event of CSFV infection. B-lymphocytes, helper T-cells and cytotoxic T-cells are all depleted to varying degrees which is due in part to apoptosis of uninfected lymphocytes. As a consequence, the cellular and humoral immune responses in CSF are delayed and virus neutralizing antibodies do not usually develop until after the third week of illness. Because of the B-cell and T-cell depletion, CSFV-infected pigs are immunocompromised, making them susceptible to secondary infections.

b. Immunization

A number of live attenuated vaccines form the basis of CSF control in many parts of the world. These include the Chinese (C) strain, the Japanese GPE$^-$ (guinea pig exaltation negative) strain, the Thiverval strain, and the Mexican PAV strain. Of these the C strain, which has been attenuated by hundreds of passages in rabbits, is the most widely used. This strain possesses many characteristics that make it suitable for use as an emergency vaccine. It has been shown to provide solid, long-lasting protection against clinical signs, virus replication, and virus excretion within a week following vaccination. Replication of the C strain is restricted primarily to lymphoid tissues and it appears to persist in pigs for no longer than 2-3 weeks. The C strain has been used to successfully control CSF in wild boar in Europe.

Despite these attributes, the C strain induces an antibody response that cannot be distinguished from that induced by infection with field virus. This has prompted the development of alternative vaccines that can be used in a DIVA (differentiating infected from vaccinated animals) strategy. Currently, 2 subunit vaccines which utilize the E2 glycoprotein of CSFV have been licensed for use in Europe. Protective immunity using these vaccines takes longer to induce compared with C strain, and the reduction of virus replication and shedding has been variable. The companion diagnostic tests that have been developed to differentiate infected from vaccinated animals in both cases are based on the detection of antibodies to another CSFV glycoprotein, E^{rns}. Early trials have indicated that these discriminatory tests are suitable for use on a herd but not on an individual animal basis. Research is under way to develop new and improved marker vaccines.

9. DIAGNOSIS

a. Field diagnosis

Due to the variability in clinical signs and pathological lesions, only a tentative diagnosis of CSF can be made in the field. A number of other septicemic diseases of pigs, including African swine fever, salmonellosis, actinobacillosis, acute pasteurellosis, and *Hemophilus suis* infections can look very much like acute CSF. Therefore, any septicemic condition in which pigs have a high fever should be carefully investigated. In addition to a systematic post-mortem involving 4-5 animals, a thorough history should also be taken with specific information gathered to determine if uncooked garbage has been fed, if unusual biological products have recently been used, or if there have been any recent additions to the herd.

b. Laboratory diagnosis

i. Samples

Necropsy specimens of particular relevance to CSF diagnosis include: tonsils, pharyngeal and mesenteric lymph nodes, spleen, kidney, and distal ileum. Tonsils are considered the best tissue to collect for the purpose of virus isolation, nucleic acid detection and antigen detection. Each specimen should be placed in a separate plastic bag, clearly identified, kept under refrigeration, and shipped to the laboratory as quickly as possible. In live animals, tonsil biopsies and whole blood (EDTA or heparin) should be collected from animals with fevers or showing other signs compatible with CSF. With the increasing use of the polymerase chain reaction (PCR) as a diagnostic tool, nasal and oropharyngeal swab specimens placed in a viral transport medium should also be given consideration.

Serum samples for antibody detection should be collected from animals that have recovered from suspected infection, from sows that are suspected of producing congenitally infected litters, and from pigs that have been in contact with infected or suspected cases. Because seroconversion is delayed and may not take place until 3 weeks after the beginning of illness, a sufficient number of samples should be collected to increase the probability of detecting seropositive animals.

All diagnostic material should be transported and stored in leak-proof containers in accordance with national regulations governing the transportation of diagnostic biologic samples.

ii. Laboratory procedures

Isolation of CSFV involves inoculation of pig kidney cell culture with subsequent virus detection by immunofluorescence or immunoperoxidase staining. Identification by this means is necessary because CSFV is a non-cytopathic Pestivirus. Anti-CSFV polyclonal conjugates are typically used to initially identify infected cultures. Because these polyclonal conjugates cross-react with BVDV and BDV, confirmatory testing employing the use of CSFV-specific monoclonal antibodies, or amplification of CSFV nucleic acid by reverse transcriptase-polymerase chain reaction (RT-PCR) and sequencing of the nucleic acid product usually follows. Definitive results for virus isolation may require a week or longer.

Assays using RT-PCR are gaining wider acceptance and use as diagnostic tools for CSF. They have been shown to be as sensitive, if not more sensitive, than virus isolation. These assays have additional advantages over virus isolation, the most important being rapid turnaround time along with high throughput capability.

The CSFV antigens can be detected in frozen tissues taken from dead pigs, in tonsil biopsy material, in impression smears, and in bone marrow aspirates. Polyclonal fluorescent conjugates have traditionally been used for this purpose in a fluorescent antibody test (FAT). The FAT can be carried out on frozen tissue sections within a matter of hours of specimen reception. Because BVDV and BDV antigens are also detected using polyclonal conjugates, confirmation using peroxidase-conjugated CSFV-specific monoclonal antibodies is highly recommended. In the past the FAT was routinely used as a rapid test for CSF, however, diagnostic sensitivity is only moderate, and is suitable for use only at the herd level, provided that an appropriate number of samples are taken per farm. Recent work on an avidin-biotin-complex (ABC) monoclonal antibody-based assay for use on frozen tissue sections has demonstrated a higher degree of sensitivity and specificity for CSFV infected tissues.

Low and moderate virulent strains of CSFV will often produce infections that are subclinical in nature. Because of this, serological tests are essential to identify and eradicate infected herds. Virus neutralization tests and enzyme-linked immunosorbent assays (ELISA) are available for detecting antibodies to CSFV.

10. PREVENTION AND CONTROL

Most countries have regulatory measures in place to prevent the introduction of CSF. These include trade restriction policies for live pigs as well as for fresh and cured pig meat. Importation of live pigs often requires serologic testing and a quarantine period. Because the clinical and pathological signs of CSF are so variable, newer surveillance programs (designed by the Dutch) are aimed at minimizing the length of the high risk period, which is defined as the time that elapses between virus introduction and the first positive diagnosis. These programs are based on routine gross pathologic examination of pig carcasses, routine virological testing of tonsils, clinical observations by the producer, planned periodic clinical inspections by veterinarians, and leukocyte counts in blood samples from diseased animals on antimicrobial group therapy. Other countries carry out lesser degrees of surveillance that may be more targeted in nature.

In the event that CSF is introduced, stamping out is the most appropriate initial response. All pigs on an affected farm are slaughtered and their carcasses, bedding, feed, etc., are appropriately destroyed. Infected, surveillance, and control zones should be established. Also, a complete epidemiological investigation with tracebacks to determine possible sources of introduction and traceforwards to assess possible spread should be initiated. Emergency vaccination can be considered should stamping out prove to be ineffective. A country can have a free OIE status for CSF even with vaccination, but only if a validated marker vaccine is used.

11. GUIDE TO THE LITERATURE

1. EDWARDS, S., MOENNIG, V. and WENSWOORT, G. 1991. The development of an international panel of monoclonal antibodies for the differentiation of hog cholera virus from other pestiviruses. *Vet. Micro*. 29:101-108.
2. RISATTI, G. R., CALLAHAN, J. D., NELSON, W.M. and BORCA, M.V. 2003. Rapid detection of classical swine fever virus by a portable real-time reverse transcriptase PCR assay. *J. Clin. Micro*. 41:500-505.
3. FLOEGEL-NIESMANN, G., BUNZENTHAL, C., FISCHER, S., and MOENNIG, V. 2003. Virulence of recent and former classical swine fever virus isolates evaluated by their clinical and pathological signs. *J. Vet. Med*. 50:214-220.
4. SUMMERFIELD, A., KNOETIG, S. M., TSCHUDIN, R. and McCULLOUGH, K.C. 2000. Pathogenesis of granulocytopenia and bone marrow atrophy during classical swine fever involves apoptosis and necrosis of uninfected cells. *Virology*. 272:50-60.
5. SUMMERFIELD, A., McNEILLY, F., WALKER, I., ALLAN, G., KNOETIG, S. M. and McCULLOUGH, K. C. 2001. Depletion of $CD4^{+}$ and $CD8^{high+}$ T-cells before the onset of viremia during classical swine fever. *Vet. Immunol. Immunopath*. 78:3-19.
6. KLINKENBERG, D., NIELEN, M., MOURITS, M. C. M. and de JONG, M. C. M. 2005. The effectiveness of classical swine fever surveillance programmes in The Netherlands. *Prev. Vet. Med*. 67:19-37.
7. VAN OIRSCHOT, J. T. 1999. Classical Swine Fever (Hog Cholera). In *Diseases of Swine,* 8th ed., B Straw, S D'Allaire, W Mengeling, D Taylor, eds., Ames, IA: Iowa State University Press, pp. 159-172.

See Part IV for photos.

John Pasick, DVM, PhD, National Centre for Foreign Animal Disease, Winnipeg, Manitoba, R3E 3M4, Canada, jpasick@inspection.gc.ca

13

CONTAGIOUS AGALACTIA OF SHEEP AND GOATS

1. NAME
Contagious agalactia

2. DEFINITION
Contagious agalactia (CA) is a serious disease syndrome of sheep and goats of both sexes characterized by fever, malaise, and septicemia followed by arthritis, keratoconjunctivitis, and in females, mastitis and agalactia.

3. ETIOLOGY
The etiologic agent of the classical disease is *Mycoplasma agalactiae (Ma)*, which, since its isolation in 1923, has been considered the main cause of the disease, especially in sheep. However, it has become evident that the "contagious agalactia" syndrome (especially in goats) can also be caused by several other mycoplasmas, notably *M. capricolum* subsp. *capricolum (Mcc), M. putrefasciens*, and the "large colony" or LC type of *M. mycoides* subsp. *mycoides (Mmm*LC*)*. Although some workers have suggested limiting the term "contagious agalactia" to the disease caused by *M. agalactiae*, the consensus of the 1999 working group on contagious agalactia of the European Cooperation in the field of Scientific and Technical Research on ruminant mycoplasmoses (EC COST Action 826) was that all 4 mycoplasmas should be considered causal agents of contagious agalactia.

Many of the routinely-used disinfectants will effectively inactivate the organisms, including sodium hypochlorite (30 ml of household bleach in 1 gallon of water), cresol, 2% sodium hydroxide (lye) (pH 12.4), formalin (1%), sodium carbonate (4% anhydrous or 10% crystalline with 1% detergent), and ionic and nonionic detergents.

4. HOST RANGE
a. Domestic animals
Goats seem to be more susceptible to the natural disease than are sheep, but *M. agalactiae* is an important pathogen of both species. Both *Mcc* and *Mmm*LC are important pathogens of goats and rarely sheep. *M. putrefaciens* causes mastitis and arthritis in goats, which is indistinguishable from that caused by *Ma*, *Mmm*LC and *Mcc*. Most outbreaks occur in the summer months and coincide with the time of births and peak lactation.

b. Humans

There is no evidence that humans are susceptible to *M. agalactiae*.

5. EPIDEMIOLOGY

a. Transmission

The disease spreads by ingestion of feed, water, or milk contaminated with infected milk, urine, feces, or nasal and ocular discharges. Transmission may also be by direct entry to the teat opening at milking or by inhalation of contaminated dust. Animals with subclinical or chronic infections can carry and shed the mycoplasmas for months, and the organisms can survive in the supramammary lymph nodes from one lactation to the next. Contaminated fomites can transmit the organisms between premises. After parturition, spread in milking animals increases and kids ingesting infected colostrum and milk becomc infected. The resulting septicemia, with arthritis and pneumonia, gives rise to high mortality in kids.

b. Incubation period

The incubation period in the natural disease varies between 1 to 8 weeks.

c. Morbidity

The economic impact of the disease lies in its high morbidity, ranging between 30%-60%, and resultant loss of milk and meat production, rather than in its mortality. The greatest number of cases develops during those periods when the young are being born and the dams are in full lactation.

d. Mortality

In most outbreaks of CA, the mortality is low, seldom exceeding 20%, but occasionally secondary bacterial pneumonia may cause a higher mortality.

6. CLINICAL SIGNS

Infection with *M. agalactiae* occurs in male and female sheep and goats and can be inapparent or can cause mild, acute, or chronic disease. Freshening female goats/sheep at the beginning of lactation are especially susceptible and often display the acute form of the disease. After an incubation period of 1-8 weeks, transient fever followed by malaise and inappetance are observed. This is followed by mastitis, polyarthritis, and keratoconjunctivitis.

The mastitis is characterized by a change in the color of the milk to greenish-yellow or grayish-blue, and in the texture of the milk to a watery and later lumpy consistency as lactation decreases and eventually ceases. The udder eventually becomes fibrosed and atrophic.

Polyarthritis is first seen as swelling of the periarticular tissues, especially of the carpal and tarsal joints, and later develops into painful chronic infection resulting in lameness and inability to stand or walk. In male goats this may be the main manifestation of the disease.

Keratoconjunctivitis is usually of short duration and is seen in about 50% of infected animals. It may develop into a chronic infection, occasionally resulting in unilateral or bilateral blindness. *Mycoplasma agalactiae* may occasionally be found in lung lesions, but pneumonia is not a consistent finding. Bacteremia is common and could account for the isolation of the organism from sites where it is only transiently present.

Abortion has been described in chronically infected animals, but its pathogenesis is not understood. *Mycoplasma agalactiae* has also been associated with granular vulvovaginitis in goats.

*Mmm*LC, one of the most widely distributed ruminant mycoplasmas, may cause mastitis, arthritis, pleurisy, pneumonia, and keratoconjunctivitis. The organism is mostly confined to goats but has occasionally been isolated from sheep with balanoposthitis and vulvovaginitis and cattle. Cases usually occur sporadically, but the disease may persist and spread slowly within a herd. After parturition, the opportunity for spread in milking animals increases, and kids ingesting infected colostrum and milk become infected. The resulting septicemia, with arthritis and pneumonia, gives rise to high mortality in kids.

Mcc is widely distributed and highly pathogenic, particularly in North Africa, but the frequency of occurrence is low. Goats are more commonly affected than sheep, and clinical signs of fever, septicemia, mastitis, and severe arthritis may be followed rapidly by death. Pneumonia may be seen at necropsy. The severe joint lesions seen in experimental infections with this organism are accompanied by intense periarticular subcutaneous edema affecting tissues some distance from the joint. In the United Kingdom, genital lesions were seen in sheep as a result of a sporadic outbreak of *Mcc* infection.

Mycoplasma putrefaciens is common in milking goat herds in Western Europe where it can be isolated from animals with and without clinical signs. It has also been associated with a large outbreak of mastitis and agalactia leading to severe arthritis in goats accompanied by abortion and death (but not pyrexia) in California. *Mycoplasma putrefaciens* was the major finding in an outbreak of polyarthritis in kids in Spain. It has been associated with genital lesions in both male and female goats in an outbreak of caprine contagious agalactia caused by both *M. agalactiae* and *M. putrefaciens* in Spain. While both species were isolated from several organs of the affected goats, only *M. putrefaciens* was isolated from the genital lesions.

7. POST- MORTEM LESIONS

The principal lesion in female animals is catarrhal mastitis with primary inflammation of the interstitial tissues followed by secondary acinar involvement. If the mastitis becomes chronic, progressive fibrosis and eventually parenchymatous atrophy will be present.

In males and females dying of the acute disease, congestion of the musculature and of the spleen and liver may be seen as a result of the septicemia. In both acute and more chronically affected animals, arthritis with periarticular edema is common and especially affects the carpal joints. Synovial membranes may be hyperemic, and joint cavities may be filled with turbid or hemorrhagic fluid. The early eye lesion is usually a serous and later a mucopurulent conjunctivitis followed by keratitis and occasionally corneal ulceration.

8. IMMUNE RESPONSE

Both live and inactivated vaccines have been used in the prevention of CA. A live-attenuated vaccine for goats and vaccine prepared from a naturally avirulent strain of mycoplasma are effective in goats. Formalin-inactivated aluminum hydroxide precipitated vaccines have been extensively used in Eastern Europe. The efficacy of inactivated vaccines is low.

9. DIAGNOSIS

a. Field diagnosis

The characteristic cluster of clinical signs of the disease, namely mastitis with loss of milk production, keratoconjunctivitis, and arthritis, all occurring at or soon after parturition, warrant a clinical diagnosis of contagious agalactia.

Several mycoplasmas (especially of goats) can cause syndromes resembling contagious agalactia. Pneumonia, mastitis, and arthritis can also be caused by *Mannheimia hemolytica*; mastitis can be caused by streptococci, staphylococci, or other bacteria; and arthritis can be caused by caprine arthritis encephalitis virus and multiple bacteria, including *Erysipelothrix rhusiopathiae*. Thus, laboratory confirmation of field diagnosis is essential.

b. Laboratory diagnosis

i. Samples

From a live animal, milk, swabs from the eyes, joint fluid, and blood all provide good samples for isolation attempts. From a dead animal that has had severe clinical disease, the best specimens to submit are blood, urine, udder and associated lymph nodes, lung lesions, liver, spleen and joint fluid from those animals with arthritis. All samples should be collected aseptically and, if possible, placed in transport medium (heart infusion broth, 20% serum, 10% yeast extract, benzylpenicillin at 250-1000 IU/ml). Samples should be kept cool and shipped on wet ice as soon as possible. If transport to the laboratory is delayed (more than a few days), samples may be frozen. Blood should be collected for serum.

ii. Laboratory procedures

Diagnosis of CA must be confirmed by isolation and immunological identification of the causative agent. Serology (complement fixation [CF], indirect hemagglutination, and enzyme-linked immunosorbent assay [ELISA]) for the detection of antibodies is useful on a herd basis after the presence of the disease has been confirmed by isolation of the organism. Several PCRs specific for *M. agalactiae* and "mycoides cluster" have been developed and have proved to be very sensitive. They can provide a rapid early warning system when used directly on milk, nasal, ocular, synovial and tissue samples enabling a full investigation to take place when results are positive. However, negative results should not be considered definitive. A positive result, particularly in an area previously free of contagious agalactia, should be confirmed by isolation and identification of the mycoplasma using standard procedures.

10. PREVENTION AND CONTROL

Because CA is a chronic disease that may exist subclinically in carrier animals, it is important to maintain sufficient regulatory restrictions to prevent its introduction in apparently healthy animals. In endemic areas, normal sanitary precautions of separating affected animals from healthy animals, separating milking animals from younger animals, cleaning and disinfection of milking utensils, practicing good hygienic principles when milking, cleaning and disinfecting stalls, and eliminating litter will reduce the incidence of disease in a flock. If possible, newborn animals should be removed from the dam immediately after birth and fed only pasteurized colostrum and then pasteurized milk.

With early antibiotic treatment (tetracyclines, microlide, florfenicol, tiamulin and fluoroquinolones) the prognosis is good, and only for those animals developing chronic arthritis or keratoconjunctivitis is recovery unlikely. Oxytetracycline does not prevent subsequent shedding of the organisms. With other drugs this still needs to be determined.

Eradication can be accomplished by slaughter of all infected and contact flocks. As noted above, live and inactivated vaccines have been used in the prevention of CA. Due to strain variation, the use of autogenous vaccines incorporating local strains of mycoplasma is recommended. The efficacy of inactivated vaccines is low. Two concerns about the use of live vaccines are 1) the vaccine organism may be shed in the milk and 2) although the vaccines may prevent the development of clinical disease, they do not prevent infection and shedding of virulent organisms.

11. GUIDE TO THE LITERATURE

1. BAR-MOSHE, B., RAPAPPORT, E. and BRENNER, J. 1984. Vaccination trials against *Mycoplasma mycoides* subsp. *mycoides* (large-colony type) infection in goats. *Israel J. Med. Sci.* 20:972–974.
2. BERGONIER, D., BERTHOLET, X. and POUMARAT, F. 1997. Contagious agalactia of small ruminants: current knowledge concerning epidemiology,
3. diagnosis and control. *Rev. sci. tech. Off. int. Epiz.* 16:848–873.
4. DAMASSA, A.J., WAKENELL, P.S., and BROOKS, D.L. 1992. Mycoplasmas of goats and sheep. *J. Vet. Diag. Invest.* 4:101-113.
5. GIL, M.C., PENA, F.J., HERMOSO DE MENDOZA, J. and GOMEZ, L. 2003. Genital lesions in an outbreak of caprine contagious agalactia caused by *Mycoplasma agalactiae* and *Mycoplasma putrefaciens*. *J. Ve.t Med. B. Infec.t Di.s Vet. Pub. Hlth.* 50:484-487.
6. GRECO,G., CORRENTE, M., MARTELLA, V., PRATELLI A. and BUONAVOGLIA, D. 2001. A multiplex-PCR for the diagnosis of contagious agalactia of sheep and goats. *Mol. Cell. Probes.* 15:21-25.
7. PEYRAUD, A., WOUBIT, S., POVEDA, J. B., DE LA FE, C., MERCIER, P. and THIAUCOURT, F. 2003. A specific PCR for the detection of *Mycoplasma putrefaciens*, one of the agents of the contagious agalactia syndrome of goats. *Mol. Cell. Probes.* 17:289-294.
8. TOLA, S., MANUNTA, D., ROCCA, S., ROCCHIGIANI, A.M., IDINI, G., ANGIOI, A., and LEORI, G. 1999. Experimental vaccination against *Mycoplasma agalactiae* using different inactivated vaccine. *Vaccine.* 17:2764–2768.

Terry McElwain, DVM, PhD and Fred Rurangirwa, DVM, PhD, Washington State University, Pullman, WA, 99165-2037, tfm@vetmed.wsu.edu, ruvuna@vetmed.wsu.edu

14

CONTAGIOUS BOVINE PLEUROPNEUMONIA

1. NAME
Contagious Bovine Pleuropneumonia

2. DEFINITION
Contagious bovine pleuropneumonia (CBPP) is an infectious bacterial disease, primarily of cattle, which can present in an acute, subacute, or chronic form. Lungs are the target organ but occasionally joints and/or kidneys will be affected as well.

3. ETIOLOGY
Contagious bovine pleuropneumonia is caused by *Mycoplasma mycoides* subspecies *mycoides* small-colony or bovine type (*MmmSC*). The organism survives well only in vivo and is quickly inactivated when exposed to normal external environmental conditions. Many of the routinely used disinfectants will effectively inactivate the organism. *MmmSC* does not survive in meat or meat products and does not survive outside the animal in nature for more than a few days. *Mycoplasma mycoides* subsp. *mycoides* large-colony type is pathogenic for sheep and goats but not for cattle.

4. HOST RANGE
a. Domestic animals
Cattle, both *Bos taurus* and *Bos indicus* are susceptible to infection, as are Asian buffalo (*Bubalus bubalis*).

b. Wild animals
Wild animals are not thought to be susceptible under natural conditions. There have been infrequent cases recorded in zoos (bison, yak), and white-tailed deer were successfully infected experimentally.

c. Humans
MmmSC does not infect humans.

5. EPIDEMIOLOGY
a. Transmission
Contagious bovine pleuropneumonia is spread by inhalation of droplets from an infected, coughing animal. Consequently, relatively close contact is required for

transmission to occur. Outbreaks usually begin as the result of movement of an infected animal into a naive herd. It is widely believed that recovered animals harboring infectious organisms within a pulmonary sequestrum may become active shedders when stressed. Although this may be a factor in some outbreaks, it has not been substantiated experimentally. There are limited anecdotal reports of fomite transmission, but this means of transmission is not generally thought to be a problem.

b. *Incubation period*

The time from natural exposure to overt signs of disease is variable but generally quite long. It has been shown that healthy animals placed in a CBPP-infected herd may begin showing signs of disease from 3 weeks to 4 months later. Experimentally, subsequent to instillation of large quantities of infective material at the tracheal bifurcation, the incubation period is 2-3 weeks.

c. *Morbidity*

The attack rate with CBPP is variable. It is not thought to be a highly contagious disease. With increased confinement of animals, morbidity rises. In a herd situation where animals have close contact, the infection rate would be 50-80%.

d. *Mortality*

The mortality rate with CBPP is quite varied and ranges from 10-70% in various outbreaks. Most of the animals that present with severe acute infection will die. Mortality in the subacute and chronic forms may depend on other intercurrent factors such as plane of nutrition, level of parasitism, and general body condition.

6. CLINICAL SIGNS

Usually the first abnormality noticed is a depressed, inappetent animal with fever. Coughing may be the next sign, followed by evidence of thoracic pain and an increased respiratory rate. When pulmonary involvement is extensive and severe, there will be labored respiration and, sometimes, open-mouthed breathing. As pneumonia progresses and animals become increasingly dyspneic, animals are inclined to stand with elbows abducted in an attempt to decrease thoracic pain and increase chest capacity. Auscultation of the lungs reveals any of a wide variety of sounds, depending on how severely the subjacent pulmonary parenchyma is affected. Crepitations, rales, and pleuritic friction rubs are all possible. Percussion over affected areas reveals dullness.

Contagious bovine pleuropneumonia often evolves into a chronic disease. This form, characterized by ill thrift and recurrent low-grade fever, may be difficult to recognize as pneumonia. Forced exercise may precipitate coughing.

In calves, the mycoplasmemia results in a polyarthritis which may be the primary presenting complaint, often without an accompanying pneumonia. Animals affected in this manner may stand stiffly with a distinctly arched back and be very reluctant to move. Getting up and down may cause obvious discomfort. Large joints may be distended and warm on palpation. If joint pain is severe, animals may be so reluctant to bend the joints that they lie in lateral recumbency with legs outstretched.

7. POST- MORTEM LESIONS

a. Gross

In the lung, gross pathologic features of CBPP are characteristic. If the animal dies, there is usually extensive and marked inflammation of the lung and associated pleurae. In severe cases there can be abundant fluid in the thoracic cavity. The inflammation is not uncommonly unilateral. The initial focus can be in any part of the lung and, in fatal cases, usually has spread locally and extensively to include a sizable segment.

The affected pulmonary parenchyma is odorless. The predominant gross change is consolidation, or thickening, of individual lobules that become encased in markedly widened interlobular septa, resulting in the characteristic marbled appearance. The organism produces a necrotizing toxin, galactan, which allows for extensive spread through septa. Interlobular septa become distended first by edema, then by fibrin, and finally by fibrosis. The overlying pleura may be thickened by an irregular layering of yellow fibrin which, with time, becomes fibrosed, often resulting in adhesions between parietal and visceral pleurae. Occasionally within an affected lung there will be a sequestrum, a focus that has undergone coagulative necrosis and is effectively avascular. Such sequestra may be found in recovered animals and it is known that *MmmSC* can survive within these sequestra for months or possibly longer, facilitating spread.

In calves, joints may be filled with abundant turbid fluid that clots on exposure to air because of the excessive amounts of fibrin. Infarcts occur in the kidney as a result of the initial mycoplasmemia and will appear as chronic fibrotic foci.

b. Key microscopic

The pulmonary histologic lesions seen with CBPP include distention of interlobular septa, coagulative necrosis, and abundant fibrin and edema. These lesions cannot reliably be distinguished from any other severe lobar pneumonia, e.g., shipping fever.

8. IMMUNE RESPONSE

a. *Natural infection*

Recovered animals are not susceptible to reinfection. In chronic cases, the antibody titer may drop and be negligible, making individual animal titers unreliable as an indicator of infection.

b. *Immunization*

A vaccine is in use in endemic areas and offers some protection (see below).

9. DIAGNOSIS

a. *Field diagnosis*

Clinical diagnosis of CBPP is difficult. At post-mortem the gross lesions of CBPP are somewhat distinct. Unlike most other pneumonias, CBPP is usually unilateral. Often there is an extensive deposition of fibrin and a large quantity of straw-colored fluid in the thoracic cavity with a prominent marbling of pulmonary parenchyma. In some chronic cases the nodules of inflammation may not be readily apparent from the pleural surface but can be palpated within the parenchyma.

b. *Laboratory diagnosis*

i. Samples

From a live animal, nasal swabs, transtracheal washes, or pleural fluid obtained by thoracic puncture all provide good samples for isolation attempts. From a dead animal that has had severe clinical disease, the best specimens to submit are affected lung, swabs of major bronchi, tracheo-bronchial or mediastinal lymph nodes, and joint fluid from those animals with arthritis. All samples should be collected aseptically and, if possible, placed in transport medium (heart infusion broth, 20% serum, 10% yeast extract, benzylpenicillin at 250-1000 IU/ml). Samples should be kept cool and shipped on wet ice as soon as possible. If transport to the laboratory is delayed (more than a few days), samples may be frozen. Blood should be collected for serum.

ii. Laboratory procedures

Isolation and identification of *MmmSC* is the preferred method of diagnosis, however the organism is notably fastidious and slow growing. There is a penside agglutination test for use in active outbreaks. Real-time PCR has been described. For serology, a complement fixation test is best for herd surveillance. A competitive ELISA has been developed and is being validated. Immunohistochemical staining of pathologic specimens has been used.

10. PREVENTION AND CONTROL

Successful control of the spread of CBPP rests on removing susceptible animals from any possible contact with CBPP-infected animals, whether they are clinically affected or only subclinical carriers. On-farm quarantine of suspicious and contact animals would be very advantageous in stemming the spread of the disease. In an outbreak situation, testing, slaughter, and quarantine would be the methods of choice.

Treatment of the disease is not recommended. Antimicrobial therapy may only serve to slow the progression of the disease or may even in some cases favor the formation of sequestra.

An attenuated strain of *MmmSC* (T1) is used as a vaccine in enzootic areas. It is delivered subcutaneously. There are frequent questions regarding the efficacy and safety of this vaccine, which may be related to the diverse nature of the laboratories that produce it as well as to genetic drift through serial passage of the cultured strain. An additional drawback of this vaccine is that it generates an unpredictable local reaction, probably due to the presence of galactic toxin. As a result, it may produce an extensive necrotic focus at the site of inoculation leading to skin necrosis and sloughing. For this reason, it is given in the tail tip rather than the neck, as the former does not have enough loose subcutaneous tissue to allow for extensive spread.

11. GUIDE TO THE LITERATURE

1. ANON. 2002. Contagious bovine pleuropneumonia. Animal diseases data, OIE, http://www.oie.int/eng/maladies/fiches/a_A060.htm.
2. BASHIRUDDIN, J.G., SANTINI, F.G., DE SANTIS, P., VISAGGIO, M.C., DI FRANCESCO, G., D'ANGELO, A. and NICHOLAS, R.A. 1999. Detection of *Mycoplasma mycoides* subspecies *mycoides* in tissues from an outbreak of contagious bovine pleuropneumonia by culture, immunohistochemistry and polymerase chain reaction. *Vet. Rec.* 145:271-274.
3. GORTON, T.S., BARNETT, M.M., GULL, T., FRENCH, R.A., LU, A., KUTISH, G.F., ADAMS, G.L. and GEARY, S.J. 2005. Development of real-time diagnostic assays specific for *Mycoplasma mycoides* subspecies *mycoides* Small Colony. *Vet. Microbiol.* 111:51-58.
4. GRIECO, V., BOLDINI, M., LUINI, M., FINAZZI, M., MANDELLI, G. and SCANZIANI, E. 2001. Pathological, immunohistochemical and bacteriological findings in kidneys of cattle with contagious bovine pleuropnumonia (CBPP). *J. Comp. Pathol.* 124:95-101.

5. THIACOURT, F., YAYA, A., WESONGA, H., HUEBSCHLE, O.J., TULASNE, J.J. and PROVOST, A. 2000. Contagious bovine pleuropneumonia. A reassessment of the efficacy of vaccines used in Africa. *Ann. N.Y. Acad. Sci.* 916:71-80.

See Part IV for photos.

Corrie Brown, D.V.M., Ph.D., Department of Pathology, College of Veterinary Medicine, University of Georgia, Athens, GA 30602-7388, corbrown@uga.edu

15

CONTAGIOUS CAPRINE PLEUROPNEUMONIA

1. NAME

Contagious caprine pleuropneumonia

2. DEFINITION

Contagious caprine pleuropneumonia (CCPP) is an acute, highly contagious disease of goats of all ages and both sexes caused by a mycoplasma and characterized by extreme fever, coughing, severe respiratory distress, and high mortality. The principal lesion at necropsy is fibrinous pleuropneumonia with massive lung consolidation, accompanied by accumulation of straw-colored pleural fluid.

3. ETIOLOGY

Classical contagious caprine pleuropneumonia (CCPP) is caused by *Mycoplasma capricolum* subsp. *capripneumoniae* (*Mccp*), originally known as the F38 biotype. *Mccp* was first isolated and shown to cause CCPP in Kenya. Disease indistinguishable from naturally occurring CCPP has been experimentally reproduced with *Mccp* by several groups of workers.

Mycoplasma capricolum subsp. *capripneumoniae* is closely related to 3 other mycoplasmas: *M. mycoides* subsp. *mycoides* large colony type (LC), *M. mycoides* subsp. *capri, and M. capricolum* subsp. *capricolum.* Unlike classical CCPP, which is confined to the thoracic cavity, the disease caused by the latter 3 mycoplasmas is accompanied by prominent lesions in other organs and/or parts of the body in addition to the thoracic cavity. For many years the causative agent of CCPP was considered to be *M. mycoides capri* (type strain PG-3). However, the infrequency with which *M. mycoides capri* has been isolated from CCPP in recent years suggests that it may be a minor cause of the disease.

M. mycoides subsp. *mycoides* (LC type) (the so-called large colony or LC variant of *M. mycoides* subsp. *mycoides*) typically produces septicemia, polyarthritis, mastitis, encephalitis, conjunctivitis, hepatitis, or pneumonia in goats. Some strains of this agent will cause pneumonia closely resembling CCPP, but the agent is not highly contagious and is not considered to cause classical CCPP. *M. capricolum* subsp. *capricolum* is commonly associated with mastitis and polyarthritis in goats, and can produce pneumonia resembling CCPP. However, it typically causes severe septicemia and polyarthritis. While *M. capricolum* subsp. *capricolum* is closely related to *M. capricolum* subsp. *capripneumonia,* they can be differentiated using monoclonal antibodies and/or polymerase chain reaction.

4. HOST RANGE

a. Domestic animals

Contagious caprine pleuropneumonia is a disease of goats. Where the classical disease has been described, only goats were involved despite the presence of sheep and cattle in the area. *M. capricolum* subsp. *capripneumoniae* does not cause disease in sheep or cattle. *M. mycoides capri* will result in a fatal disease in experimentally inoculated sheep and can spread from goats to sheep. It is, however, not recognized as a cause of natural disease in sheep.

b. Humans

Human infection with *Mccp* or *M. mycoides capri* has not been reported.

5. EPIDEMIOLOGY

a. Transmission

Contagious caprine pleuropneumonia is transmitted by direct contact through inhalation of infective aerosols. Of the 2 known causative agents, *Mccp* is far more contagious. Outbreaks of the disease often occur after heavy rains (e.g., after the monsoons in India), after cold spells or after transportation over long distances. This is probably because recovered carrier animals begin shedding the infectious agent after the stress of sudden climatic or environmental changes. It is believed that a long-term carrier state may exist.

b. Incubation period

The incubation period can be as short as 6-10 days but may be very prolonged (3-4 weeks) under natural conditions.

c. Morbidity

Morbidity can be 100%. Gathering or increased confinement of animals facilitates the spread of the disease.

d. Mortality

Mortality may be in the range of 70-100%.

6. CLINICAL SIGNS

The classical disease caused by *M. capricolum* subsp. *capripneumoniae* is limited to the respiratory tract. It is characterized by a fever as high as 106°F (41°C), coughing, and a distinct loss of vigor. Affected goats have labored breathing;

later they may grunt or bleat in obvious pain. Frothy nasal discharges and stringy salivation are often seen shortly before death. In the acute disease, which occurs in fully susceptible populations of goats, death occurs within 7-10 days of the onset of clinical signs. A more chronic form of the disease is often seen in endemic areas and may lead to recovery of a higher percentage of infected animals, many of them carriers of the mycoplasmas.

M. mycoides capri tends to cause a more generalized infection in which septicemia is frequently seen. An acute or peracute septicemic form of the disease involving the reproductive, respiratory and alimentary tracts has been described. In addition, thoracic and reproductive forms of the disease have been attributed to this agent. The disease is considerably less contagious than *M. capricolum* subsp. *capripneumoniae*-induced disease, and the mortality and morbidity rates are lower.

7. POST- MORTEM LESIONS

Gross lesions in classical CCPP are confined to the thoracic cavity. Pea-sized yellowish nodules are present in the lungs in early cases; in more established cases there is marked congestion around the nodules. The lesions may be confined to one lung or involve both, and an entire lobe may become solidified. The pulmonary pleura becomes thickened, and there may be adhesions to the chest wall. Straw-colored pleural fluid may accumulate.

In sharp contrast, *M. mycoides capri* has been reported to cause lesions in a wide variety of organ systems and to produce lung lesions closely resembling those seen in CBPP. The generalized lesions described include encephalitis, meningitis, lymphadenitis, splenitis, genitourinary tract inflammation, and intestinal lesions, none of which are a feature of classical CCPP. The lung lesions, which resemble those seen in CBPP, are usually confined to one lung and reflect various stages of fibrinous pneumonia. Extensive pleuritis is usually present, and various stages of pulmonary hepatization with marked dilation of interlobular septa are commonly seen. The cardiac and diaphragmatic lung lobes are most commonly involved. Some describe this as a mild form of CCPP; others argue that it is not CCPP.

8. IMMUNE RESPONSE

Vaccines to *M. mycoides capri* have been used, but with little success. This is probably because classical CCPP is usually caused by *Mccp* F-38 biotype. Since that time both live attenuated and inactivated *Mccp* F-38 biotype vaccines have been tested with varying degrees of success. The most promising of the experimental vaccines is the lyophilized saponin-inactivated *Mccp* F-38 vaccine shown in field tests to confer 100% protection to contact exposure. This vaccine has been in use in Kenya for several years.

9. DIAGNOSIS

a. Field diagnosis

A highly contagious disease occurring in goats and characterized by pyrexia of 106°F (41°C) and above, severe respiratory distress, high mortality, and postmortem lesions of fibrinous pleuropneumonia with pronounced hepatization and pleural adhesions warrants a field diagnosis of CCPP.

b. Laboratory diagnosis

i. Samples

From a dead animal that has had severe clinical disease, the best specimens to submit are affected lung, swabs of major bronchi, and tracheobronchial or mediastinal lymph nodes. All samples should be collected aseptically and if possible, placed in transport medium (heart infusion broth, 20% serum, 10% yeast extract, benzylpenicillin at 250-1000 IU/ml). Samples should be kept cool and shipped on wet ice as soon as possible. If transport to the laboratory is delayed (more than a few days), samples may be frozen. Blood should be collected for serum.

ii. Laboratory procedures

Diagnosis must be confirmed by isolation of the agent (*Mccp*). The causative agent, once isolated, can be identified by immunofluorescence or by growth or metabolic inhibition tests. Several serological tests can be used in the laboratory for the detection of antibodies to *Mccp*. These include complement fixation (CF), passive hemagglutination (PH), and enzyme-inked immunosorbent assay (ELISA). Latex agglutination tests are very convenient field tests for detecting antibodies to a carbohydrate antigen in whole blood or serum, or to detect carbohydrate antigen in serum. Polymerase chain reaction may also be used.

10. PREVENTION AND CONTROL

Sufficient regulatory restrictions should be maintained to prevent introduction of CCPP into apparently healthy animals. Serologic testing of susceptible animals for importation is a recommended safeguard.

Successful control of the spread of CCPP rests on removing susceptible animals from any possible contact with CCPP-infected animals which are clinically affected or are subclinical carriers. On-farm quarantine of suspicious and contact animals would be very advantageous in stemming the spread of the disease.

The mycoplasmas are sensitive to several broad-spectrum antibiotics (notably the tetracyclines, tylosin, and tiamulin). *Mccp* is particularly sensitive to streptomycin. Although early treatment can be effective, chemotherapy and chemoprophylaxis have not played important roles in CCPP control programs.

As noted above, both live attenuated and inactivated F-38 biotype vaccines have been tested with varying degrees of success.

11. GUIDE TO THE LITERATURE

1. MACOWAN, K.J. 1984. Role of Mycoplasma strain F-38 in contagious caprine pleuropneumonia. *Isr. J. Med. Sci.* 20:979-981.
2. MACOWAN, K.J. and MINETTE, J.E. 1977. Contact transmission of experimental contagious caprine pleuropneumonia (CCPP). *Trop. Anim. Health Prod.* 9:185–188.
3. MARCH, J.B., GAMMACK, C. and NICHOLAS, R. 2000. Rapid detection of contagious caprine pleuropneumonia using a *Mycoplasma capricolum* subsp. *capripneumoniae* capsular polysaccharide-specific antigen detection latex agglutination test. *J. Clin. Microbiol.* 38:4152–4159.
4. RURANGIRWA, F.R., MCGUIRE, T.C., KIBOR, A. and CHEMA, S. 1987. A latex agglutination test for field diagnosis of caprine pleuropneumonia. *Vet. Rec.* 121:191–193.
5. RURANGIRWA, F.R., MCGUIRE, T.C., KIBOR, A. and CHEMA, S. 1987. Aninactivated vaccine for contagious caprine pleuropneumonia. *Vet. Rec.* 121:397–402.
6. RURANGIRWA, F.R., MCGUIRE, T.C., MUSOKE A.J. and KIBOR, A. 1987. Differentiation of F38 mycoplasmas causing contagious caprine pleuropneumonia with a growth-inhibiting monoclonal antibody. *Infect. Immun.* 55:3219–3220.
7. THIAUCOURT, F. and BOLSKE, G. 1996. Contagious caprine pleuropneumonia and other pulmonary mycoplasmoses of sheep and goats. *Rev. sci. tech. Off. Int. Epiz.* 15:1397–1414.
8. WOUBIT, S., LORENZON, S., PEYRAUD, A., MANSO-SILVAN, L., and THIAUCOURT, F. 2004. A specific PCR for the identification of *Mycoplasma capricolum subsp. capripneumoniae*, the causative agent of contagious caprine pleuropneumonia (CCPP). *Vet. Microbiol.* 104:125-132.

See Part IV for photos.

Terry McElwain, DVM, PhD and Fred Rurangirwa, DVM, PhD, Washington State University, Pullman, WA, 99165-2037, tfm@vetmed.wsu.edu, ruvuna@vetmed.wsu.edu

16

CONTAGIOUS EQUINE METRITIS

1. NAME

Contagious equine metritis

2. DEFINITION

Contagious equine metritis (CEM) is a highly contagious venereal disease characterized by acute purulent metritis and a shift to copious mucopurulent vaginal discharge that develops 10-14 days postbreeding to a stallion infected with *Taylorella equigenitalis*. The first exposure usually results in temporary infertility in the mare. Infection may become chronic and so the causal organism may be carried for several months or longer.

3. ETIOLOGY

T. equigenitalis is a microaerophilic gram-negative non-motile coccobacillus that may be either sensitive or resistant (MIC 128 µg/ml) to streptomycin. Isolates of varying geographic origin share 99.5% or greater sequence identity of their 16S rDNA. A second species, *T. asinigenitalis*, shares 26% DNA homology with *T. equigenitalis*, is phenotypically similar, but does not cause disease.

T. equigenitalis is susceptible to most commonly used disinfectants including sodium hypochlorite at 400 parts per million, 2% chlorhexidine, and ionic and nonionic detergents.

4. HOST RANGE

a. Domestic animals

Only the equine species appear to be natural hosts for *T. equigenitalis*. Thoroughbreds appear to be more severely affected than other breeds. A related species, *T. asinigenitalis*, has been isolated from male donkeys, and on one occasion from a stallion.

b. Humans

There is no evidence that *T. equigenitalis* can infect humans.

5. EPIDEMIOLOGY

a. Transmission

T. equigenitalis is naturally transmitted by mating, but may also be transmitted indirectly to mares and stallions via contaminated instruments and equipment as well as through the use of artificial insemination. Carrier mares and stallions may be sources of infection for acute outbreaks. However, the carrier stallion is often epizootiologically more significant since it may infect many mares before the disease is suspected and diagnosed. High risk mares and stallions are those which were previously infected, are from a premise where the disease was diagnosed during the previous year, or are from countries which do not enforce high standards of CEM control. Asymptomatic stallions may carry *T. equigenitalis* on their external genitalia for years. The primary site of localization is the urethral fossa. Foals may become infected at birth and remain infected until mature.

b. Incubation period

The naturally occurring disease does not become evident until 10-14 days postbreeding and is signalled when the mare short-cycles and shows signs of estrus. An inflammatory reaction begins 24 hours after exposure and reaches maximum intensity 10-14 days later.

c. Morbidity

Morbidity is high in animals following venereal exposure to the organism.

d. Mortality

Fatal infections have not been observed.

6. CLINICAL SIGNS

A copious mucopurulent vaginal discharge occurs 10-14 days postbreeding to an infected stallion. The first indication of infection is short-cycling and return to estrus. At this time, a mucopurulent vaginal discharge or a dried vaginal discharge may be found on the tail and inside the thighs. The discharge subsides after a few days, but the mare may remain chronically infected for several months. In experimental infections in ponies and horses, a mucopurulent vaginal discharge was evident 24-48 hours postinfection, and lasted for 2-3 weeks. Most infected mares will not conceive. In the unlikely event they conceive, they either abort or carry to term. The newborn foal may then become a carrier.

7. POST- MORTEM LESIONS

The lesions of contagious equine metritis are not pathognomic. Lesions are most severe in the uterus, but salpingitis, cervicitis, and vaginitis also occur. The most severe lesions occur about 14 days postinfection and gradually decrease in severity over the next several weeks as the disease becomes chronic. There is edema and swelling of the uterine endometrial folds with a mucopurulent exudate between the folds at the height of the infection. The cervix is edematous and hyperemic, and covered with a mucopurulent exudate.

8. IMMUNE RESPONSE

a. *Natural infection*

Natural infection confers some immunity. The first exposure causes a severe metritis, which usually results in temporary infertility. Subsequent exposure results in less severe disease and conception may ensue. However, the carrier state may result. This suggests that vaccination may not be a useful strategy since the resulting antibodies would have to act locally to abolish the carrier state.

b. *Immunization*

Currently there is no vaccination for *T. equigenitalis.*

9. DIAGNOSIS

a. *Field diagnosis*

Mares with a copious mucopurulent vaginal discharge 10-14 days postbreeding are suspect cases of CEM. Chronically infected mares and stallions may not show clinical signs of the disease.

b. *Laboratory diagnosis*

i. Samples

Isolation of *T. equigenitalis* is necessary for a diagnosis of both the acute and chronic forms of the disease. Mares suspected as carriers should be cultured during estrus, preferably during the first part of the cycle. Culture sites are the uterus, clitoral fossa, and clitoral sinuses and the urethra, urethral fossa, diverticulum, and sheath of the stallion. Culture swabs should be placed immediately in Amies transport media and maintained at 4°C or lower to maintain viability and to prevent overgrowth by contaminating bacteria. Should the culture swabs not be cultured within a few hours, specimens in transport media should be frozen. The frozen organism remains viable, and has been cultured from Amies transport media stored for 18 years at -20°C.

ii. Laboratory procedures

Smears of the uterine exudate during the acute phase of the disease are helpful in making a presumptive diagnosis of CEM. Gram and Giemsa stained smears of exudate may reveal many inflammatory cells, mainly neutrophils. Numerous gram-negative coccobacillary bacteria may be demonstrated free in mucus and in the cytoplasm of neutrophils. The organisms are usually seen individually or in pairs arranged end to end.

The polymerase chain reaction (PCR) utilizing primers from 16S ribosomal DNA sequence or from a cloned DNA fragment unique to *T. equigenitalis* may be used for detection in specimens and in cultures. Real time Tag Man® PCR will detect *T. equigenitalis* and distinguish it from the closely related *T. asinigenitalis*. PCR is rapid, specific and more sensitive than culture. However, since it detects DNA of both live and dead organisms, a positive PCR does not necessarily signify the presence of live, infectious *T. equigenitalis*. It follows, therefore, that the gold standard for laboratory diagnosis of CEM is isolation of *T. equigenitalis* on culture media. Nevertheless, results of PCR testing have very significant diagnostic value since culture attempts may fail because of overgrowth of contaminants or presence of too few viable organisms in the specimen. A positive PCR test and negative culture result require that additional samples be cultured. A negative PCR test is more definitive diagnostically given the much greater sensitivity of PCR relative to culture.

PCR is effective for differentiating *T. equigenitalis* from *T. asinigenitalis*.

Various tests have been used to detect serum antibodies to *T. equigenitalis*, including rapid plate agglutination (RPA), antiglobulin, fluorescent monoclonal antibody enzyme-linked immunosorbent assay (ELISA), passive hemagglutination (PHA), complement fixation (CF), and agar-gel diffusion. The CF test has proved unreliable during the chronic disease because of increase in titers of non-complement fixing antibody and relative decrease in complement fixing IgM. Most acute and chronic CEM infections are detected with the RPA, ELISA, and PHA tests. Infected stallions do not develop detectable antibodies to *T. equigenitalis*.

10. PREVENTION AND CONTROL

It is uncertain whether antibiotic treatment eliminates or hastens elimination of *T. equigenitalis*. Natural clearance may take several months. Disinfectants and antibiotics may be applied to the external genitalia of the mare and stallion. A standard treatment is thorough washing of the external genitalia with soap and water followed by chlorhexidine surgical scrub daily for 5 days. After application, chlorhexidine is removed by rinsing the external genitalia with warm water and applying nitrofurazone-containing ointment. A disadvantage of this treatment is destruction of the normal flora and potential overgrowth of opportunistic pathogens such as *Pseudomonas* spp. and *Klebsiella* spp. The clitoral sinuses are sites of persistence in carrier mares and are difficult to expose for cleaning and treatment. Surgical excision usually rids the mare of infection. However, repeated washing with chlorhexidine solution and packing of the clitoral sinuses and fossa with nitrofurazone ointment may be effective in eradicating *T. equigenitalis*.

Inapparent infections in carrier mares and stallions make CEM difficult to control. Prevention of spread requires detection and treatment of the infection in the mare and stallion before breeding. Suspect carrier mares should be cultured to ensure they do not infect the stallion which subsequently has the potential to quickly transmit to many more mares. Suspect stallions should be cultured and may also be bred to susceptible test mares, which are then cultured for *T. equigenitalis*.

Small-colony types of *T. equigenitalis* are less virulent and may account for the gradual decrease in number of naturally infected horses showing typical clinical signs of CEM in the field. These variants present special challenges, both clinically and in laboratory testing, since they have no distinguishing cultural characteristics other than the small size and transparency of their colonies. In addition, their slow growth, the presence of contaminating bacteria, and occurrence of streptomycin-sensitive strains reduce the success rate of bacteriologic culture.

Since bacterial isolation of streptomycin-sensitive strains may be difficult, serologic testing is a valuable aid in detecting mares that have previously been exposed to *T. equigenitalis*. The CF test becomes positive about 2 weeks after infection and remains positive for the next 3 weeks or so. ELISAs and other serologic tests have been shown to detect mares that have previously been exposed. Serologic testing is of no value in stallions because they do not produce detectable antibodies.

11. GUIDE TO THE LITERATURE

1. ACLAND, H.M. and KENNEY, R.M. 1983. Lesions of contagious equine metritis in mares. *Vet. Pathol.* 20:330-341.
2. ANZAI, T., EGUCHI, M., SEKIZAKI, T., KAMADA, M., YAMAMOTO, K. and OKUDA, T. 1999. Development of a PCR test for rapid diagnosis of contagious equine metritis. *J. Vet. Med. Sci.* 61:1282-1292.
3. ANZAI, T., WADA, R., OKUDA, T. and AOKI, T. 2002. Evaluation of the field application of PCR in the eradication of contagious equine metritis from Japan. *J. Vet. Med. Sci.* 64:999-1002.
4. ARATA, A.B., COOKE, C.L., JANG, S.S. and HIRSH, D.C. 2001. Multiplex polymerase chain reaction for distinguishing *Taylorella equigenitalis* from *T. equigenitalis*-like organisms. *J. Vet. Diagn. Invest.* 13:263-264.

5. BAVERUD, V., NYSTROM, C. and JOHANSSON, K.E. 2006. Isolation and identification of *Taylorella asinigenitalis* from the genital tract of a stallion; first case of a natural infection. *Vet. Microbiol.* 116:294-300.
6. BLUEMINK-PLUYM, N.M.C., WERDLER, M.E.B., HOUWERS, D.J., PARLEVLIET, J.M., COLENBRANDER, B. and VAN DER ZEIJST, B.A.M. 1994. Development and evaluation of PCR test for detection of *Taylorella equigenitalis*. *J. Clin. Microbiology*. 32:893-896.
7. MATSUDA, M. and MOORE, J.E. 2003. Recent advances in molecular epidemiology and detection of *Taylorella equigenitalis* associated with contagious equine metritis, (CEM). *Microbiol. 97*:111-122.
8. MATSUDA, M., TAZUMI, A., KAGAWA, S., SEKIZUKA, T. and MURAYAMA, O. 2006. Homogeneity of the 16 S r DNA sequence among geographically disparate isolates of *Taylorella equigenitalis*. *BMC Vet. Res. 2*:1.
9. SWERCZEK, T.W. 1979. Contagious equine metritis – outbreak of the disease in Kentucky and laboratory methods for diagnosing the disease. *J. Reproduc. Fertil.* (Suppl). 27:361-365.
10. TIMONEY, P.J., MCARDLE, J.F., O'REILLY, P.J. and WARD, J. 1978. Infection patterns in pony mares challenged with the agent of contagious equine metritis 1977. *Equine Vet. J.* 10:148-152.
11. WAKELY, P.R., ERRINGTON, J. HANNON, S., ROEST H.I. CARSON, T. and HUNT, B. 2006. *Vet. Microbiol. 118*:247-54.

See Part IV for photos.

John Timoney, DVM, PhD, Gluck Equine Research Center, University of Kentucky, Lexington, KY 40546-0099, jtimoney@uky.edu

17

DOURINE

1. NAME
Dourine, Slapsiekte, Morbo Coitale maligno, Beschalseuche

2. DEFINITION
Dourine is a chronic trypanosomal disease of Equidae. The disease is transmitted almost exclusively by coitus, is characterized by edematous lesions of the genitalia, nervous system involvement, and progressive emaciation, and is ultimately fatal in most cases.

3. ETIOLOGY
Dourine is caused by *Trypanosoma equiperdum*, a protozoan parasite related morphologically and serologically to *T. brucei*, *T. rhodesiense*, *T. evansi* and *T. gambiense* (of the subgenus *Trypanozoon* of the *Salivarian* section of organisms of the pathogenic genus *Trypanosoma*). Different strains of the parasite vary in pathogenicity.

4. HOST RANGE
a. Domestic animals
Dourine is typically a disease of horses, donkeys and mules. The organism has been adapted to a variety of laboratory animals.

b. Wild animals
Positive compliment-fixation (CF) tests have been obtained from zebras, although it has not been shown that zebras can be infected with *T. equiperdum* or transmit the disease. Improved breeds of horses seem to be more susceptible to the disease. The disease in these animals often progresses rapidly and involves the nervous system. In contrast, native ponies and donkeys often exhibit only mild signs of the disease. Infected male donkeys, which may be asymptomatic, are particularly dangerous in the epidemiology of the disease, for they may escape detection as carriers.

c. Humans
Humans are not susceptible to infection with *T. equiperdum*.

5. EPIDEMIOLOGY

a. Transmission

This venereal disease is spread almost exclusively by coitus. Organisms are present in the urethra of infected stallions and in vaginal discharges of infected mares. The organism may pass through intact mucous membranes to infect the new host. Infected animals do not transmit the infection with every sexual encounter, however. As the disease progresses, trypanosomes periodically disappear from the urethra or vagina; during these periods, the animals are noninfective. Noninfective periods may last for weeks or months and are more likely to occur in the later stages of the disease. Thus, transmission is most likely early in the disease process.

It is possible for mares to become infected and pregnant after mating with an infected stallion. Foals born to infected mares may be infected. It is unclear if this occurs in utero, during birth or via suckling. Because trypanosomes may occur in the milk of infected mares, these foals may be infected per os during birth or by ingestion of infected milk. Foals infected in this way may transmit the disease when mature and have a lifelong positive CF titer. This method of disease transmission is rare, however. Some foals may acquire passive immunity from colostrum of infected mares without becoming actively infected; in such foals, the CF titer declines, and the animal becomes seronegative by 4-7 months of age. Although the possibility of noncoital transmission remains uncertain, it is supported by sporadic infections in sexually immature equids.

b. Incubation period

The incubation period is highly variable. Clinical signs usually appear within a few weeks of infection but may not be evident until after several years.

c. Morbidity

Because this is a venereal disease, morbidity is associated with coitus.

d. Mortality

Although the course of the disease may be long, it is usually fatal. Uncomplicated dourine does not appear to be fatal unless the nervous system is involved. The progressive debilitation associated with the neurological manifestation of the disease predisposes infected animals to a variety of other conditions. Because of the long survival time in some experimental cases, reports of recovery from dourine should be regarded with skepticism.

6. CLINICAL SIGNS

Clinical signs vary considerably, depending on the virulence of the infecting strain, the nutritional status of the infected animal, and the presence of other stress factors. The strain prevalent in southern Africa (and formerly in the Americas) is apparently less virulent than the European, Asian, or north African strains and produces an insidious, chronic disease. In some animals, clinical signs may not be apparent for up to several years (so-called latent infection). Clinical signs may be precipitated by stress in these animals.

In mares, the first sign of infection is usually a small amount of vaginal discharge, which may remain on the tail and hindquarters. Swelling and edema of the vulva develop later and extend along the perineum to the udder and ventral abdomen. There may be vulvitis and vaginitis with polyuria and other signs of discomfort such as an elevated tail. Abortion is not a feature of infection with mild strains, but significant abortion losses may accompany infection with a more virulent strain.

In stallions, the initial signs are variable edema of the prepuce and glans penis, spreading to the scrotum and perineum and to the ventral abdomen and thorax. Paraphimosis may be observed. The swelling may resolve and reappear periodically. Vesicles or ulcers on the genitalia may heal and leave permanent white scars (leukodermic patches). Transient, circular cutaneous plaques are a feature of the disease in some locations and strains but not others. These are sometimes called “silver dollar plaques,” and, when they occur, are pathognomonic.

Conjunctivitis and keratitis are often observed in outbreaks of dourine and may be the first signs noted in some infected herds.

Nervous disorders may be seen soon after the genital edema or may follow by weeks or months. Initially these signs consist of restlessness and the tendency to shift weight from one leg to another followed by progressive weakness and incoordination and ultimately by paralysis and recumbency. Anemia and emaciation sometimes accompany development of clinical signs even though the appetite remains unaffected.

Dourine is characterized by stages of exacerbation, tolerance, or relapse that may vary in duration and occur several times before death or recovery. The course of the disease may last several years after infection with a mild strain. Experimentally, horses have survived for up to 10 years after infection. The course is apparently more acute in the European and Asian forms of the disease in which the mortality rate is higher.

7. POST- MORTEM LESIONS

Anemia and cachexia are consistent findings in animals that have succumbed to dourine. Edema of the genitalia and ventral abdomen become indurated later in the course of the disease. Chronic lymphadenitis of most lymph nodes may be evident. Perineural connective tissue becomes infiltrated with edematous fluid in animals with nervous signs, and a serous infiltrate may surround the spinal cord, especially in the lumbar or sacral regions.

8. IMMUNE RESPONSE

Trypanosomes are covered by a layer of a single protein known as a variablc surface glycoprotein (VSG) which is strongly antigenic and evokes production of protective antibodies. However, the parasite population evolves to evade these antibodies by producing a different VSG, a process which is repeated throughout the long period of natural infection. The sequence of specific VSGs seems to be consistent in the course of infection. This phenomenon means that antibodies are useful for diagnostic purposes but are not protective. This has also prevented the development of effective vaccines to date.

9. DIAGNOSIS

a. Field diagnosis

Diagnosis on physical signs is unreliable because many animals develop no sign. When signs are present, however, they are suggestive of a diagnosis of dourine. If "silver dollar plaques" occur, they are considered pathognomonic for dourine.

b. Laboratory diagnosis

i. Samples

Detection of trypanosomes is highly variable and is not a reliable means for diagnosis of dourine. The following specimens should be submitted: serum, whole blood in EDTA, and blood smears.

ii. Laboratory procedures

A reliable CF test has been the basis for the successful eradication of dourine from many parts of the world. Thc antigen used in the CF test is group-specific, leading to cross-reactions with sera of horses infected with *T. brucei*, *T. rhodesiense*, or *T. gambiense*. The test is therefore most useful in areas where these parasites do not occur. Indirect fluorescent antibody, card agglutination, and an enzyme-linked immunosorbent assay test (ELISA) have also been developed for dourine but have not replaced the CF test.

10. PREVENTION AND CONTROL

Although there are reports of successful treatment with trypanocidal drugs (e.g., suramin at 10 mg/kg IV, quinapyramine dimethylsulfate at 3-5 mg/kg SC), treatment is more successful when the disease is caused by the more virulent (European) strains of the parasite. In general, treatment is not recommended for fear of continued dissemination of the disease by treated animals. Treatment may result in inapparent disease carriers and is not recommended in a dourine-free territory. Immunity to trypanosomiasis is complicated. *T. equiperdum* has the ability periodically to replace major surface glycoprotein antigens, which is a strategy supporting chronic infections. No method of immunization against dourine exists at present. The most successful prevention and eradication programs have focused on serologic identification of infected animals. Infected animals should be humanely destroyed or castrated to prevent further transmission of the disease. Some geldings may still show service behavior and constitute a risk. All equids in an area where dourine is found should be quarantined and breeding should be stopped for 1-2 months while testing continues.

Sanitation and disinfection are ineffective means of controlling the spread of dourine because the disease is normally spread by coitus.

11. GUIDE TO THE LITERATURE

1. BUCK, G.A., LONGACRE, S., RALBAUD, A., HIBNER, U., GIRAUD, C., BALTZ T., BALTZ, D. and EISEN, H. 1984. Stability of expression-linked surface antigen gene in *Trypanosoma equiperdum*. *Nature.* 307:563-566.
2. HERR, S., HUCHZERMEYER, H.F., TE BRUGGE, L.A., WILLIAMSON, C.C., ROOS, J.A., and SCHIELE, G.J. 1985. The use of a single complement fixation test technique in bovine brucellosis, Johne's disease, dourine, equine piroplasmosis and Q-tever serology. *Onderstepoort J.Vet.Res*. 52:279-282.
3. LOSOS, G.J. 1986. *Infectious Tropical Diseases of Domestic Animals*. New York: Churchill Livingstone, Inc. pp. 182-318.
4. LUCKINS, A.G., BARROWMAN, P.R., STOLTSZ, W.H. AND VAN DER LUGT, J.J. 2005. Dourine. In: *Infectious Diseases of Livestock,* 2nd ed., JAW Coetzer, RC Tustin, eds, Oxford University Press, pp. 297-304.
5. McENTEE, K. 1990. *Reproductive Pathology of Domestic Animals*. New York: Academic Press, pp. 204-205, 267-268.

6. WILLIAMSON, C.C., STOLTSZ, W.H., MATTHEUS, A., and SCHIELE, G.J. 1988. An investigation into alternative methods for the serodiagnosis of dourine. *Onderstepoort J. Vet. Res*. 55:117-119.

See Part IV for photos.

R.O. Gilbert, BVSc, MMedVet, MRCVS, College of Veterinary Medicine, Cornell University, Ithaca, NY 14853-6401, rog1@cornell.edu

18

DUCK VIRUS HEPATITIS

1. NAME
Duck virus hepatitis

2. DEFINITION
Duck virus hepatitis (DVH) is a highly contagious and fatal disease of young ducklings characterized by rapid spread, high mortality, and characteristic liver hemorrhages. The disease first appeared in 1945 in pekin ducklings on Long Island, New York.

DVH is caused by at least 3 different viruses that produce similar clinical signs and lesions. Type 1 causes the highest mortality and is the most widely prevalent. Type 2 has only been reported in ducks from England in 2 outbreaks occurring in 1964-68 and 1983-84. Type 3 has only been reported in the U.S. from 1969-78.

3. ETIOLOGY
DVH type 1 is caused by a picornavirus that can produce mortality as high as 100% in young ducklings. The virus contains RNA and is less than 50nm in size. The virus is very resistant to physical and chemical agents including chloroform and trypsin, and is stable at 50°C and pH 3. It is inactivated when treated with 5% phenol, undiluted Wescodyne (an iodine-containing disinfectant) or 5.25% sodium hypochlorite solution (bleach). Generally, DVH refers to an infection caused by the type 1 virus. A variant of type 1 (1a) was isolated from ducks in the U.S. and was identified as partially related to type 1 by cross-neutralization and cross-protection tests. DVH type 2 is caused by a totally different virus identified as an astrovirus that seems to be different from astrovirus isolates of chickens and turkeys. DVH type 3 was reported in ducklings immune to DVH type 1,and was later identified as a picornavirus antigenically unrelated to type 1.

4. HOST RANGE
a. Domestic animals and wild animals
Natural infection with DVH type 1 occurs in young domestic ducklings. Pekin ducklings are highly susceptible. Natural infection has also been reported in domestic mallards. Ducklings under 3 weeks of age are highly susceptible, but low mortality has been observed in ducklings up to 6 weeks of age. Experimental infection has resulted in mortality of young goslings, mallards, guinea fowl,

pheasants, quails and turkey poults, but clinical disease was not seen in young chickens and Muscovy ducklings.

DVH types 2 and 3 have only been seen in pekin ducklings. Type 2 has been reported in ducklings up to 6 weeks of age, while type 3 rarely occurs in ducklings over 2 weeks of age.

b. Humans

There is no evidence of human infection with DVH.

5. EPIDEMIOLOGY

a. Transmission

DVH is transmitted through contact with infected ducklings or contaminated environment. In natural infections, the disease spreads very rapidly and is highly contagious. The route of transmission seems to be both oral and respiratory, although oral administration of virus by pipetting or in a gelatin capsule placed in the esophagus failed to reproduce the disease. Aerosol or intratracheal administration has resulted in infection, suggesting that the natural route of infection may be pharynx or upper respiratory tract. There is no evidence of vertical egg transmission. Recovered ducklings remain as carriers and shed infective virus in their droppings for 8-10 weeks. No biological vector or wildlife reservoir has been detected.

b. Incubation period

The incubation period for DVH type 1 infection is 18-48 hours. Almost all of the mortality occurs within 3-5 days after the onset of an outbreak.

c. Morbidity

Morbidity in ducklings under 3 weeks of age can be as high as 100% with DVH type 1 infection.

d. Mortality

Mortality ranges between 70-90%. The morbidity and mortality in 4- to 6- week-old ducklings is lower.

DVH type 2 caused 10-25% losses in 3- to 6-week-old ducklings and up to 50% in ducklings below 3 weeks of age. Type 3 virus infection produced 50-60% morbidity and 10-30% mortality.

6. CLINICAL SIGNS

The disease has a sudden onset and runs an acute course. Affected ducklings become listless, reluctant to move with the brood and develop spasmodic contractions of legs. Sick ducklings lose their balance, fall on their side and die within 1-2 hours with their head drawn backward as in an opisthotonos position. Mortality is highest on the second and third days of clinical disease, and is over by 4-5 days. Clinical signs in ducklings affected with DVH types 2 and 3 are similar to type 1.

7. POST- MORTEM LESIONS

a. Gross

Most characteristic lesions are seen in the liver, which is enlarged and shows petechial or ecchymotic hemorrhages against a light-colored background. In older birds the hemorrhages may be more diffuse. The spleen is slightly enlarged and mottled. The kidneys are congested. Types 2 and 3 produce similar gross lesions. Catarrhal to hemorrhagic enteritis and petechial hemorrhages on the myocardium have been seen in DVH type 2.

b. Key microscopic

There is extensive hepatic cell necrosis and bile duct hyperplasia. Varying degrees of inflammatory cell infiltrates and hemorrhages may also be present. There are scattered heterophils in the hepatic sinusoids. Spleen and kidneys show degenerative changes.

8. IMMUNE RESPONSE

a. Natural infection

Ducklings that have recovered from DVH become immune to infection with a homologous type virus.

b. Immunization

Ducklings can be passively immunized by parenteral administration of serum or egg yolk obtained from recovered or immunized birds. The progeny of immunized breeder ducks are resistant to a virulent infection due to maternal antibodies passed through the egg yolk. This immunity lasts up to 2-3 weeks of age. The immune response of susceptible ducklings to experimental infection or immunization results in the development of 19S antibodies, but in ducklings over the age of 30 days, initial immune response of 19S antibodies is followed by production of 7S antibodies.

9. DIAGNOSIS

a. Field diagnosis

A presumptive diagnosis can be made on the basis of presence of characteristic signs, high mortality and typical lesions in the liver.

b. Laboratory diagnosis

i. Samples

The optimal sample to submit from a dead bird is liver. From a live bird, serum can be submitted.

ii. Laboratory procedures

The liver is the ideal organ to use for virus isolation. Virus can be isolated by inoculation of liver homogenate into 1- to 7-day-old susceptible pekin ducklings resulting in death within 24-48 hours with typical signs and liver lesions. Duck hepatitis virus can also be isolated by allantoic sac inoculation of 10- to 12-day-old duck embryos or 8- to 10-day-old chicken embryos. Duck embryos usually die within 24-72 hours while chicken embryos die within 5-8 days. Affected chicken embryos are stunted and show cutaneous hemorrhages, edema and enlarged greenish livers with necrotic foci. DVH type 2 virus grows erratically in the amnion of chicken embryos but is diagnosed by the presence of astrovirus particles on EM examination of liver homogenate. Type 3 virus can only be isolated in duck embryos or duck kidney cells. The isolated virus can be further identified by virus-neutralization test using a specific antiserum or by immunofluorescent test to demonstrate viral antigens.

10. PREVENTION AND CONTROL

Raising ducks in strict isolation along with strict biosecurity has kept some duck farms DVH-free in areas where there is no disease, but this approach has not worked in endemic areas.

DVH is prevented and controlled through immunization. Subcutaneous injection of immune serum obtained from recovered ducklings has been successfully used to control this disease even in the face of an outbreak. A similar approach was practiced using yolk antibody preparation from eggs laid by vaccinated breeder ducks. At the Cornell University Duck Research Laboratory in Eastport, New York, this procedure has been modified by hyperimmunizing chickens with a virulent DHV type 1 to produce yolk antibody preparation. Adult chickens are not susceptible to DVH.

The method that has proven most successful in the control and eradication of DVH is by immunization of breeder ducks to provide protection in ducklings through maternal immunity. A modified live DVH type 1 vaccine is used for vaccination. Breeder ducks are vaccinated 3 times at 12, 8 and 4 weeks before coming into lay. Re-vaccination is recommended every 3 months thereafter to maintain adequate protection in progeny. Immunity in ducklings usually lasts up to 2-3 weeks of age. Inactivated oil-emulsified vaccines when inoculated after an initial sensitization with a live modified vaccine have been reported to be effective.

DVH type 1 can also be prevented by active immunization of young ducklings. The same modified live virus vaccine is used for active immunization of 1- to 2-day-old susceptible ducklings. Since maternal antibodies interfere with active immunization, ducklings should originate from unexposed and unvaccinated parent flocks for this program to be successful.

Experimental live modified virus vaccines have been used to provide protection in ducklings against DVH types 2 and 3, but no commercial vaccines are available.

11. GUIDE TO THE LITERATURE

1. FABRICANT, J., RICKARD, C.G. and LEVINE, P.P. 1957. The pathology of duck virus hepatitis. *Avian Dis.* 1: 256-275.
2. GOUGH, R.E. and STUART, J.C. 1993. Astoviruses in ducks (Duck virus hepatitis type II). In: *Virus Infections of Birds*, J. B. McFerran, M.S. McNulty, eds., Amsterdam: Elsevier Science Publishers, B.V. pp. 505-508.
3. HAIDER, S.A. and CALNEK, B.W. 1979. In vitro isolation, propagation and characterization of duck hepatitis virus type III. *Avian Dis.* 23:715-729.
4. SANDHU, T.S., CALNEK, B.W. and ZEMAN, L. 1992. Pathologic and serologic characterization of a variant of duck hepatitis type 1 virus. *Avian Dis.* 36:932-936.
5. TOTH, T.E. 1969. Studies of an agent causing mortality among ducklings immune to duck virus hepatitis. *Avian Dis.* 13: 834-846.
6. WOOLCOCK, P.R. 1998. Duck Hepatitis. In: *A Laboratory Manual for the Isolationand Identification of Avian Pathogens*, 4th ed. D.E. Swayne, et al., eds., Kennett Square, PA: American Association of Avian Pathologists. pp. 200-204.
7. WOOLCOCK, P.R. 2003. Duck Hepatitis. In: *Diseases of Poultry*, 11th ed. Y. M. Saif et al., eds. Ames, IA: Iowa State Press, A Blackwell Publishing Company, pp. 343-354.

8. ZHAO, X., PHILLIPS, R.M., LI, G., and ZHONG, A. 1991. Studies on the detection of antibody to duck hepatitis virus by enzyme-linked immunosorbent assay. *Avian Dis*. 35:778-782.

See Part IV for photo.

Tirath S. Sandhu, DVM, MS, PhD, Cornell University Duck Research Laboratory, Eastport, NY, tss3@cornell.edu

19

EAST COAST FEVER

1. NAME

East coast fever, Theileriasis, Theileriosis, Zimbabwean tick fever, African Coast fever, Corridor disease, January disease

2. DEFINITION

East Coast fever (ECF), a form of bovine theileriosis, is a tick-transmitted protozoal disease of cattle characterized by high fever and lymphadenopathy. The disease causes high mortalities in breeds non indigenous to the endemic areas, and is confined to eastern, central, and parts of southern Africa.

3. ETIOLOGY

The causative agent of classical ECF is *Theileria parva parva*. Some previously recognized separate species and subspecies have been combined with *T. parva* as a result of recent studies on their DNA.

The life cycle of *T. parva* is complex in its tick and mammalian hosts. Sporozoite stages, produced in large numbers in the acinar cells of the salivary glands of the infected tick vector, *Rhipicephalus appendiculatus* are inoculated along with saliva during feeding and rapidly enter target lymphocytes, which become transformed after the development of the macroschizont stage. The infected lymphocyte is transformed into a lymphoblast and divides in conjunction with the schizont, giving rise to 2 infected daughter cells. This process has been termed "parasite-induced reversible transformation" because, if the cells are treated with anti-theileria drugs, the transformed cells revert to quiescent lymphocytes. Clonal expansion of infected cells occurs with an approximate tenfold increase of schizonts every 3 days.

From day 14 after tick infection of cattle, individual schizonts undergo merogony to produce merozoites (traditionally called microschizonts). Merozoites invade the erythrocytes to become piroplasms, which may subsequently undergo limited division also by merogony, Piroplasm-infected erythrocytes are ingested by ticks of the larval or nymphal stages and undergo a sexual cycle in the gut of the replete tick to produce zygotes, which in turn develop into motile kinete stages that infect the salivary glands of the next instar, the nymph or adult.

4. HOST RANGE

a. *Domestic animals*

Cattle in endemic areas, particularly the zebu type (*Bos indicus*), appear less susceptible to ECF, as do young animals. In addition, introduced cattle, whether of a taurine, zebu, or sanga breed, are much more susceptible to theileriosis than cattle from endemic areas. The Indian water buffalo (*Bubalus bubalis*) is as susceptible to *T. parva* infection as cattle.

b. *Wild animals*

The African buffaloes (*Syncerus caffer*) are reservoirs of *T. parva* infection, and it has recently been demonstrated that waterbucks (*Kobus* spp.) also are reservoirs. Buffaloes may suffer clinical disease from *T. parva* infection, but its effects on waterbuck are unknown. Piroplasms can be demonstrated in most wild antelopes in east Africa, but the relationship of most of them to *T. parva* is unclear.

c. *Humans*

Theileria parva parva does not infect humans.

5. EPIDEMIOLOGY

a. *Transmission*

Rhipicephalus appendiculatus is the main field vector of ECF, although in certain areas other field vectors occur, such as *R. zambeziensis* in drier areas of southern Africa and *R. duttoni* in Angola. East Coast Fever is not maintained in the absence of these field vectors. The rhipicephalid vectors are three-host ticks, and transmission occurs from stage to stage; transovarian transmission does not occur. Ticks can remain infected on the pasture for up to 2 years depending on the climatic conditions. The parasite dies out faster in hot climates and in nymphs compared with adults. Normally, for transmission to occur, the infected tick has to attach for several days to enable sporozoites to mature and be emitted in the saliva of the feeding tick. However, under high ambient temperatures, ticks on the ground may develop infective *Theileria* sporozoites, which can be transmitted to cattle within a few hours after attachment.

b. *Incubation period*

Under experimental conditions, using either ticks of known infection or sporozoite stabilate, the incubation period has a medium range of 8-12 days. The incubation period may be much more variable in the field owing to differences in challenges experienced by the cattle and may extend to beyond 3 weeks after attachment of infected ticks.

c. *Morbidity*

Morbidity and mortality depend on, among other factors, the magnitude of the infected tick challenge and susceptibility of the host and strain of parasite. Animals that recover are often unthrifty and sickly. Zebu cattle residing for many generations in endemic areas become infected (100% morbidity), but only a minor proportion succumb. However, many become carriers, and early infection with *T. parva* can affect their growth and productivity.

d. *Mortality*

Host susceptibility is a factor in mortality rates. East Coast fever in susceptible cattle, which are not indigenous to the enzootic area, is very severe with a mortality approaching 100%.

6. CLINICAL SIGNS

The first clinical sign is usually a swelling of the draining lymph node, usually the parotid, for the ear is the preferred feeding site of the vector. This is followed by a generalized lymphadenopathy in which superficial lymph nodes such as the parotid, prescapular, and prefemoral lymph nodes, can easily be seen and palpated. Fever ensues and continues throughout the course of infection. This rise in temperature is rapid and is usually in excess of 103°F (39.5°C) but may reach 106°F (42°C). There is marked petechial and ecchymotic hemorrhage on most mucous membranes of the conjuctiva and the buccal cavity. Anorexia develops, and loss of condition follows. Other clinical signs may include lacrimation, corneal opacity, nasal discharge, terminal dyspnea, and diarrhea.

Before death the animal is usually recumbent, the temperature falls, and there is a severe dyspnea due to pulmonary edema that is frequently seen as a frothy nasal discharge. Mortality in fully susceptible cattle can be nearly 100%. The severity and time course of the disease depend on, among other factors, the magnitude of the infected tick challenge (ECF is a dose-dependent disease), and on the strain of parasites. Some stocks of parasites cause a chronic wasting disease. A fatal condition called "turning sickness" is associated with the blocking of brain capillaries by infected cells and results in neurological signs.

In recovered cattle, chronic disease problems can occur that result in stunted growth in calves and lack of productivity in adult cattle. However, this syndrome tends to be in the minority of recovered clinical cases. In a majority of cases, asymptomatic carriers can be recognized with apparently little or no effect on their productivity.

7. POST- MORTEM LESIONS

a. Gross

A frothy exudate is frequently seen around the nostrils of an ECF-infected animal. Signs of diarrhea, emaciation, and dehydration may be seen. Lymph nodes are greatly enlarged and may be hyperplastic, hemorrhagic, and edematous. In acute cases of ECF, lymph nodes are edematous and hyperemic but often become necrotic and shrunken in more chronic disease. Generally, muscles and fat appear normal but, depending on relative acuteness of infection, fat may become greatly depleted. Serosal surfaces have extensive petechial and ecchymotic hemorrhages, and serous fluids may be present in body cavities. Hemorrhages and ulceration may be seen throughout the gastrointestinal tract — particularly in the abomasum and small intestine, where necrosis of Peyer's patches can be observed. Lymphoid cellular infiltrations appear in the liver and kidney as white foci. The most striking changes are seen in the lungs. In most cases of ECF, interlobular emphysema and severe pulmonary edema appear, the lungs are reddened and filled with fluid, and the trachea and bronchi are filled with fluid and froth.

b. Key microscopic

One method of diagnosis involves finding schizonts in affected tissues. Although these are more readily detectable in Giemsa-stained smears, their presence in affected lymph nodes examined histologically is also diagnostic.

8. IMMUNE RESPONSE

a. Natural infection

Recovery from ECF usually results in an excellent immunity to homologous or related stocks of the parasite, lasting for over 3.5 years in the absence of reinfection.

b. Immunization

There are several methods of immunization using live parasites. The most successful involves an "infection and treatment method" using oxytetracycline or comparable drugs. Animals are inoculated with a potentially lethal dose of infective sporozoite stabilate prepared from ticks and treated with a drug either simultaneously (tetracyclines, buparvaquone) or subsequently (parvaquone and halofuginone). Problems do occur in the recognition of suitable antigenic stocks for immunization, and any vaccination scheme can only follow after a careful assessment of the local complex of *T. parva* parasites. In any population of *T. parva* parasites in the field, an isolate may constitute several strains. This method of immunization requires a reliable cold chain and extensive monitoring.

Protective immunity involves generation of cytotoxic T lymphocytes (CTL) targeting the schizont-infected lymphocytes. Hence, immunity is induced by schizont antigens and there are a number that are being evaluated in imunogenicity and efficacy trials for development of recombinant vaccines. Neutralizing antibodies to sporozoites have been recognized and related to a 67Kda antigen on the surface of the sporozoites. This has been synthesized by recombinant technology and has been shown to provide a degree of protection to cattle immunized with it.

9. DIAGNOSIS

a. *Field diagnosis*

East Coast Fever is only found in association with its known tick vectors, *Rhipicephalus appendiculatus*, *R. zambeziensis* and possibly *R. duttoni* and *R. nitens*. A febrile disease with signs of enlarged lymph nodes associated with infestation by tick vectors is suggestive of ECF. An acute disease with high mortality on farms, where tick control is not effectively applied, also is suggestive of ECF. In the field, diagnosis is usually achieved by finding *Theileria* parasites in Giemsa-stained blood smears and lymph node needle biopsy smears.

b. *Laboratory diagnosis*

i. Samples

Specimens consisting of buffy coat smears air-dried and fixed in methanol, lymph node impressions air-dried and fixed in methanol, lymph nodes, spleen, lung, liver, and kidney samples for histopathology, and serum should be collected.

ii. Laboratory procedures

The demonstration of schizont-infected cells in Giemsa-stained blood smears, lymph node impression smears, or histologic sections, is diagnostic of ECF. Small piroplasms in erythrocytes are suggestive of ECF, but diagnosis must be confirmed by the detection of schizonts. Schizonts can be detected in sections but are best seen in smears of lymph node biopsies. Since there is considerable similarity between schizonts of other theileria parasites (*T. mutans*, *T. velifera*, *T. taurotragi* and *T. buffeli*), which may co-infect an animal, it is important to differentiate the infecting species. This can be done by using serological and DNA-based assays.

A number of PCR methods (targeting sequences TpR, p104, p67) can be used to detect *T. parva parva.* These are specific and detect infected and carrier animals.

Antibody in the mammals can be detected by a variety of serological tests of which the most widely used is the indirect fluorescent antibody test (IFAT) employing cell culture schizont antigen. Though informative and widely used, the IFAT does not discriminate between *T. parva parva* and *T. taurotragi* antibodies. More specific enzyme-linked assays have been developed using whole parasite lysates or specific antigens isolated by monoclonal antibodies. The serological test of choice because of its high specificity for *T. parva parva* antibodies is the sub-unit ELISA targeting the polymorphic immunodominant molecule (PIM) antigen. This ELISA has high sensitivity and does not cross-react with *T. mutans*, *T. taurotragi* or *T. buffeli*. Because of the often acute nature of the ECF, serological tests are useful in detecting a changed immune status of recovered animals within an exposed herd. An antigen capture *T. mutans* specific ELISA (based on the 32kDA antigen), is available which does not detect the other theileria species.

10. PREVENTION AND CONTROL

There are currently 3 effective drugs for the treatment of ECF: parvaquone (Clexon), buparvaquone (Butalex), and halofuginone lactate (Terit). Of these, Butalex is used most commonly. The availability of a therapeutic means of controlling ECF is a significant development.

The current primary method of controlling ECF in cattle is immunization and treatment of cattle with chemical acaricides. A number of acaricides, mainly organochlorides and organophosphorus compounds but recently synthetic pyrethroids and amidens, are applied in dips, spray races, or by handspraying. More recently, "pour on" or "spot on" formulations have been introduced. The application is usually on a weekly basis, but this rate has to be increased when the challenge is high. Very high levels of acaricide exposure lead to resistance of vectors, residues in milk and meat, and, often, creation of an epidemic instability with a large proportion of the cattle population becoming susceptible. Integrated control measures are recommended including effective fencing, pasture management, rotational grazing to reduce the level of challenge, selection of tick resistant cattle, and new methods of immunization. These are all combined with strategic acaricide application.

11. GUIDE TO THE LITERATURE

1. NORVAL, R.A.I., PERRY, B.D. and YOUNG, A.S. 1992. *The Epidemiology of Theileriosis in Africa*. London:Academic Press, 481 pp.
2. ole-MOIYOI, O.K. 1989. Theileria parva: An intracellular parasite that induces reversible lymphocyte transformation. *Exptl. Parasitol.*, 69: 204-210.
3. MORRISON, W.I., BOSCHER, G., MURRAY, M., EMERY, D.L., MASAKE, R.A., COOK, R.H. and WELLS, P.W. 1981. *Theileria parva*: Kinetics of infection in lymphoid system of cattle. *Exptl. Parasit.*, 52:248-260.
4. BISHOP, R., SOHANPAL, B., KARIUKI, D.P., YOUNG, A.S., NENE, V., BAYLIS, H., ALLSOPP, B.A., SPOONER P.R., DOLAN, T.T., and MORZARIA, S. P. 1992. Detection of a carrier state in *Theileria parva* infected cattle using the polymerase chain reaction. Parasitology, 104:215-232.
5. MOLL, G., LOHDING, A., YOUNG, A.S., and LEITCH, B.L. 1986. Epidemiology of theileriosis in calves in an edemic area of Kenya. *Vet. Parasitol.*, 19:255-273.
6. MORZARIA, S.P., KATENDE, J., MUSOKE, A.J., NENE, V., SKILTON, R. and BISHOP, R. 1999. Development of sero-diagnostic and molecular tolls for the control of important tick-borne pathogens of cattle in Africa. *Parasitologia*. 41:73-80.
7. CUNNINGHAM, M.P. 1977. Immunization of Cattle Against Theileria parva. In *Immunity to Blood Parasites of Animals and Man.* L.H. Miller, J. A. Pino, and J.J. McKelvey, Jr., eds., New York: Plenum Press, pp. 189-207.
8. GRAHAM, S.P., PELLE, R., HONDA, Y., MWANGI, D.M., TONUKARI, N.J., YAMAGE, M., GLEW, E.J., de VILLIERS, E.P., SHAH, T., BISHOP, R., ABUYA, E., AWINO, E., GACHANGA J, LUYAI, A.E., MBWIKA, F., MUTHIANI, A.M., NDEGWA, D.M., NJAHIRA, M., NYANJUI, J.K., ONONO, F.O., OSASO, J., SAYA, R.M., WILDMANN, C., FRASER, C.M., MAUDLIN I, GARDNER, M.J., MORZARIA, S.P., LOOSMORE, S., GILBERT, S.C., AUDONNET, J.C., van der BRUGGEN, P., NENE, V. and TARACHA, E.L. 2006. *Theileria parva* candidate vaccine antigens recognized by immune bovine cytotoxic T lymphocytes. *Proc. Nat. Acad. Sci.* 103:3286-3291.
9. MUSOKE, A.J., MORZARIA, S. P., NKONGE, C., JONES, E. and NENE, V. 1992. A recombinant sporozoite surface antigen of *Theileria parva* induces protection in cattle. *Proc. Nat. Acad. Sci.* 89:514-519.
10. YOUNG, A.S., GROOCOCK, C.M. and KARIUKI, D.P. 1988. Integrated control of ticks and tick-home diseases of cattle in Africa. *Parasitology.* 96:403-441.

11. COETZER, J.A.W., and TUSTIN, R.C. 2005. Theilerioses. In *Infectious Diseases of Livestock* 2nd ed. JAW Coetzer, RC Tustin, eds., Oxford University Press. pp. 447-466.

See Part IV for photos.

Suman M Mahan, BVM, MSc PhD, Pfizer Animal Health, Kalamazoo, Michigan, 49001, suman.mahan@pfizer.com

20

EPIZOOTIC LYMPHANGITIS

1. NAME
Epizootic lymphangitis, Pseudoglanders, Histoplasmosis farciminosi, Equine Blastomycosis, Equine Histoplasmosis, Equine Cryptococcosis, African Farcy

2. DEFINITION
Epizootic lymphangitis is a chronic, infectious granulomatous disease of the skin, lymph vessels, and lymph nodes of the neck and legs of horses caused by *Histoplasma capsulatum* var. *farciminosum*, a thermally dimorphic fungus previously known as *Histoplasma farciminosum.*

3. ETIOLOGY
Epizootic lymphangitis is caused by a dimorphic fungus, *Histoplasma capsulatum* var. *farciminosum*, formerly known as *Histoplasma farciminosum*, *Cryptococcus farciminosis*, *Zymonema farciminosa*, or *Saccharomyces farciminosus.* In tissue, the organism is present in a yeast form; it forms mycelia in the environment, has a saprophytic phase in the soil, and is relatively resistant to ambient conditions, which allows it to persist many months in warm, moist conditions.

4. HOST RANGE
a. *Domestic animals*
The natural host range seems to be limited to horses, donkeys, and occasionally mules.
b. *Wild animals*
There are no reports of the disease in wild animals.

c. *Humans*
Rare cases of human infection have been reported, but identification of the causative organism has not been substantiated.

5. EPIDEMIOLOGY
a. *Transmission*
H. farciminosum is introduced via open wounds. Transmission generally involves infection of wounds by flies contaminated by feeding on the open wounds of infected animals. The organism has been isolated from the gastrointestinal tract of flies.

b. *Incubation period*

The incubation period is variable and is usually several weeks.

c. *Morbidity*

Morbidity is high only when large numbers of animals are collected together such as in military situations, for racing, or on village commonages.

d. *Mortality*

Mortality is low.

6. CLINICAL SIGNS

There is no breed, sex, or age predilection in epizootic lymphangitis. This disease most typically involves the skin and associated lymph vessels and nodes. In addition, the conjunctiva and nictitating membrane may be involved. Occasionally there is involvement of the respiratory tract. The body temperature and general demeanor of the animal are not changed. The initial lesion is a painless cutaneous nodule about 2 cm in diameter. This nodule is intradermal and is freely moveable over the subcutis. Lesions are most commonly found on the skin of the face, forelimbs, thorax, and neck or (less often) the medial aspect of the rear limbs. The subcutaneous tissue surrounding the nodule becomes diffusely edematous. The nodule gradually enlarges and ultimately bursts. Some cases do not progress beyond small, inconspicuous lesions that heal spontaneously. More typically, resultant ulcers increase in size and undergo cycles of granulation and partial healing followed by renewed eruption. The surrounding tissues become hard, variably painful, and swollen.

The infection spreads along lymph vessels and causes cord-like lesions, leading to diffuse and irregular involvement of an area of skin. After initial increase in lesion size, additional cycles of eruption and granulation lead to progressively smaller areas of ulceration until eventually only a scar remains. The scar is usually stellate in shape. The development and regression of a lesion takes about 3 months. Where lesions overlie joints, involvement may extend to synovial structures and produce severe arthritis.

Conjunctivitis or keratoconjunctivitis may occur — usually in conjunction with skin lesions. A serous or purulent nasal discharge containing abundant organisms may be observed. Although respiratory lesions are described as common in older literature, this form of the disease appears to be rare in more recent outbreaks.

7. POST- MORTEM LESIONS

The affected skin and subcutaneous tissue is thickened, fibrous, and firm. Several purulent foci may be apparent on cut section. Lymphatic vessels are distended with pus. Regional lymph nodes are swollen, soft, and reddened and may contain purulent foci. Arthritis, periarthritis, and periostitis have been described. The nasal mucosa may have multiple, small gray-white nodules or ulcers with raised borders and granulating bases. Nodules and abscess may occur in internal organs, including the lungs, spleen, liver, and testes. A pyogranulomatous reaction with fibroplasia and a predominance of macrophages is the characteristic histopathological lesion.

8. IMMUNE RESPONSE

a. Natural infection

Horses that recover from clinical infection are immune to reinfection.

b. Immunization

Although promising results have been obtained with experimental vaccines, a vaccine is not commercially available.

9. DIAGNOSIS

a. Field diagnosis

Although the clinical presentation of the disease may lead to a presumptive diagnosis of epizootic lymphangitis, the similarity of this disease to glanders makes laboratory confirmation essential.

b. Laboratory diagnosis

i. Samples

A whole or section of a lesion and a serum sample should be collected aseptically. The samples should be kept cool and shipped on wet ice as soon as possible. Sections of lesions in 10% buffered formalin and air-dried smears of exudate on glass slides should be submitted for microscopic examination.

ii. Laboratory procedures

Demonstration of the yeast in tissue sections or smears of lesions is considered the most reliable means of diagnosis. Attempts to culture the organism fail in up to half of cases. The organism in the tissues is in its yeast form. It may be stained with Giemsa, Diff-Quik, or Gomori methenamine silver. In addition, an indirect fluorescent antibody technique for demonstration of the organism has been developed. Rabbits, mice and guinea pigs may be experimentally infected with exudate from suspect animals. Affected animals do mount a humoral immune response to the infection; an enzyme-linked immunosorbent assay

(ELISA), direct and indirect fluorescent antibody tests, and a passive hemagglutination test have been developed for the diagnosis of epizootic lymphangitis. Attempts have also been made to utilize intradermal skin testing (with histoplasmin or histofarcin) with encouraging results.

10. PREVENTION AND CONTROL

Successful treatment with intravenous administration of sodium iodide, oral administration of potassium iodide, and surgical excision of lesions where possible have been reported, but recurrences of clinical signs months later are possible. In vitro sensitivity of the organism to amphotericin B, nystatin, and clotrimazole has been reported. In most areas, epizootic lymphangitis is a reportable disease, and treatment is not allowed. To prevent contamination of the environment, affected animals must be destroyed.

Strict hygienic precautions are essential to prevent spread of epizootic lymphangitis. Great care should be taken to prevent spread by grooming or harness equipment. Contaminated bedding should be burned. The organism may persist in the environment for many months.

Epizootic lymphangitis is a chronic disease. Many mildly affected horses recover. Those that do are reputedly immune for life, a belief that has led to a premium being placed in endemic areas on horses with characteristic scars. In most areas of the world, however, this is a reportable disease. Treatment of clinical cases is not permitted, and destruction of affected horses is usually mandatory. In most areas, epizootic lymphangitis has been eradicated by a strict policy of slaughter of infected animals.

11. GUIDE TO THE LITERATURE

1. AL-ANI, F.K. and AL-DELAIMI, A.K. 1986. Epizootic lymphangitis in horses: Clinical, epidemiological and haematological studies. *Pakistan Vet. J.*, 6:96-100.
2. CHANDLER, F.W., KAPLAN, W. and AJELLO, L. 1980. *Color Atlas and Text of the Histopathology of Mycotic Diseases*. Chicago:Year Book Medical Publishers, pp.70-72, 216-217.
3. GABAL, M.A., BANNA, A.A. and GENDI, M.E. 1983. The fluorescent antibody technique for diagnosis of equine histoplasmosis (epizootic lymphangitis). Zbl.Vet. Med. (B), 30:283-287.
4. GABAL, M.A. and KHALIFA, K. 1983. Study on the immune response and serological diagnosis of equine histoplasmosis (epizootic lymphangitis). *Zbl. Vet. Med. (B)*, 30:317-321.

5. GABAL, M.A. and MOHAMMED, K.A. 1985. Use of enzyme-linked immunosorbent assay for the diagnosis of equine histoplasmosis farciminosi (epizootic lymphangitis). *Mycopathologia*. 91:35-37.
6. MORROW, A.N., and SEWELL, M.M.H.1990. Epizootic Lymphangitis, in *Handbook on Animal Diseases in the Tropics* 4th ed, Sewell, M.M.H. and Brocklesby, D.W. eds, London: Bailliere Tindall, pp.364-367.
7. SELIM, S.A., SOLIMAN, R., OSMAN, K., PADHYE, A.A. and AJELLO, L. 1985. Studies on histoplasmosis farciminosi (epizootic lymphangitis) in Egypt. Isolation of *Histoplasma farciminosum* from cases of histplasmosis farciminosi in horses and its morphological characteristics. *Eur. J. Epidemiol*. 1:84-89.
8. SOLIMAN, R., SAAD, M.A. and REFAI, M. 1985. Studies on histoplasmosis farciminosi (epizootic lymphangitis) in Egypt. 111. Application of a skin test ("Histofarcin") in the diagnosis of epizootic lymphangitis in horses. *Mykosen*. 28:457-461.

R.O. Gilbert, BVSc, MMedVet, MRCVS, College of Veterinary Medicine, Cornell University, Ithaca, N Y 14853-6401, rog1@cornell.edu

21

EQUINE ENCEPHALOSIS

1. NAME
Equine encephalosis

2. DEFINITION
Equine encephalosis virus, transmitted by *Culicoides*, is the cause of a subclinical to mild disease in horses, characterized by fever, congestion or icterus of mucous membranes, and occasionally neurologic signs.

3. ETIOLOGY
The causative agent, equine encephalosis virus (EEV) is an orbivirus in the Family Reoviridae. As such, it is closely related to the viruses of bluetongue, epizootic hemorrhagic disease, and African horse sickness. EEV consists of ten segments of dsRNA, surrounded by a protein coat. There are seven antigenically distinct serotypes of EEV.

4. HOST RANGE
a. Domestic animals
Equine encephalosis is a disease of horses. Donkeys and mules may be productively infected, with resulting seroconversion, but there is no clinical disease.
b. Wild animals
Seropositivity rates in zebra in endemic areas are about 25%, although there is no clinical disease recorded. Similarly, EEV-specific antibodies have been detected in several elephants, but no clinical disease. Other wildlife species tested have been negative for the presence of antibodies to EEV. Zebra are probably the reservoir of EEV, although this has not been definitely proven, as it has been for the closely related agent, African horse sickness virus.

c. Humans
There is no indication that EEV infects humans.

5. EPIDEMIOLOGY

a. Transmission

Culicoides imicola is the main vector in the endemic region but other species of biting midges can be competent vectors. Infection usually occurs during late summer and fall, which are the times for maximal vector numbers and activity.

b. Incubation period

Horses show clinical signs 2-6 days after experimental inoculation of the virus. Subsequent to natural infection (by *Culicoides*), incubation is slightly longer.

c. Morbidity

Serologic surveys indicate that infection among equids is extensive in the endemic regions, with up to 75% of horses in an area being seropositive.

d. Mortality

Mortality with EEV is infrequent. Early reports in the literature described a peracute neurologic disease with death and virus isolated from tissues of dead animals. However, it is unknown if this is unique to a certain serotype, and/or if the EEV infection was responsible for the clinicopathologic syndrome.

6. CLINICAL SIGNS

Infected horses usually experience the disease subclinically. For those that are ill, there is fever, inappetence, congestion or icterus of mucous membranes, and occasionally neurologic dysfunction. Older literature describes myocardial degeneration and fibrosis but it is not certain that this is due to EEV rather than another disorder. Some reports connect the presence of the virus with abortion but, similarly, the role of EEV in these cases is unclear.

7. POST- MORTEM LESIONS

Gross

In horses that are necropsied during the period of clinical disease, there is localized enteritis, cerebral edema, and fatty degeneration of the liver.

8. IMMUNE RESPONSE

a. Natural infection

After natural infection, horses are resistant to challenge with the same serotype. There is limited cross-protection among serotypes.

b. Immunization

No vaccine exists for EEV.

9. DIAGNOSIS

a. Field diagnosis

Icterus and neurologic disease in an area endemic for the disease, and when *Culicoides* insects are active, would raise suspicion, prompting collection of samples. Laboratory testing is essential for diagnosis of the disease.

b. Laboratory diagnosis

i. Samples

Blood, serum, and spleen should be collected.

ii. Laboratory procedures

Serum neutralization, ELISA (specific for EEV, does not cross-react with other orbiviruses), and virus isolation in embryonated eggs, suckling mouse brain, or cell culture may all be used to diagnose EEV.

10. PREVENTION AND CONTROL

There is no vaccine for this disease. There are no known treatments. Preventing contact with *Culicoides* is the most effective means to limit transmission from one equid to another.

11. GUIDE TO THE LITERATURE

1. BARNARD, B.H.J. 1997. Antibodies against some viruses of domestic animals in southern African wild animals. *Onderstepoort Journal of Veterinary Research.* 65:95-110.
2. CRAFFORD, J.E., GUTHRIE, A.J., VAN VUUREN, M., MERTENS, P.P.C., BURROUGHS, J.N., HOWELL, P.G. and HAMBLIN, C. 2003. A group-specific, indirect sandwich ELISA for the detection of equine encephalosis virus antigen. *Journal of Virological Methods.* 112:129-135.
3. HINCHCLIFF, K.W. 2007. Equine encephalosis. Chapter 26, Miscellaneous viral diseases. In: *Equine Infectious Diseases*, D. Sellon and M. Long, eds., Saunders, St. Louis, MO. p. 233
4. HOWELL, P.G., GROENEWALS, D., VISAGE, C.W., BOSMAN, A.M., COETZER, J.A. and GUTHRIE, A.J. 2002. The classification of seven serotypes of equine encephalosis virus and the prevalence of homologous antibody in horses in South Africa. *Onderstepoort Journal of Veterinary Research*, 69:79-93.
5. LORD, C.C., VENTER, G.J., MELLOR, P.S., PAWESKA, J.T. and WOOLHOUSE, M.E. 2002. Transmission patterns of African horse sickness and equine encephalosis viruses in South African donkeys. *Epidemiology and Infection*, 128:265-275.

6. PAWESKA, J.T. and VENTER, G.J. 2004. Vector competence of *Culicoides* species and the seroprevalence of homologous neutralizing antibody in horses for 6 serotypes of equine encephalosis virus (EEV) in South Africa. *Medical and Veterinary Entomology,* 18:398-407.

Corrie Brown, DVM, PhD, Department of Pathology, College of Veterinary Medicine, University of Georgia, Athens, GA 30602-7388, corbrown@uga.edu

22

FOOT-AND-MOUTH DISEASE

1. NAME

Foot-and-Mouth Disease
Fiebre Aftosa (Sp), Fievre Aphteuse (Fr), Maul-und-Klauenseuche (Gr)

2. DEFINITION

Foot-and-Mouth Disease (FMD) is an extremely contagious viral disease, primarily of domestic cloven-hoofed animals and many wild animals. FMD is characterized by fever, vesicular lesions, and subsequent erosions of the epithelium of the mouth, tongue, nares, muzzle, feet, or teats.

3. Etiology

FMD virus (FMDV) belongs to the genus *Aphthovirus* within the family Picornaviridae. FMDV was for a long time the only member of the *Aphthovirus* genus. However, recently this genus included the Equine Rhinovirus A virus (previously known as the Equine Rhinovirus Type 1).

There are 7 completely immunologically distinct FMDV types: A, O, C (or the so called European types); SAT-1, SAT-2, SAT-3 (South African Territories types) and Asia 1. In addition, within a given type there are many immunologically related sub-types, particularly among the A, O and C (i.e. A_5, A_{24}, C_1, O_1, etc), totaling more than 60 type-subtype known combinations. The high number of FMDV sub-types is the result of error-prone replication of RNA viruses. The implication is that in a given infection there are a great number of mutants being produced (quasispecies concept), many of which are selected out due to host immunological pressures. New variants of FMDV can also be the result of homologous recombination between 2 different strains of FMDV. The combined result of the genetic variation of FMDV through mutations, recombinations and selection is that new FMDV variants are constantly being generated with important implication for the selection of FMDV vaccine strains.

As a *picornavirus*, FMDV is a small (27-28 nm) icosahedral non-enveloped virus that contains a single positive-sense strand of RNA. A large open reading frame codes for a single large polyprotein that is cleaved into 12 different proteins by viral proteases. Some of these proteins are incorporated into the virus capsid (structural proteins) and they carry the unique antigens that induce protective immunity and codify for its type and sub-type characteristics. Other proteins are produced during the virus replication in the infected animal, but are not incorporated into the virus capsid (non-structural proteins or NSPs). Some of these NSPs are common among all types and do not induce protective immunity in the infected host.

Members of the genus *Aphthovirus* differ from other members of the Picornaviridae family in their susceptibility to low pH. By their nature, most enteric Picornaviruses are quite resistant to the low pH of the stomach which allows them to reach the intestinal tract, their target organ. However, FMD viruses are typically denatured at pH below 6.0 or above 9.0. This characteristic does not allow the FMDV to survive in muscle tissue when its pH is <6.0 after *rigor mortis*. Still, FMDV can persist for longer periods of time after death in lymph nodes and bone marrow that maintain a close to neutral pH at *post mortem* conditions. Depending on the temperature, ultraviolet radiation (sunlight), humidity and pH conditions, FMDV can persist in the environment for up to one month.

FMDV can be inactivated by low pH solutions (2% acetic acid or straight vinegar, or 0.2% citric acid solutions); or alkaline solutions (2% sodium hydroxide or 4% sodium carbonate). FMDV is somewhat resistant to iodophores, quaternary ammonium compounds, hypochlorite and phenol, especially in the presence of organic matter.

4. HOST RANGE

a. Domestic animals:

While all cloven-hoofed domestic animals are susceptible to FMDV infection, the clinical manifestations of the disease are more evident in cattle (both *Bos indicus* and *Bos taurus*), water buffalo (*Bubalus bubalis*), yaks (*Bos grunniens*), and pigs (*Suidae* sp.). The clinical manifestation of FMD in sheep (*Ovis aries*) and goats (*Capra aegagrus hircus*) is quite mild or not apparent in many cases. FMD infection in sheep and goats tends to be self-limiting in the absence of continual challenge from cattle or other domestic animals. Proof of this has been seen in the experience in South America where sheep have not been vaccinated against FMD even when they commingle with vaccinated cattle. In those situations, once FMD is eliminated from the cattle, FMD infection disappears in sheep and goats.

Old World camelidae (camels – *Camelus bactrianus* and dromedaries – *Camelus dromedarius*) may be affected by some FMDV strains. In contrast, New World South American camelids (llamas – *Lama glama*, alpacas – *Vicugna pacos*, and vicuñas – *Vicugna vicugna*) have low susceptibility.

b. Wild animals:

It is considered that all cloven-hoofed wild animals are susceptible to FMDV infection. However, their clinical manifestation can be quite variable, from severe clinical and pathological manifestations to inapparent infections. The most important hosts of FMDV in Africa are the African or Cape buffalo (*Syncerus caffer*) and the impala (*Aepyceros melampus*). High mortality was observed in Israel in free-ranging mountain gazelles (*Gazella gazelle*). Other susceptible wild animals include hedgehogs (*Erinaceinae* sp.), armadillos (*Dasypodidae* sp.), nutrias (*Myocastor coypus*), elephants (*Elephantidae* sp.), capybaras (*Hydrochoerus hydrochoeris*), rats (*Rattus* sp.) and mice (*Mus* sp.).

Except for a report of an FMD outbreak in mule deer (*Odocoileus hemionus*) in California in 1924, and an experimental FMD infection of white-tailed deer (*Odocoileus virginianus*) in the 1970's, relatively little is known regarding the effects of FMDV infection in the major North American wildlife species. Experimental infections of North American bison (*Bison bison*), elk (*Cervus elaphus nelsoni*), pronghorn antelope (*Antilocapra americana*), and mule deer have been conducted recently by the USDA at the Plum Island Animal Disease Center. It was observed that all inoculated animals as well as all contact-exposed bison, pronghorn, and mule deer developed clinical vesicular disease manifestations. However, no clinical vesicular disease was observed in contact-exposed elk or in cattle exposed to inoculated elk. Only one contact elk showed serologic evidence of FMD virus infection. Mild oral lesions were observed in pronghorn antelope and elk, and severe lesions were present in bison cattle, and mule deer. Mild foot lesions were observed in elk, while severe foot lesions were present in the other species. Intra and interspecies transmission occurred in all species except elk.

The role of wildlife in FMD spread and persistence has been a matter of discussion for years. However, it is a fact that countries that have eradicated FMD from domestic animals did nothing systematic to control FMD in wildlife with no recurrence of the disease. The only exception to this has been the documented experiences in South Africa with carrier African buffalo (*Syncerus caffer*) in the Kruger National Park where FMDVs (SAT types) have been transmitted from the buffalo to domestic cattle and perhaps to other wildlife. There is good evidence that most FMD infections in wildlife in many parts of the world have been the result of transmission from livestock, and, once FMD is eliminated from domestic animals, FMD disappears from wildlife.

c. Humans:

FMDV infections in humans are rare and of little consequence. For that reason, FMD is not considered to be a zoonotic disease. A review of human infections from 1921-97 revealed slightly over 40 human cases, in all continents, mostly due to type O viruses, followed by type C and rarely by type A. However, FMD is considered by the Pan American Health Organization as a public health issue due to its impact on availability of quality protein of animal origin for human nutrition and its effects on mental health (depression, high rate of suicides, and post-traumatic mental disturbances) as a result of activities during the control of serious outbreaks requiring massive depopulations.

5. EPIDEMIOLOGY

a. Transmission:

Transmission of FMDV occurs by direct contact between infected and susceptible animals; direct contact of susceptible animals with contaminated inanimate objects (hands, footwear, clothing, vehicles, etc.); consumption (primarily by pigs) of contaminated meat products; ingestion of contaminated milk (by calves); artificial insemination with contaminated semen; and, by the inhalation of infectious aerosols. This latter mode of transmission can occur in close quarters during the viremic period of FMDV infection, or by wind-borne aerosols from far distances (up to 60 km overland and 300 km by sea). In these cases, there is a need for proper environmental conditions of high humidity with decreased UV radiation (foggy conditions) and low temperatures. These conditions have been observed in northern portions of Europe but not in the tropics due to low humidity, higher temperatures and abundant sunlight.

The potential for humans to harbor the FMDV in their respiratory tract has been investigated. It was noted that FMDV can be carried for a short period in the throats of people. Sampling of human subjects who had been in isolation rooms containing FMDV-infected animals demonstrated that virus could be recovered from the nose, throat, and saliva from 7 of 8 people up to 24 hours immediately after exiting the room. However, nasal swabs from one person yielded FMDV after 24 hours but not after 48 hours. These findings have led to the common practice of 3-5 days of personal quarantine on personnel exposed to isolation rooms containing FMD infected animals at FMD research sites around the world. This personal quarantine can be reduced to an overnight period of time through change of clothing and complete shower with shampoo and expectoration during an active outbreak situation.

FMDV is shed in large amounts from animals from as early as 1-2 days before the appearance of clinical signs, and for 7-10 days after the onset of clinical signs. Virus is excreted in all body fluids and excretions, as well as in exhaled air. Pigs are notorious for excreting large amounts of FMDV (estimated at 10 million to 10 billion infective doses per day). Cattle are not that far apart in the amount of virus shed, primarily in the form of sloughed oral and pedal epithelium. There is also significant amount of FMDV shed in the milk of lactating cows reaching levels of 5 million infective doses per milliliter of milk.

It should be pointed out that there have been naturally-occurring strains of FMDV that have atypical affinity for some animal species otherwise susceptible to this virus. The best recent example of this phenomenon was the emergence of the Taiwan FMD type O/97 as a natural mutant that lost its ability to infect cattle or other ruminants while retaining a high virulence for pigs. Because of the presence of SVD in that country, the wrong assumption was made that what was observed in the field was SVD (only pigs affected) and could not be FMD (cattle were not affected). By the time that FMD was properly diagnosed, it was too late to prevent the wide spread of FMD throughout the island, and there were severe economic, political and social consequences.

b. *Incubation period*

FMD has one of the shortest incubation periods of any major infectious disease known. In most situations, the incubation period for clinical signs is usually 3-5 days. However, once an active outbreak of FMD occurs and there is a large amount of FMDV in the environment, incubation periods of as short as 24-36 hours are seen. Experimentally, clinical signs can be seen as early as 12 hours post exposure

c. *Morbidity*

Morbidity is usually very high (close to 100%) in fully susceptible cloven-hoofed domestic animals. Morbidity in susceptible wildlife is quite variable from high to very low depending on the FMDV sub-type and the species involved.

d. *Mortality*

Mortality due to FMDV infections in general is very low in adult animals (1-5%) but higher in young calves, lambs and piglets (20% or higher).

e. *FMDV carriers*

An FMDV carrier animal is defined as a convalescent or sub-clinically infected animal in which FMDV persists in the pharyngeal region for more than 28 days (4 weeks) after infection. The rates of carriers in cattle vary from 15-50%. The mechanism for developing a carrier state is still not well understood as it happens regardless of the vaccination status of the animal prior to infection. The carrier state in cattle has been reported to be as long as 3.5 years. The rate of carriers among the African buffalo (*Syncerus caffer*) with SAT types is much higher (50-70%) with durations up to 5 years. Sheep and goats can also become FMDV carriers. Ultimately, the number of carriers in a herd and the duration of the carrier state depend on the incidence of the disease, the herd immunity, and the type of virus.

The role of FMD carrier animals in the infection of animals in the field is quite controversial. Evidence exists of transmission of FMD from African buffalo carriers to cattle during outbreaks in southern Africa. However, the evidence of carrier bovine being the source of an outbreak is lacking. The South American experience supports the notion that FMDV carrier cattle are not a source of infection to susceptible cattle even under close contact.

The FMDV carrier state in sheep and goats is less understood and appears to be less frequent and of shorter duration (1-5 months). Other susceptible wildlife species, apart from the African buffalo, could also become FMDV carriers. Carrier conditions have been observed in experimentally infected white-tailed deer (*Odocoileus virginianus*), kudu (*Tragelaphus strepsiceros*), wildebeest (*Connochaetes taurinus*), and, sable antelope (*Hippotragus niger*), but not in impala (*Aepyceros melampus*) which happens to be one of the most frequently affected species in Africa.

6. CLINICAL SIGNS

a. Cattle

The clinical signs of FMD are indistinguishable from those of other vesicular diseases. Thus clinical signs of FMD in cattle are identical to those of Vesicular Stomatitis.

FMDV infected cattle will develop fever of 103°–105°F (39.4°–40.6°C) accompanied by dullness, anorexia and drop in milk production. Cattle develop excessive drooling of saliva, and the presence of excess nasal mucous secretions. The saliva may look thicker with long strands, and more foaming due to the dehydration of the febrile animal. These signs are not necessarily resulting from excess production of saliva, but rather from the inability of the animals to swallow the saliva or clean the nostrils with the tongue, due to painful mouth lesions. It is fairly common to observe cattle "smacking" their lips and grinding their teeth, with accompanied noticeable sounds.

Vesicles are formed in the nares, lips, gums, tongue, teats, coronary bands, and interdigital spaces. As vesicles rupture, erosions form in the same locations in the mouth, as well as in the hard palate. As these lesions progress there may be more marked drooling and lameness. Cattle may develop erosions on the rumen pillars. The course of FMD in cattle may last for 2-3 weeks or longer if there are serious secondary bacterial infections. Pregnant cows may abort and young calves may die without developing any vesicular lesions. Lactating animals may not recover production of milk to pre-infection levels. Other long term sequelae include mastitis, unthriftiness, chronic breeding problems and chronic panting.

b. Swine

Clinical signs of FMD in pigs are identical to those caused by other vesicular diseases: vesicular stomatitis, swine vesicular disease or vesicular exanthema of swine.

FMDV-infected pigs will develop fever of 104°–105°F (40.0°–40.6°C), and anorexia. Fever is followed by "blanching" of the coronary bands (a white discoloration of the coronary bands) due to the formation of vesicles in all toes including the dew claws. There is a considerable painful lameness and reluctance to move due to lameness. This is accompanied by strong vocalization when the pigs are forced to move. There is no drooling in pigs. Foot lesions may lead to the loss of claws and to other secondary skin abrasions as pigs try to walk on the knuckles or knees as a way to relieve the pressure on the painful feet. Vesicles are formed in the snout and muzzle, gums, coronary bands, and interdigital spaces. As vesicles rupture, erosions form. While pigs do not develop extreme tongue vesicles (due to the thinness of the tongue epithelium) erosive lesions covered with fibrin could be observed on the surfaces of the tongue. It is also possible to observe teat lesions in sows and gilts. Pregnant sows or gilts may abort and piglets may die without showing any clinical sign.

c. Sheep & goats

FMD lesions in sheep and goats are very mild or non-apparent. They may include fever, dullness, and occasional small vesicles or erosions on the dental pad, lips, gums or tongue. Mild lameness may be the only observable sign. In these animals there may be vesicles or erosions on the coronary band or in the interdigital spaces. Agalactia in lactating sheep and goats is a frequent consequence of FMD in these animals. Pregnant animals may abort and nursing lambs may die without showing any clinical signs.

7. POST-MORTEM LESIONS

a. Cattle

The diagnostic lesions are single or multiple vesicles ranging from 5mm to 10cm. These can occur at all sites of predilection.

Gross lesions on the tongue:

Lesions usually progress in the following manner:

- A small blanched whitish area develops in the epithelium.
- Fluid fills the area, and a vesicle (blister) is formed.
- Vesicle enlarges and may coalesce with adjacent ones.
- Vesicle ruptures.
- Vesicular covering sloughs leaving an eroded (red) area.
- Gray fibrinous coating forms over the eroded area.
- Coating becomes yellow, brown or green.
- Epithelium is restored, but line of demarcation remains; line then gradually fades with time.

Occasionally "dry" FMD lesions develop in the tongue. Instead of forming a vesicle, the fluid is apparently lost as it forms and the upper layers of the epithelium become necrotic and discolored. The lesion therefore appears necrotic rather than vesicular.

Gross lesions on the feet:

The vesicle in the interdigital space is usually large because of the stress on the epithelium caused by movement and weight. The lesion at the coronary band at first appears blanched; then, there is separation of the skin and horn. When healing occurs, new horn is formed, but a line resulting from the coronitis is seen on the wall of the hoof.

Gross cardiac and other lesions:

Animals that die may have grayish or yellowish streaking in the myocardium due to degeneration and necrosis. These findings are known as "tiger heart", but are histologically no different from any other acute myodegeneration or viral inflammation of heart muscle. Skeletal muscle lesions occur but are rare. Occasional erosions are observed on the epithelium of the rumen pillars.

b. Swine

Vesicles on the snout can be large and filled with clear or bloody fluid. Mouth lesions are usually the "dry" type and appear as necrotic epithelium. Feet lesions are usually severe, and the hoof can become detached. Animals that die may have grayish or yellowish streaking in the myocardium with degeneration and necrosis ("tiger heart").

c. Sheep

Lesions in the mouth and vesicles on the coronary band may be few, small, and difficult to find. Young animals that die may have grayish or yellowish streaking in the myocardium with degeneration and necrosis ("tiger heart").

8. IMMUNE RESPONSE

a. Natural infection

Animals infected by FMDV develop a humoral antibody that is transient in duration and specific for the infective sub-type virus. Protective antibodies are developed against FMDV structural proteins present in the virus capsid between 7 to 14 days post infection. The predominant immunoglobulin type is IgG_2 as well as IgA in nasal secretions. These protective antibodies have the ability to neutralize the FMDV. There is also the possibility of macrophages clearing the virus from the infected animal by phagocytosis of opsonized virus. The role of cell-mediated immunity against FMDV in the infected animal is still largely unknown.

b. Immunization

Immunization of cattle against FMDV has been practiced for over 50 years with good results. Current vaccines are prepared from killed whole-virus preparations in emulsions with single or double oil adjuvants. These vaccines have been effective not only in preventing clinical disease but also in arresting the spread of FMDV when used in the face of an ongoing outbreak. Despite that, the current oil-adjuvated vaccines have many limitations, including:

- Vaccine production requires the propagation of large amounts of live virulent FMDV. This requirement poses significant risks and requires high levels of biocontainment for vaccine producing plants present in FMD-free areas of the world.
- Vaccines need to match the FMDV type- and sub-type present in the area to be vaccinated. This requirement causes the problem of not being able to rapidly vaccinate against a new emerging sub-type and on continuously monitoring for antigenic variations in the field FMDV to be sure that the vaccine antigen matches the antigenic makeup of the circulating field strain.
- The duration of immunity of FMDV killed-oil adjuvated vaccines is rather short. Primo-vaccinated animals are expected to be protected

for about 6 months. After animals receive 2-3 doses, the vaccination schedule can be spaced to once a year.

- Depending on the production technologies, killed FMDV vaccines may be contaminated with NSP, creating difficulties in the differential serological diagnosis of vaccinated (having only antibodies to FMDV capsid or structural proteins) from infected animals (having both antibodies to capsid and NSP fractions).
- Vaccines do not induce rapid protection against the challenge FMDV, thus creating a window of susceptibility of vaccinated animals prior to the development of an active immune response.
- Vaccinated animals could become carriers upon infection by the challenge FMDV.

There are new technologies that are likely to enhance the availability of FMD vaccines in the future. They include the development of "empty capsid" vaccines that contain all the necessary immunogenic proteins but lack the infectious nucleic acid and thus cannot replicate in the vaccinated host, nor do they contain NSPs. Another promising technology is the use of live virus vectors, particularly a human Adenovirus, for the expression of FMDV capsid proteins. These live FMDV-adenovirus constructs cannot replicate in the animal host; they can only replicate in artificial cell culture systems used for vaccine production. Preliminary results indicate that these adenovirus-vectored vaccines could be effective in providing immunity to animals after a single injection even in the presence of an outbreak.

9. DIAGNOSIS

a. Field diagnosis

The presence of a vesicular disease should be suspected any time that there are vesicular lesions in cloven-hoofed domestic or wild animals. Due to the inability to distinguish clinically among the different vesicular diseases, it is absolutely necessary to submit samples to a competent laboratory. The fact that FMD virus spreads extremely fast and that there are serious trade consequences if FMD is diagnosed in any country in the world means that prompt submission of quality samples for laboratory diagnosis is a must. It is also important to note that there havc been outbreaks of FMD in sheep where the only clinical sign has been the mortality of lambs with myocardial necrosis ("tiger heart" lesions).

b. Laboratory diagnosis

i. Samples

The ideal samples for FMD (or any other vesicular disease condition) are: vesicular fluid, vesicular epithelium, scrapings or deep swabs of erosions, nasal swabs (for RRT-PCR), and acute and convalescent serum samples. Esophageal-pharyngeal fluid collected by means of a probang cup is used for the detection of FMDV carrier animals.

ii. Laboratory procedures

Laboratory procedures for the diagnosis of FMD or other vesicular diseases include tests for the detection of virus (virus isolation), detection of viral antigens (Antigen-capture ELISA, Complement Fixation), detection of antibodies (virus neutralization, agar-gel immunodiffusion, antibody detection ELISA), or the new technology of nucleic acid detection (real-time reverse-transcriptase polymerase chain reaction – RRT-PCR). There are a number of new tests developed for the differentiation of animals infected by FMDV from those vaccinated but not infected. While the technologies vary, the basis for the test is the fact that infected animals (vaccinated or not) develop antibodies to the viral protein NSPs. Conversely, animals just vaccinated against FMD but never infected will not develop antibodies to the NSPs that are absent in most commercial vaccines.

c. DIFFERENTIAL DIAGNOSIS

i. Cattle

The fact that FMD in cattle is clinically identical to Vesicular Stomatitis requires that any bovine with lesions compatible to vesicular diseases must be tested for VS and FMD. FMD in cattle also has some common features with a number of other diseases including: rinderpest, mucosal disease – bovine viral diarrhea, infectious bovine rhinotracheitis, bluetongue, bovine mammillitis, bovine papular stomatitis, and other chemical and/or traumatic diseases of the oral cavity and feet. There is a condition in Uruguay and southern Brazil known as Bocopa (from Spanish **bo**ca = mouth; **co**la = tail; **pa**ta = foot) that has some similarities to FMD due to massive erosions on the mouth and feet. Bocopa is caused by an alkaloid toxin of the fungus *Ramaria flavo-brunnescens* that is seasonally present (February to June) in association with eucalyptus (*Eucaliptus* spp.) forests.

ii. Swine

Pigs are the only species susceptible to all vesicular diseases, therefore there is a need to test any pig with vesicular lesions for at least FMD, VS and SVD. (vesicular exanthema of swine –VES – is mostly a historical disease and tests should only be performed if the history may indicate contact of swine with marine mammals or fish.) This is particularly important in countries where SVD is endemic.

10. PREVENTION AND CONTROL

A review of the literature of 627 documented outbreaks of FMD from 1870 through 1963 revealed that the majority of these outbreaks (>68 %) were caused by the legal or illegal importation of infected animals or infected animal products. Therefore, a first line of defense against the introduction of FMD into a free area is to have adequate import controls and quarantine procedures for live animals, as well as to establish proper risk analysis of the hazards associated with the importation of animal products from FMD affected areas of the world. The practice in the U.S. and several other countries of collecting and destroying foreign garbage at the ports of entry, along with the controls or prohibition of feeding meat-containing garbage to pigs, has proven to be an effective means of preventing the introduction into the U.S. of not only FMDV but also of many other serious diseases of domestic animals, primarily those affecting pigs.

Prevention interventions also include education and training on the recognition and diagnosis of FMD, as well as the maintenance of competent laboratories that use validated and controlled testing procedures. The development of RRT-PCR techniques for diagnosing FMD and other vesicular diseases is being adopted at a number of laboratories as part of the National Animal Health Laboratory Network (NAHLN). This is an exceptional asset for our country.

Due to the fast spread and high contagiousness of FMDV, the control of FMD outbreaks requires several simultaneous actions. First, there is a need for prompt and accurate diagnosis, followed by prompt quarantines of affected index farms and aggressive trace back of animal movements within the last 3-4 weeks of the event. Most effective control strategies will call for the depopulation of affected index farms and their contact animals and the establishment of control and surveillance zones following the guidelines promulgated by the World Organization for Animal Health (OIE).

The steps used in the UK and other European countries to control an FMD outbreak in 2001 through massive depopulation of affected, contact, and healthy non-exposed at risk animals (pre-emptive slaughter) are (for the most part) neither socially nor politically acceptable.

The procedures taken by the Netherlands, Uruguay and Argentina, also in 2001, to control FMD outbreaks through massive vaccination demonstrated that it is possible to control an FMD outbreak faster and with lower costs through the use of strategic vaccination campaigns, even with the current vaccines which are less than optimal.

Control and eradication programs in South America have led to the implementation of twice per year vaccination practices where animals (only cattle, not sheep) are vaccinated every 6 months up to 2 years of age and then once annually.

There are encouraging data that indicate that current FMD vaccines can induce protection even in infected animals by reducing the amount of virus being shed and the number of days of shedding. To prepare for this control strategy, many countries and/or groups of countries have developed FMD vaccine banks. One of the oldest and more developed FMD vaccine banks is the North American FMD Vaccine Bank (NAFMDVB) established in 1982 as a consortium among the governments of Canada, Mexico and the United States. The NAFMDVB is composed of 2 elements: vaccine antigen concentrates that can be reconstituted rapidly into a final product, and vaccine seeds for the production of additional doses over longer periods of time.

The OIE has established guidelines for a country to be declared free of FMD and for the safe trade of animals and animal products from FMD affected areas.

These OIE Guidelines can be summarized as follows:

For non-vaccinated areas

FMD-free without vaccination

- 12 months with no outbreaks
- 12 months with no evidence of infection
- 12 months with no vaccination

Recovery of FMD-freedom

- 3 months after last case with stamping-out
- 3 months after slaughter of all vaccinates, or
- 6 months after last vaccination (no slaughter)

For vaccinated areas

FMD-free with vaccination

- 24 months with no outbreaks
- 12 months with no evidence of infection
- Routine vaccination program maintained

Recovery of FMD-freedom with vaccination

- 6 months after last case with stamping-out
- 18 months after last vaccination (no slaughter)

It is important to note that while many countries have endemic presence of one or more FMD virus types, they are as concerned as we are in the U.S. about having the incursion of a foreign FMD type or the introduction of any type of FMDV. For example, many South American countries that have only FMDV types A and O treat any incursion of Type C as an exotic animal disease with aggressive depopulation and /or vaccination campaigns.

Geographic Distribution

There are several areas of the world that are historically free of FMD. Those areas include New Zealand, all Central America (Guatemala, Honduras, Nicaragua, El Salvador, Costa Rica and Panama) as well as all Caribbean islands. Other countries have been free for a long time, including the U.S. (1929), Australia, Canada (1952), and Mexico (1946 -1954). During the last 2 decades of the 20th century, many countries gained FMD-freedom status, including most Western European countries and countries in the southern cone of South America (Chile, Argentina, and Uruguay). Many of these countries, including the UK, France, the Netherlands, Ireland, Argentina and Uruguay, lost the FMD freedom due to outbreaks of FMD during 2001 (with type O in Europe and type A in South America). Today many of these nations have regained their FMD freedom without vaccination or with vaccination as per the OIE guidelines. A complete list of FMD-free countries is available at the OIE web pages (www.oie.int).

11. GUIDE TO THE LITERATURE

1. BAUER, K. 1997. Foot-and-Mouth disease as zoonoses. *Ann Rev. Microbiol.* 22:201-244.
2. BEARD, C. W. and P. MASON. 2000. Genetic determinants of altered virulence of Taiwanese foot-and-mouth disease virus. *Journal of Virology*, 74:987-991.
3. GRUBMAN, M.J. and BAXT, B. 2004. Foot-and-Mouth Disease. *Clinical Microbiology Reviews.* 17:465–493.
4. KEANE, C. 1927. The outbreak of foot-and-mouth disease among deer in the Stanislaus National Forest. *Monthly Bulletin of the California State Dept. of Agriculture*. 16:216-226.
5. McVICAR, J.W. et. al. 1974. Foot-and-mouth disease in white-tailed deer: clinical signs and transmission in the laboratory. *Proc of 78th Annual Meeting U.S. Anim. Health Assoc*. pp. 169 – 180.
6. MEBUS, C., et al. 1997. Survival of several porcine viruses in different Spanish dry-cured meat products. *Food Chem.* Vol. 59 (4):555-559.

7. RHYAN, J. et. al. Susceptibility of North American wild ungulates to foot-and-mouth disease virus: initial findings. http://www.fao.org/AG/AGAInfo/commissions/en/documents/reports/paphos/App23RhyanShalev.pdf
8. SUTMOLLER, P., BARTELING, S.S., OLASCOAGA, R.C. and SUMPTION, K.J. 2003. Control and eradication of foot-and-mouth disease. *Virus Research*. 91:101-144.
9. THOMPSON, G.R., VOSLOO, W. and BASTOS, A.D. 2003. Foot-and-Mouth Disease in wildlife. *Virus Research*. 91:145-161.
10. USDA:APHIS:VS: Centers for Epidemiology and Animal Health. Foot-and-Mouth Disease: Sources of outbreaks and hazard categorization of modes of virus transmission. December 1994.
11. WOODBURY, E. L., SAMUEL, A.R. and KNOWLES, N.J. 1995. Serial passage in tissue culture of mixed foot-and-mouth disease virus serotypes. *Archives of Virology*, 140:783-787.
12. WOOLDRIDGE, M., HARTNETT, E., COX, A. and SEAMAN. M. 2006. Quantitative risk assessment case study: smuggled meats as disease vectors, Rev. sci. tech. Off. int. Epiz. 25:105-117.

See Part IV for photos.

Alfonso Torres, DVM, MS, PhD, Associate Dean for Public Policy, College of Veterinary Medicine, Cornell University, Ithaca, NY, 14852. at97@cornell.edu

23

GETAH

1. NAME
Getah virus

2. DEFINITION
Getah virus, transmitted by mosquitoes, causes a transient febrile illness in horses that is characterized by fever, hindlimb edema, and a papular rash.

3. ETIOLOGY
Getah virus is a member of the *Alphavirus* genus within the Semliki Forest complex of the Family Togaviridae. It is a small, enveloped, single-stranded positive sense RNA virus.

4. HOST RANGE
a. *Domestic animals*
Although Getah virus may infect many species of mammals, only horses and swine show signs of clinical disease. In the laboratory, mice, hamsters, guinea pigs, and rabbits can all be infected experimentally.

b. *Wild animals*
Antibodies have been seen in a wide range of animals, including wild birds, reptiles, and marsupials, but no disease has been recorded in any species other than horses.

c. *Humans*
Although antibodies have been found in humans, there is no evidence to suggest that Getah virus is a human pathogen.

5. EPIDEMIOLOGY
a. *Transmission*
Transmission from one animal to another is through the bite of an infected mosquito. Known species that can vector the virus include *Culex tritaeniorhynchus* and *Aedes vexans nipponii*. Virus is maintained in the mosquito-vertebrate-mosquito cycle typical of many arboviruses. Swine may be an amplifying host. There is some evidence to suggest that direct horse-to-horse transmission may occur during outbreaks via nasal discharges that contain considerable amounts of virus. Vertical transmission through milk is known to

happen in experimentally infected mice, but there is no data as to whether or not this might be occurring in horses.

b. Incubation period

Disease may begin 2-9 days after exposure to the virus.

c. Morbidity

Clinical disease is much less frequent than infection. In an endemic area, there may be extensive seropositivity in equine populations, but only few clinical cases. During an epizootic, morbidity may approach 40%.

d. Mortality

Mortality with Getah virus infection is negligible.

6. CLINICAL SIGNS

a. Horses

There can be anorexia, fever, serous nasal discharge, submandibular lymphadenopathy, edema of the limbs (especially the hind limbs), and there may be an urticarial, papular rash. Animals walk in a stiff manner. The period of illness may be as long as 7-14 days. The syndrome could probably best be characterized as mild and self-limiting, without any apparent sequelae.

b. Swine

Abortion and illness or death of newborn piglets is the main form the disease takes in this species. Experimentally infected piglets showed depression and transient diarrhea.

7. POST- MORTEM LESIONS

a. Gross

The disease does not usually result in death. Consequently, lesion descriptions are taken from experimentally inoculated animals euthanized during the period of clinical disease. There is hyperplasia of lymphoid tissues such as might be seen in any acute viral infection.

b. Key microscopic

Histologically, there may be a perivascular dermatitis and mild cuffing of blood vessels in the cerebrum.

8. IMMUNE RESPONSE

a. Natural infection

After natural infection, if the SN titer is greater than 1:4, horses are considered resistant to challenge.

b. Immunization

A killed vaccine is available for use in areas where prevention is desired. Vaccinated horses are protected against challenge.

9. DIAGNOSIS

a. Field diagnosis

Clinical signs consistent with the disease should prompt collection of samples for laboratory testing.

b. Laboratory diagnosis

i. Samples

Blood and nasal discharge should be collected.

ii. Laboratory procedures

Virus isolation, serum neutralization, hemagglutination inhibition, and complement fixation can all be used to diagnose Getah virus.

10. PREVENTION AND CONTROL

If the disease is occurring in an area, it is important to take measures to decrease contact with mosquitoes and to minimize congregation of horses. An inactivated vaccine is available.

11. GUIDE TO THE LITERATURE

1. BROWN, C.M. and TIMONEY, P.J. 1998. Getah virus infection of Indian horses. *Tropical Animal Health and Production*, 30:241-252.
2. FUKUNAGA, Y., KUMANOMIDO, T., and KAMADA, M. 2000. Getah virus as an equine pathogen. *Veterinary Clinics of North America: Equine Practice.* 16:605-617.
3. HINCHCLIFF, KENNETH W. 2007. Getah and Ross River Viruses, Miscellaneous viral diseases. In: *Equine Infectious Diseases*, D. Sellon and M. Long, eds., Saunders, St. Louis, MO. pp. 233-235.
4. KAMADA, M., KUMANOMIDO, T., WADA, R., FUKUNAGA, Y., IMAGAWA, H., and SUGIURA, T. 1991. Intranasal infection of Getah virus in experimental horses. *J Vet Med Sci.* 53(5):855-858.

5. KUMANOMIDO, T., WADA, R., KANEMARU, T., KAMADA, M., HIRASAWA, K., and AKIYAMA, Y. 1988. Clinical and virological observations on swine experimentally infected with Getah virus. *Veterinary Microbiology,* 16:295-301.

Corrie Brown, DVM, PhD, Department of Pathology, College of Veterinary Medicine, University of Georgia, Athens, GA 30602-7388, corbrown@uga.edu

24

GLANDERS

1. NAME
Droes, Farcy, Malleus

2. DEFINITION
Glanders is a highly contagious disease of solipeds caused by *Burkholderia mallei* and characterized by nodular lesions of the lungs and other organs as well as ulcerative lesions of the skin and mucous membranes of the nasal cavity and respiratory passages. The disease typically has a progressive course and poses a significant human health risk.

3. ETIOLOGY
Glanders is caused by the bacteria *Burkholderia mallei*, a gram-negative, non-spore-forming, aerobic coccobacillus. Former names of this pathogen include *Pseudomonas mallei, Loefflerella mallei, Pfeifferella mallei, Malleomyces mallei, Actinobacillus mallei, Corynebacterium mallei, Mycobacterium mallei*, and *Bacillus mallei*. In experimental infection of guinea pigs *B. mallei* produces a tenacious capsule that may serve to protect it from phagocytosis. The organism is closely related to *B. pseudomallei*, the cause of melioidosis, and is serologically indistinguishable in some cases. Genetic homology between *B. mallei* and *B. pseudomallei* approaches 70%. Because of this, many consider them to be biotypes or isotypes. The organism is destroyed by direct sunlight and is sensitive to desiccation. It is readily killed by common disinfectants. It may survive for up to 6 weeks in infected stables.

4. HOST RANGE

a. Domestic animals
Glanders is primarily a disease of solipeds – particularly horses, donkeys, and mules. Traditionally, donkeys are regarded as most likely to experience the acute form of the disease and horses a more chronic form, with mules intermediate in susceptibility. Recent reports suggest that chronic and even latent infections are equally likely in mules. Carnivores are susceptible to disease if they consume glandered meat; felids appear to be more susceptible than canids, and several laboratory animals including hamsters and guinea pigs are susceptible to infection. The susceptibility of the latter species formed the basis of the Strauss reaction in the diagnosis of the disease. Swine and cattle are resistant to infection with *B. mallei*, but goats can be infected.

b. Wild animals
Outbreaks of glanders in captive wild felids have been reported.

c. Humans
Humans also are susceptible to infection with glanders and it is an important occupational disease of veterinarians, farriers, and other animal workers. The human form of the disease is painful and frequently fatal. Laboratory workers and animal attendants are most at risk. Symptoms of glanders in people include nodular eruption on the face, legs, and arms; involvement of the nasal mucosa; and later pyothorax and metastatic pneumonia. Human glanders may be confused with a variety of other diseases, including typhoid fever, tuberculosis, syphilis, erysipelas, lymphangitis, pyothorax, yaws, and melioidosis. The diagnosis can be confirmed by serology and by isolation of the causative organism.

5. EPIDEMIOLOGY

a. Transmission
The disease is introduced into horse populations by diseased or latently infected animals. Ingestion of the pathogen, present in secretions from infected animals, constitutes the major route of infection in glanders. Experimental evidence suggests that inhalation of the organism is less likely to result in typical cases of the disease. Although invasion by way of skin lesions is possible, it is regarded as being of minor importance in the natural spread of the disease. Close proximity alone does not usually result in transmission of glanders; transmission is facilitated if the animals share feeding troughs or watering facilities or if they nuzzle each other. Underfed and overcrowded horses are especially susceptible.

b. Incubation period
After artificial (intra-tracheal) infection, fever, 104°F (40°C), develops in 1-2 days, respiratory distress is evident by 3 or 4 days, and purulent nasal discharge is conspicuous by 5 days after inoculation. After natural infection, weeks or months may elapse before manifestations of the disease are apparent. Such latent infections are a feature of the epidemiology of glanders.

c. Morbidity
When horses, donkeys, and mules are concentrated, the morbidity can be high.

d. Mortality
Mortality is high but figures may be confounded due to euthanasia because of public health or trade concerns.

6. CLINICAL SIGNS

Classical descriptions of glanders distinguish between cutaneous, nasal, and pulmonary forms of the disease, but in most outbreaks these forms are not clearly distinct and may occur simultaneously in an animal. Chronic infections with slow progression of an insidious disease are more common than the acute form of glanders. The acute form (more common in donkeys and mules than in horses) typically progresses to death within about a week.

The nasal form of glanders is characterized by unilateral or bilateral nasal discharge. The yellowish-green exudate is highly infectious. The nasal mucosa has nodules and ulcers. These ulcers may coalesce to form large ulcerated areas, or they may heal as stellate scars of the mucosa. In some cases the septum may even be perforated. Nasal lesions are accompanied by enlargement and induration, or sometimes rupture and suppuration, of regional lymph nodes.

In the cutaneous form of glanders, multiple nodules may develop in the skin of the legs or other parts of the body. These nodules may rupture, leaving ulcers that discharge a yellow exudate to the skin surface and heal slowly. Cutaneous lymphatic vessels in the region become involved. They become distended and firm by being filled with a tenacious, purulent exudate. (These may be referred to as "farcy pipes.") In the pulmonary form of glanders, lesions in the lungs develop in concert with nasal and cutaneous lesions or may be the sole manifestation of the disease (typical of latent cases). The lung lesions begin as firm nodules or as a diffuse pneumonic process. The nodules are gray or white and firm, surrounded by a hemorrhagic zone, and may become caseous or calcified. Clinical signs in animals with lung lesions may only range from inapparent infection to mild dyspnea, or severe coughing and obvious lower respiratory tract involvement.

Lesions may also occur in the liver or spleen and, in male animals, glanderous orchitis is a common lesion.

7. POST- MORTEM LESIONS

Nodular lesions of glanders are most consistently found beneath the pleura of the lung. In some acute cases, however, a more diffuse form of lobular pneumonia may be present. The nodular lesions, typically about 1 cm in diameter, consist of a gray or white core of necrotic material that may become calcified and are surrounded by a zone of hyperemia and edema. Similar lesions may be found in other viscera. Glanderous orchitis may be seen in intact males.

Nasal lesions consist of submucosal nodules surrounded by a small zone of hyperemia. These nodules may rupture, leaving exudative ulcers. As new lesions develop it is not unusual to find small nodules, ulcers, and scars side by side. Lymphadenitis of associated lymph nodes is a consistent finding. In some cases laryngeal lesions similar to the nasal lesions may be found.

Cutaneous lesions consist of cord-like thickening of subcutaneous lymphatics along which are distributed chains of nodules, some of which are ulcerated.

8. IMMUNE RESPONSE

Horses may recover spontaneously from glanders, but are frequently susceptible to recurrence of disease. It is therefore not surprising that no successful vaccine has been developed.

9. DIAGNOSIS

a. Field diagnosis

Typical nodules, ulcers, scars, and a debilitated condition can be sufficient to diagnose glanders. Unfortunately, many cases of glanders are latent and clinically inapparent. Therefore, systematic testing is essential to identify all infected animals in an outbreak. The mallein test has been the mainstay of field diagnosis. Mallein is a lysate of *B. mallei* containing both endotoxins and exotoxins elaborated by the organism. Infected animals mount an allergic response to mallein and exhibit local and systemic hypersensitivity after mallein inoculation similar to that exhibited in tuberculin testing. Inoculation with mallein may trigger a humoral serologic reaction to the complement fixation test. This seroconversion is thought to be transient but may be permanent if the animal undergoes repeated mallein testing. This is extremely important to consider if animals are destined for export to countries that depend on the complement fixation test.

The preferred method of application of mallein is intrapalpebral. The mallein (0.1 ml) is injected into the dermis of the lower eyelid. In positive cases marked edema of the eyelid, purulent conjunctivitis, photophobia, pain, and depression may be observed within 12-72 hours. The test is usually read 48 hours after injection.

The ophthalmic mallein test consists of the instillation of mallein into the conjunctival sac. A positive reaction is characterized by development of severe purulent conjunctivitis within 6-12 hours. A larger volume of dilute mallein (2.5 ml) may be injected subcutaneously, causing fever, local swelling, and pain in positive animals.

b. Laboratory diagnosis

i. Samples

A whole or section of a lesion and a serum sample should be collected aseptically. The samples should be kept cool and shipped on wet ice as soon as possible. Sections of lesions in 10% buffered formalin and air-dried smears of exudate on glass slides should be submitted for microscopic examination.

ii. Laboratory procedures

The causative organism may be cultured from fresh lesions or lymph nodes. It may also be demonstrated microscopically in films made from this material.

The Strauss reaction is observed when infectious material from glanders patients is injected intraperitoneally into male guinea pigs. In positive cases, the guinea pig develops localized peritonitis involving the scrotal sac. Glanderous orchitis follows with painful enlargement of the testes. The testis becomes enlarged and painful and ultimately necrotic and is discharged through the scrotal skin.

A variety of serologic tests for glanders has been developed. These are superior to mallein testing in sensitivity and specificity. The complement fixation test is widely used and is reported to have an overall accuracy of 95%. A counter-immunoelectrophoresis test has been described. Recently a dot enzyme-linked immunosorbent assay has been developed and found to be superior to all previously described tests in its sensitivity. This test is inexpensive, rapid, and easy to perform and is not influenced by anticomplement activity. Cross-reactions with *B. pseudomallei,* the cause of melioidosis, are features of all of the serological tests for glanders. Therefore, these tests will result in false positive reactions in animals from areas where melioidosis is endemic.

10. PREVENTION AND CONTROL

B. mallei is sensitive to many antimicrobials, but the risk of spreading infection to other equids or to people dictates that infected animals be destroyed. This policy has successfully eradicated glanders from most parts of the world. Sulfonamides have traditionally been used for the treatment of human infection.

Protective vaccines have not been developed.

In endemic areas, routine testing and destruction of positive animals have proven successful in the eradication of the disease. Particular care is required where animals are congregated — most often for military purposes. In endemic areas, communal feeding and watering sites should be avoided.

B. mallei is quite sensitive to heat, desiccation, and common disinfectants. In warm, moist environments, however, it may remain viable for several months. In outbreaks it is important to bury or burn all contaminated bedding and foodstuffs to prevent infection of susceptible animals. Stalls and harness equipment should be thoroughly disinfected. Removal of susceptible species from contaminated premises for a period of months is advisable.

11. GUIDE TO THE LITERATURE

1. ALIBASOGLU, M., YESILDERS, T., CALISLAR, T., INAL, T. and CALSIKAN, U. 1986. Malleus-Ausbruch bei Lowen im Zoologischen Garten Istanbul. *Berl. Munch. Tierarztl. Wochenschr.* 99:57-63.
2. AL-IZZI, S.A. and AL-BASSAM, L.S. 1990. In vitro susceptibility of *Pseudomonas mallei* to antimicrobial agents. Comp. *Immunol. Microbiol. Infect*. Dis. 13:5-8.
3. JANA, A.M., GUPTA, A.K., PANDYA, G., VERMA, R.D. and RAO, K.M. 1982. Rapid diagnosis of glanders in equines by counterimmuno-electrophoresis. *Indian Vet. J.* 59:5-9.
4. RAY, D.K. 1984. Incidence of glanders in the horses of mounted platoon of 4th A.P. Bn. Kahilipara, Gauhati-19 - - a case history. *Indian Vet. J.* 61:264.
5. VAID, M.Y., MUNEER, M.A. and NAEEM, M. 1981. Studies on the incidence of glanders at Lahore. *Pakistan Vet. J.* 1:75
6. VAN DER LUGT, J.J. and BISHOP, G.C. 2005. Glanders. In: *Infectious Diseases of Livestock*, 2nd ed., JAW Coetzer, RC Tustin, eds., Oxford University Press, pp. 1500-1504.
7. VERMA, R.D. 1981. Glanders in India with special reference to incidence and epidemiology. *Indian Vet. J.* 58:177-183.
8. VERMA, R.D., SHARMA, J.K., VENKATESWARAN K.S. and BATRA, H.V. 1990. Development of an avidin—biotin dot enzyme-linked immunosorbent assay and its comparison with other serological tests for diagnosis of glanders in equines. *Vet. Microbiol.* 25:77-85.

See Part IV for photos.

R.O. Gilbert, BVSc, MMedVet, MRCVS, College of Veterinary Medicine, Cornell University, Ithaca, NY 14853-6401, rog1@cornell.edu

25

HEARTWATER

1. Name

Heartwater, Cowdriosis

2. Definition

Heartwater (HW) or cowdriosis is an acute noncontagious infectious disease of ruminants affecting cattle, sheep, goats, and certain antelope species caused by the rickettsial organism *Ehrlichia ruminantium.* Heartwater is transmitted by ticks of the genus *Amblyomma.* The disease is characterized by fever, dyspnea, nervous system signs, hydropericardium, hydrothorax, ascites, edema of the lungs, and high mortality. The disease occurs in several forms: peracute, acute, sub-acute and mild. The name "heartwater" is derived from the hydro-pericardium, which is often seen with this disease.

3. Etiology

Heartwater is caused by the obligate intracellular rickettsial agent *Ehrlichia ruminantium* which infects endothelial cells, macrophages and neutrophils. *E. ruminantium* was previously classified as *Cowdria ruminantium.* It has been recently renamed after reclassification was considered necessary based on the comparison of several gene sequences, e.g. the 16S RNA, the *groESL* and genes encoding the major outer membrane proteins. These *E. ruminantium* genes have high sequence homology with genes of other ehrlichial agents, namely *E. chaffeensis, E. canis, E. ewingii* and *E. muris*. Hence in the order Rickettsiales, all the erhlichial agents including *E. ruminantium* are classified under the family Anaplasmaceteae and the tribe Ehrlichia. The genus *Cowdria* has been abolished.

E. ruminantium multiplies in vascular endothelial cells throughout the body and hence causes severe vascular compromise. The agent is pleomorphic, usually coccoid, occasionally ring-formed, and measures from 400 to over 1,000nm in diameter. It usually occurs in clumps of from less than 5 to several thousand organisms within the cytoplasm of infected capillary endothelial cells, and is easily detected in the brain. The HW organism is extremely fragile and cannot persist outside of a host for more than a few hours. Because of its fragility, the organism must be stored in dry ice or liquid nitrogen to preserve its infectivity. *E. ruminantium* strains vary in virulence, and this influences the outcome of clinical infection.

4. Host range

a. Domestic animals and wild animals

Heartwater causes severe disease in cattle, sheep, and goats, with milder disease in some indigenous African breeds of sheep and goats, and inapparent disease in several species of antelope indigenous to Africa. Heartwater has caused mortality in the African buffalo (Syncerus caffer) in some situations. Other species that have shown to be susceptible are the blesbok (*Damaliscus albifrons*), the black wildebeest (*Connochaetes gnu*), the eland (*Taurotragus oryx oryx*), giraffe (*Giraffe camelopardalis*), greater kudu (*Tragelaphus strepsiceros*), sable antelope (*Hippotragus niger*), sitatunga (*Tragelaphus spekii*), steenbok (*Raphicerus campestris)*, and lechwe (*Kobus leche kafuensis*). It is believed that these species serve as reservoirs of HW and naturally the disease in these animals is usually mild or undetectable. Deaths in springbok (*Antidorcas marsupialis*) in South Africa have been attributed to HW. A number of non-African ruminants are susceptible to HW, shown by experimental infection, include the Timor deer (*Cervus timorensis*) and chital (*Axis axis*) of southern Asia. There is one report of an African elephant dying of heartwater, but this animal was also infected with anthrax. Other species suspected to be susceptible HW but lacking in definitive proof are nilgai (*Boselaphus tragocamelus*), fallow deer (*Dama dama*), Himalayan tahr (*Hemitragus jemlahicus*), barbary sheep (*Ammotragus lervia*), mouflon (*Ovis aries*), blackbuck (*Antilope cervicapra*), white and black rhinoceros.

It was once believed that the guinea fowl and leopard tortoises were the nonruminant hosts of *E. ruminantium*, but recent data have confirmed that these species are not susceptible and do not transmit to vector ticks that usually feed on them. The scrub hare's susceptibility to infection is also not fully substantiated. Although the striped mouse and the multimammate mouse have been shown to be susceptible to *E. ruminantium*, they are not hosts of the vector ticks and are not believed to play a role in the epizootiology of HW. Some laboratory inbred strains of mice have been shown to be susceptible to *E. ruminantium* and have assisted in defining disease and immune mechanisms, but these are not indicated as important in disease maintenance.

In the U.S. the most common deer species, *Odocoileus virginianus* (the white-tailed deer), has been shown by experimental inoculation to be susceptible to *E. ruminantium* and would suffer high mortality. Severe clinical signs were noted along with typical post-mortem lesions. *Amblyomma maculatum* and *A. cajennense* ticks are experimentally proven vectors of HW, and are common parasites of the white-tailed deer in the southern United States.

b. Humans

Recently, there have been 2 reports of suspected HW infections in humans in South Africa. This was suggested because *E. ruminantium* sequences were detected by PCR in the samples of both patients, but so far no isolation of the agent has been made.

5. EPIDEMIOLOGY

a. Transmission

Heartwater is transmitted transstadially and intrastadially by ticks of the genus *Amblyomma*. *Amblyomma* ticks are biological vectors of heartwater. Of the 13 species capable of transmitting the disease, *A. variegatum* (tropical bont tick) is by far the most important because it is the most widespread. Other major vector species are the bont tick *A. hebraeum* (in southern Africa), *A. gemma* and *A. lepidum* (in Somalia, East Africa and the Sudan). *A. hebraeum* is a more efficient vector compared to *A. variegatum* and is susceptible to infection (develops high infection levels and rates) with diverse *E. ruminantium* strains, whereas *A. variegatum* is susceptible to infection with strains that occur within its distribution and less susceptible to infection with strains that occurred in the distribution of *A. hebraeum*.

Amblyomma astrion (mainly feed on buffalo) and *A. pomposum* (distributed in Angola, Zaire and Central African Republic) are also natural vectors of the disease. Four other African ticks–*A. sparsum* (feed on reptiles and buffalo mainly), *A. cohaerans* (feed on African buffalo), *A. marmoreum* (adults occur on tortoises and immature stages on goats) and *A. tholloni* (adults feed on elephants) –have experimentally been shown to be capable of transmitting HW.

Three North American species of *Amblyomma* ticks have been shown to be capable of transmitting the disease experimentally. They are *A. maculatum* (the Gulf Coast tick), *A. cajennense* (the Cayenne tick) and *A. dissimile*, but none of these ticks has been incriminated so far in natural transmission of HW. *A. maculatum* is widely distributed in the eastern, southern, and western U.S., and feeds on ungulates (cattle, sheep, goats, horses, pigs, bison, donkeys, mules, white tailed deer, sambar deer and axis deer), various carnivores, rodents and lagomorphs, marsupials, birds, and reptiles. *A. maculatum* was shown to be as efficient a vector as one of the principal African vectors, *A. hebraeum*, and was found to be susceptible to a wide range of *E. ruminantium* strains. *A. cajennense* has host preference similar to *A. maculatum* but is not as widely distributed and is a less efficient HW vector. *A. dissimile* feeds on reptiles and amphibians.

Amblyomma ticks are three-host ticks whose life cycles may take from 5 months to 4 years to complete. Because the ticks may pick up the infection as larvae or nymphs and transmit it as nymphs or as adults, the infection can persist in the tick for a long time (at least 15 months). The infection does not pass transovarially. Whilst transmission of heartwater can be by adult and nymphal ticks in the field, in general, adult ticks prefer to feed on cattle rather than goats and sheep. Hence, in the field, an abundance of adult ticks are found on cattle and few nymphal ticks in comparison, whereas transmission of infection to sheep and goats is mainly by nymphal stages.

Besides tick transmission, heartwater can also be transmitted naturally by vertical transmission. There is also a possibility that horizontal transmission may occur because in the colostrum of carrier dams, infected cells exist which may cause exposure to infection in neonates, especially during the first 48 hours of life through intake of colostrum. Infection can also be initiated artificially by needle injection of infectious blood, tick homogenates or cell culture material containing *E. ruminatium.*

b. Incubation period

Under field conditions, susceptible animals can be expected to show signs of the disease from 14 to 28 days (average 18 days) after introduction into an HW-endemic area.

c. Morbidity

The morbidity is highly variable and depends on the degree of tick infestation, the previous exposure of the animals to infected ticks, and the level of acaricide protection.

d. Mortality

Once signs of the disease have developed, the prognosis is poor for nonnative and exotic sheep, goats, and cattle. The mortality rate in sheep and goats may be 80% or higher, in contrast to 6% mortality observed in native breeds. Angora goats are extremely susceptible to HW. In cattle, mortality of 60-80% is not uncommon. Recovery from HW infection usually results in complete immunity from disease development although the animals remain carriers of infection.

6. CLINICAL SIGNS

Heartwater occurs in 4 different clinical forms determined by variations in susceptibility of the hosts and the virulence of various strains of the HW agent.

The relatively rare peracute form of the disease is usually seen in Africa in European (nonnative) breeds of cattle, sheep, and goats introduced to a HW endemic area. In this form, sudden death occurs, usually preceded only by a high fever, respiratory distress, and terminal convulsions. Severe diarrhea may also be present.

The acute form of the disease, by far the most commonly observed syndrome, is seen in nonnative and indigenous domestic ruminants. A fever develops followed by a decrease to normal levels for a day or 2, followed by an escalation which can last for up to 3-6 days. The body temperature rises as high as 107°F (42°C), and then is followed by inappetance, depression, listlessness, and rapid breathing. Nervous signs may then develop, the most prominent being chewing movements, twitching of the eyelids, protrusion of the tongue and circling, often with high stepping gait. The animal may stand with legs apart and head lowered. The nervous signs increase in severity, and the animal convulses. Galloping movements and opisthotonos are commonly seen before death. Hyperesthesia is often observed in the terminal stages of the disease, as is nystagmus and frothing at the mouth. Diarrhea is occasionally seen, especially in younger animals. The acute disease is usually fatal within a week of the onset of signs.

Rarely, the disease may run a subacute course characterized by prolonged fever, coughing (a result of lung edema), and mild incoordination; recovery or death occurs in 1-2 weeks. A mild or subclinical form of the disease can also occur in partially immune cattle or sheep, in calves less than 3 weeks old, in antelope, and in some indigenous breeds of sheep and cattle with high natural resistance to the disease. The only clinical sign in this form of the disease is a transient febrile response.

7. POST- MORTEM LESIONS

a. Gross

The gross lesions in cattle, sheep, and goats are very similar. Heartwater derives its name from one of the prominent lesions observed in the disease, namely pronounced hydropericardium. The accumulation of straw-colored to reddish fluid in the pericardium is more consistently observed in sheep and goats than in cattle. Ascites, hydrothorax, mediastinal edema, and edema of the lungs, all resulting from increased vascular permeability with consequent transudation, are frequently encountered. Often hydrothorax will be present without a major increase of fluid collection in the pericardial sac. Froth in the trachea is often seen, reflecting terminal dyspnea due to pulmonary edema. Subendocardial petechial hemorrhages are usually present and submucosal and subserosal hemorrhages may occur elsewhere in the body. Gross brain lesions can be remarkably few when one considers the severity of the nervous signs observed in this disease, and are usually absent except for subtle swelling of the brain which may result in conal herniation.

b. Key microscopic

Visualizing the organisms in brain smears or in histologic sections of brain is the time-tested method for making a diagnosis of heartwater infection.

8. IMMUNE RESPONSE

a. Natural infection

Animals recovering from the natural disease or from artificial exposure to the organism are solidly immune for a variable period ranging from 6 to 18 months. Animals exposed to reinfection during this period of resistance will have their immunity reinforced and will remain immune as long as they are periodically reinfected.

b. Immunization

Calves of less than 4 weeks of age, and lambs in the first week of life can be immunized by intravenous inoculation of HW-infected blood. The infection that follows is usually asymptomatic or mild, and results in protective immunity which gets continuously stimulated by natural exposure to the organism. Older animals or very valuable calves should be examined daily after inoculation with the infected blood and should be treated with antibiotics as soon as the febrile response commences (ideally on the second day of febrile reaction). A subcutaneous implant of doxycycline at the time of infection may eliminate the labor-intensive tetracycline treatment method. The subsequent immunity that develops is not affected by the antibiotic treatment because this treatment does not usually clear infection. Flock immunization of sheep and goats can be accomplished by inoculation followed by mass treatment on the second day of fever.

An inactivated vaccine based on the whole cell-cultured organism has been developed and can be applied in all species affected by HW. This vaccine does not prevent infection but prevents death of the vaccinated animal when challenged with live virulent HW. The inactivated vaccine has been field tested in southern Africa and is currently being commercialized.

9. DIAGNOSIS

a. Field diagnosis

To diagnose heartwater in the field, it is necessary to establish some basic history of the animals that are affected. Heartwater has most severe consequences in animals (cattle, sheep and goats or susceptible deer species) that have never been exposed previously, e.g. when animals are introduced from a heartwater-free area to a heartwater-endemic area, or, when there is a breakdown of tick control (no acaricide applications) on a farm where the disease is prevented by inhibiting transmission. Since heartwater can cause very high levels of mortality in such epidemiological situations, many animals are likely to be affected. The presence of *Amblyomma* ticks (natural or proven experimental vectors), plus the rather characteristic signs and lesions of heartwater allows tentative field diagnosis of the disease.

b. Laboratory diagnosis

i. Samples

From live animals, collect 10 ml of blood using heparin as an anticoagulant and add sufficient DMSO to make a 10% concentration; freeze on dry ice. This sample can be used to infect naïve sheep or goats to reproduce the disease (xenodiagnosis). Collect an additional 50 ml of heparinized blood and 10ml of serum. The heparinsed blood can be used to isolate *E. ruminantium* in vitro in endothelial cells and for detection of *E. ruminantium* by molecular methods. The serum may be used for detection of antibodies but this is not a reliable method of HW diagnosis and there may not be any antibody present if the sampling time is not ideal (see below). From a dead animal, submit smears of cerebral cortex or half of the brain unpreserved and a set of tissues in 10% buffered formalin. Examine the animal for presence of *Amblyomma* ticks (adults and nymphal ticks) and if present, place in clean tubes either containing 70% alcohol for detection of *E. ruminantium* DNA sequences or in ventilated top tubes to preserve the ticks' viability for xenodiagnosis.

ii. Laboratory procedures

1. Demonstration of the Organism: Definitive diagnosis of HW is by demonstration of the organism in the animal tissues. *E. ruminantium* stains purplish-blue with Giemsa stain and can be seen by microscopic examination of brain smears prepared as follows: A small piece of cerebrum, cerebellum, hippocampus, or other well vascularized portion of the brain is macerated between 2 microscope slides. The resultant pulp is then drawn across a slide with varying pressure, which results in "ridges and valleys" on the slide. The slide is then air-dried, fixed with methanol, and stained with Giemsa. Under low magnification, the capillaries will be found extending from the "thick" areas of the slide. Examination of the capillary endothelial cells under oil immersion will reveal the blue to reddish-purple clumps of organisms.

Although microscopic examination of Giemsa-stained brain smears is still widely employed in HW diagnosis, a PCR that detects *E. ruminantium* DNA in tissues of infected livestock (blood particularly or other target organs e.g. brain, lungs, kidneys thoracic fluids), and ticks has been evaluated and has high sensitivity and specificity.

2. Antibody Detection: Serological diagnosis of heartwater poses a problem because of lack of specificity of all available tests. The indirect fluorescent antibody test was previously considered the best test for detection of antibodies to *E. ruminantium.* However, due to lower sensitivity, specificity, and limitations of its applications to all species, several other diagnostic assays have been developed. These include: the competitive enzyme-linked immunosorbent assays (cELISA); the immunoblotting assay targeting responses to the immunodominant MAP-1 antigen of *E. ruminantium;* and, the indirect ELISAs based on whole organisms or on recombinant antigens (MAP-1, MAP-2 or antigenic regions of MAP-1), namely the MAP-1A and -1B. At present, the indirect MAP-1B ELISA is considered to be the most specific serological test for heartwater. However, cross-reactions still exist between *E. ruminantium* and known ehrlichial agents such as *E. canis, E. chaffeensis, E. ewingii* and agents which are yet to be fully characterized by isolation and genotyping.

Serological diagnosis of heartwater is subjective and should be used only as a tool of investigation rather than for definitive diagnosis. Definitive diagnosis should be by demonstration of the organism on a smear, or by PCR amplification using the pSC20 nested-PCR assay and corroborated by isolation of *E. ruminantium* in endothelial cell culture.

10. PREVENTION AND CONTROL

The HW organism is extremely fragile and cannot persist outside of host cells for more than a few hours. The principal mode of bringing the disease into an area is thus through introduction of infected ticks or carrier animals. It is not known for how long wild or domestic ruminants can be a source of infection for ticks in nature, but it may be many months. Ticks are a robust reservoir of *E. ruminantium* and infection can persist in them for at least 15 months. Careful dipping and hand-dressing followed by inspection to ensure the absence of ticks is recommended for animals in transit to HW free areas.

Vector control measures aimed at eradication of *Amblyomma* ticks by dipping of cattle have failed principally because the vector is a multihost tick with a high rate of reproduction. In endemic areas of Africa, tick levels are now allowed to remain at levels high enough to permit reinfection of immune animals to boost the immunity and develop endemic stability.

Cattle, sheep, and goats moving into an HW-endemic area can be protected from HW by prophylactic treatment with tetracycline (short- or long-acting) either by feeding or by inoculation. However, they should be kept under surveillance and individually treated if clinical signs are seen.

Tetracycline antibiotics (especially oxytetracycline) can be effective in the treatment of HW, especially when animals are treated early in the course of the disease. Tetracycline antibiotics administered before signs appear will suppress the disease entirely, but will allow immunity to develop if the timing of the treatment is optimal.

11. GUIDE TO THE LITERATURE

1. CAMUS, E., BARRE, N., MARTINEZ, D. and UILENBERG, G. 1996. *Heartwater (cowdriosis: A Review*, 2nd ed., Office International des Epizooties, Paris.
2. ANDREW, H.R. and NORVAL, R.A.I. 1989. The carrier status of sheep, cattle and African buffalo recovered from heartwater. *Vet. Parasitol.* 34:261-266.
3. BOWIE, M.V., REDDY, G.R., SEMU, S.M., MAHAN, S.M. and BARBET, A.F. 1999. Potential value of Major Antigenic Protein 2 for serological diagnosis of heartwater and related ehrlichial infections. *Clin. Diag. Lab. Immun.* 6:209-215.
4. COLLINS, N.E., PRETORIUS, A., VAN KLEEF, M., BRAYTON, K.A., ZWEYGARTH, E. and ALLSOPP, B.A. 2003. Development of improved vaccines for heartwater. *Annal NY Acad Sci.* 990:474-484.
5. JONGEJAN, F. 1991. Protective immunity to heartwater (*Cowdria ruminantium* infection) is acquired after vaccination with in vitro-attenuated rickettsiae. *Infect. Immun.* 59:729-731.
6. KATZ, J.B., BARBET, A.F., MAHAN, S.M., KUMBULA, D., LOCKHART, J,M., KEEL, M.K., DAWSON, J.E., OLSON, J.G. and EWING, S.A. 1996. A recombinant antigen from the heartwater agent (*Cowdria ruminantium*) reactive with antibodies in some southeastern U.S. white-tailed deer (*Odocoileus virginianus*), but not cattle, sera. *J. Wild. Dis.* 32:424-430.
7. MAHAN, S.M., KUMBULA, D., BURRIDGE, M.J. and BARBET, A.F. 1998. The inactivated *Cowdria ruminantium* vaccine for heartwater protects

against heterologous strains and against laboratory and field tick challenge. *Vaccine*. 16:1203-1211.

8. MAHAN, S.M., KUMBULA, D., BURRIDGE, M.J., and BARBET, A.F. 2001. An inactivated *Cowdria ruminantium* vaccine for heartwater protects cattle, sheep and goats against field *Amblyomma* tick challenge in four southern African countries. *Vet. Parasitol*. 97:295-308.
9. MBOLOI, M.M., BEKKER, C.P.J., KRUITWAGEN, C., GREINER, M. and JONGEJAN, F. 1999. Validation of the indirect MAP1-B enzyme-linked immunosorbent asssay for diagnosis of experimental *Cowdria ruminantium* infection in small ruminants. *Clin. Diag. Lab. Immun*. 6:66-72.
10. SUMPTION, K.J., PAXTON E.A. and BELL-SAKYI, L. 2003. Development of a polyclonal competitive enzyme-linked immunosorbent assay for detection of antibodies to *Ehrlichia ruminantium*. Clin. Diag. Lab. Immun. 10: 910-916.
11. ZWEYGARTH, E., JOSEMANS, A.I., VAN STRIJP, M.F., LOPEZ-REBOLLAR, L., VAN KLEEF, M. and ALLSOPP, B.A. 2005. An attenuated *Ehrlichia ruminantium* (Welgevonden stock) vaccine protects small ruminants against virulent heartwater challenge. *Vaccine*. 23:1695-1702.

See Part IV for photos.

Suman M Mahan BVM, MSc, PhD, Pfizer Animal Health, Kalamazoo, Michigan, 49001, suman.mahan@pfizer.com

26

HEMORRHAGIC SEPTICEMIA

1. NAME
Hemorrhagic septicemia (HS)

2. DEFINITION
Classical hemorrhagic septicemia is a disease of high mortality affecting cattle and buffaloes. The causative agent is a bacterium, *Pasteurella multocida*, serotypes 6:B and 6:E. Disease is peracute; case fatality is 100% if not treated and death is presumed due to endotoxic shock.

3. ETIOLOGY
Hemorrhagic septicemia is caused by 2 serotypes of *Pasteurella multocida*, 6:B and 6:E. The letter denotes the capsular antigen and the number stands for the somatic antigen. The 6:B organism is found predominantly in Asia and 6:E is the more commonly isolated causative bacteria in Africa. These 2 serotypes were formerly known as B:2 and E:2, respectively, in the previously utilized Carter-Heddleston system of classification. The newer designations are based on the classification system known as Namioka-Carter.

4. HOST RANGE
a. Domestic animals
Cattle and water buffaloes (*Bubalus bubalis*) are the principal hosts of hemorrhagic septicemia, and it is widely considered that buffaloes are the more susceptible. Although outbreaks of hemorrhagic septicemia have been reported in sheep and swine, it is not a frequent or significant disease.

b. Wild animals
Infrequent cases have been reported in deer, elephants, and yaks. The disease is thought to be endemic in one large herd of North American range bison.

c. Humans
There are no reported cases of human infection.

5. EPIDEMIOLOGY
a. Transmission
Transmission occurs through exposure to infected animals, carrier animals, or fomites. The bacteria do not survive well in soil. The route of entry is presumed

to be oronasal. Shortly after an outbreak, as many as 20% of the surviving animals may be carriers; after 6 months, survivor carrier rate is probably less than 5%. Close contact of animals in situations that induce crowding facilitates spread.

b. Incubation period
The incubation period is short and varies between 3 and 5 days.

c. Morbidity
Morbidity varies according to the level of immunity in the herd and the degree of association among the animals. Close herding and wetness predispose to more animals being infected.

d. Mortality
Most animals that develop clinical signs die.

6. CLINICAL SIGNS
Clinical disease usually lasts less than 72 hours. First signs may be dullness and reluctance to move. There may be respiratory distress, with frothing at the mouth, and recumbency. Edematous swellings can be seen in the submandibular region, jugular groove, and brisket. The disease tends to be more severe in buffaloes than in cattle.

7. POST- MORTEM LESIONS
a. Gross
Lesions seen are those of severe sepsis, with extensive damage to the capillary bed. Widely distributed hemorrhages and edema are the most obvious tissue changes observed in infected animals. In almost all cases there is an edematous swelling of the head, neck, and brisket region. Incision of the edematous swellings reveals a coagulated, serofibrinous mass with straw-colored or blood-stained fluid. Petechiation is present in multiple organs and serosal surfaces. There may be serosanguinous effusions in body cavities. There is an interstitial reaction in the lung, typical of a toxic state, and represented by a diffuse congestion and a rubbery feel to the lungs grossly. Occasionally, some atypical cases occur in which throat swelling is absent or pneumonia is more extensive and severe.

b. Key microscopic
There are no microscopic features that are specific for hemorrhagic septicemia – all of the lesions are consistent with severe endotoxic shock and massive capillary damage.

8. IMMUNE RESPONSE

a. *Natural infection*

Animals that are exposed to *Pasteurella multocida* serotypes 6:B and 6:E and that survive are considered solidly immune.

b. *Immunization*

Vaccination is routinely practiced in endemic areas. Two preparations are used – bacterins combined with either alum adjuvant or oil adjuvant. The oil adjuvant bacterin is thought to provide protection for up to one year and the alum bacterin for 4-6 months. A live vaccine for the Asian strain (6:B) has been in use in some countries in Asia.

9. DIAGNOSIS

a. *Field diagnosis*

The peracute nature of the disease and the extensive edema and hemorrhage make it difficult to differentiate this disease from others with similar clinicopathologic findings, e.g., lightning strike, blackleg, and anthrax.

b. *Laboratory diagnosis*

i. Samples

As with many septicemias, a number of body tissues could be useful for diagnosis. Spleen and bone marrow provide excellent samples for the laboratory, as these are contaminated relatively late in the post-mortem process by other bacteria.

ii. Laboratory procedures

Aerobic culture and subsequent serotype identification is the preferred procedure. PCR can be used to determine serogroup. Serology is not usually done for HS.

10. PREVENTION AND CONTROL

Vaccination is routinely practiced in endemic areas. Avoiding crowding, especially during wet conditions, will also reduce the incidence of disease.

11. GUIDE TO THE LITERATURE

1. ANONYMOUS. 2005. Hemorrhagic septicemia (chapter 2.3.12), In: *Manual of Diagnostic Tests and Vaccines for Terrestrial Animals*, OIE World Organization for Animal Health, www.oie.int/eng/normes/mmanual/A_0063.htm
2. DE ALWIS, M.C. 1992. Haemorrhagic septicaemia – a general review. *Br. Vet. J.* 148:99-112.

3. MYINT, A., JONES, T.O. and NYUNT, H.H. 2005. Safety, efficacy and cross-protectivity of a live intranasal aerosol hemorrhagic septicaemia vaccine. *Vet. Rec.* 156:41-45.
4. VERMA, R. and JAISWAL, T.N. 1998. Haemorrhagic septicaemia vaccines. *Vaccine.* 16:1184-1192.

See Part IV for photos.

Corrie Brown, DVM, PhD, Department of Pathology, College of Veterinary Medicine, University of Georgia, Athens, GA 30602-7388, corbrown@uga.edu

27

HENDRA

1. NAME

Hendra virus disease; previously known as morbillivirus pneumonia

2. DEFINITION

Infection with Hendra virus (HeV) causes an acute febrile respiratory infection of horses characterized by fever, increased respiratory rate, profuse nasal discharge, and, less frequently, jaundice or neurological disease with a uniformly fatal outcome. Infection with this virus has also resulted in the death of 2 of 4 humans known to have been infected with the virus.

3. ETIOLOGY

Hendra virus is a newly recognized virus in the family Paramyxoviridae within the subfamily Paramyxovirinae. Initially known as equine morbillivirus, subsequent genetic analysis led to reclassification of the virus into a new genus, *Henipavirus,* along with the only other known member of that group, Nipah virus.

4. HOST RANGE

a. Domestic animals

Horses are the only domestic species naturally infected with Hendra virus.

b. Wild animals

Large fruit bats (*Pteropodidae*) also known as flying foxes are the virus reservoir of HeV. Hendra virus antibodies are present in all 4 species of *Pteropodidae* bats in Australia – *Pteropus alecto, P. poliocephalus, P. scapulatus,* and *P. conspicillatus*. In addition, 6 species in Papua New Guinea, *P neohibernicus, P. capistratus, P. hypomelanus, P. admiralitatum, Dobsonia moluccense,* and *D. andersoni* have demonstrated serologic positivity.

c. Humans

Hendra virus has resulted in 4 human cases, with death of 2 of these patients. The route of infection of the human cases is unknown.

5. EPIDEMIOLOGY

a. Transmission

Hendra virus has been isolated from Australian flying foxes and has been present in the population for at least 20 years. The virus is shed in placental fluids of flying foxes and probably also in urine. In the original outbreak the trainer, due to his close contact with his horses, may have inadvertently spread the virus to his horses and himself. In the other human cases the route of infection is unknown. Experimentally HeV infections have been initiated in horses by nasopharynx inoculation, but not by aerosol transmission. The virus is shed from horses by the urinary tract without evidence of excretion from the conjunctiva, nasal cavity or feces. The low number of field infections and the low transmission rate experimentally suggests that the virus does not persist for long periods in the environment.

b. *Incubation period*

The incubation period in experimentally infected horses ranges from 5 to 10 days. Typically clinical disease lasts only 2 days.

c. *Morbidity*

Hendra virus disease in horses is of low morbidity.

d. *Mortality*

Mortality in horses with Hendra virus disease is high.

6. CLINICAL SIGNS

Affected horses rapidly develop increased respiratory rate and dyspnea, fever, increased heart rate, lethargy and, terminally, a frothy nasal discharge. Facial edema was observed in the initial outbreak and jaundice has been observed. Neurological signs were seen in: 2 horses from the initial outbreak; 1 experimentally infected horse; and, 1 of the human fatalities. In subsequent disease reports the clinical signs in affected horses have included severe respiratory distress, ataxia, muscle trembling, hemorrhagic nasal discharge, and swelling of the face, lips, neck and supraorbital fossa.

7. POST- MORTEM LESIONS

a. *Gross*

The most significant gross lesions in horses are diffuse pulmonary edema and congestion with marked dilation of the pulmonary lymphatics. Less frequently there is hydrothorax and hydropericardium, edema of the mesentery, and congestion of lymph nodes. In field but not experimental cases, thick froth in the trachea and extending from the nares was observed.

b. *Key microscopic*

Histologically, there is severe diffuse interstitial pneumonia with necrosis of alveolar walls in the lungs, fibrinoid necrosis and thrombosis of blood vessels, alveolar edema, and increased alveolar macrophages. Vascular lesions can be found most commonly in the kidney but also in the heart, gastrointestinal tract, lymphoid tissue and brain. Syncytial cells within blood vessels are a unique histological feature.

8. IMMUNE RESPONSE

a. Natural infection

Antibodies postinfection are detectable but because of the reportable and zoonotic nature of the disease, no information is available on whether or not these antibodies are protective against challenge.

b. Immunization

No vaccine is available.

9. DIAGNOSIS

a. Field diagnosis

In geographically affected areas, Hendra virus should be suspected as a cause of sudden death in horses when there are severe clinical respiratory illness and post-mortem lesions of pulmonary edema. Protective clothing, gloves, and protection of mucous membranes during handling of infected horses and during necropsies are recommended as a minimum standard of personal protection.

b. Laboratory diagnosis

i. Samples

Hendra virus is classified at Biosecurity Level 4 (BSL-4) and samples should be submitted to laboratories with BSL4 containment facilities. At necropsy, specimens of lung, kidney, spleen, and, if neurological signs are present, brain should be collected aseptically and sent chilled for virus isolation. A complete set of tissues should be submitted to the laboratory in 10% formalin. Serum samples from acute and convalescent stages of the disease should be submitted as well, for serology.

ii. Laboratory procedures

Hendra virus can be isolated in a variety of tissue culture cells, although Vero and RK-13 cells have been used most successfully. A syncytial cytopathic effect typically develops within 3 days and the virus identity can be confirmed by immunofluorescence, electron microscopy and molecular testing. Blood from acute and convalescent stages of the disease can be tested for serum antibody by virus neutralization or an indirect ELISA.

10. PREVENTION AND CONTROL

The geographic distribution of flying foxes in Australia is known and horses in these areas are presumed to be at increased risk of exposure to the virus. Hendra virus infected horses should be handled carefully due to the potential for zoonotic infection. Outbreaks of HeV disease in Australia have been handled by slaughter of infected horses and movement controls of other in-contact animals. Flying foxes are the known reservoir of HeV, and this species should be handled with great caution by experienced field personnel. Field observations and experimental studies indicate the virus is not highly contagious.

11. GUIDE TO THE LITERATURE

1. DANIELS, P., KSIAZEK, T. and EATON, B.T. 2001. Laboratory diagnosis of Nipah and Hendra virus infections. *Microb. Infect*. 3:289-295.
2. HALPIN, K., YOUNG, P.L., FIELD, H.E. and MACKENZIE, J.S. 2000. Isolation of Hendra virus from pteropid bats: a natural reservoir of Hendra virus. *J. Gen. Virol*. 81:1927-1932.
3. HANNA J.N., McBRIDE, W.J., BROOKES, D.L., SHIELD, J., TAYLOR, C.T., SMITH, I.L., CRAIG, S.B. and SMITH G.A. 2006. Hendra virus infection in a veterinarian. *Medical Journal of Australia*. 185:562-564.
4. HOOPER, P.T., KETTERER, P.J., HYATT, A.D. and RUSSELL, G.M. 1997. Lesions of experimental equine morbillivirus pneumonia in horses. *Vet. Pathol*. 34:312-322.
5. HOOPER, P.T. and WILLIAMSON, M.M. 2000. Hendra virus and Nipah virus infections. In: *Veterinary Clinics of North America: Equine Practice. Emerging Infectious Diseases*, P. Timoney, ed.; pp. 597-603.
6. MURRAY, P.K., SELLECK, P., HOOPER, P.T., HYATT, A., GOULD, A., GLEESON, L., WESTBURY, H., HILEY, L., SELVEY, L., RODWELL, B, and KETTERER, P. 1995. A morbillivirus that caused fatal disease in horses and humans. *Science*, 268:94-97.
7. WILLIAMSON, M. 2004. Hendra virus. In *Infectious Diseases of Livestock*. Vol 2, 2nd ed. J.A.W. Coetzer, R.C. Tustin, eds., Oxford University Press, pp. 681-691.
8. WILLIAMSON, M.M., HOOPER, P.T., SELLECK, P.W., GLEESON, L.J., DANIELS, P.W., WESTBURY, H.A. and MURRAY, P.K. 1997. Transmission studies of Hendra virus (equine morbillivirus) in fruit bats, horses and cats. *Aust. Vet. J*. 76:813-818.
9. WILLIAMSON, M.M., HOOPER, P.T., SELLECK, P., WESTBURY, H. and SLOCOMBE, R. 2000. The effect of Hendra virus on pregnancy in fruit bats and guinea pigs. *J. Comp. Pathology*. 122:201-207.
10. YOUNG, P.L., HALPIN, K., SELLECK, P, W., FIELD, H., GRAVEL, J, L., KELLY, M.A. and MacKENZIE, J.S. 1996. Serologic evidence for the presence in *Pteropus* bats of a paramyxovirus related to equine morbillivirus. *Emerg. Infect. Dis*. 2:239-240.

See Part IV for photo.

Mark M Williamson BVSc (Hons), MVS, MACVSc, PhD, Diplomate ACVP,
Veterinary Pathologist, Gribbles Veterinary Pathology, The Gribbles Group,
1868 Dandenong Rd., Clayton, Victoria, Australia, 3168.
mark.williamson@gribbles.com.au

28

INFECTIOUS SALMON ANEMIA

1. NAME

Infectious salmon anemia, also known formerly as Hemorrhagic Kidney Syndrome (HKS)

2. DEFINITION

Infectious salmon anemia (ISA) is an important, OIE-notifiable, infectious disease of Atlantic salmon that has been reported since 1984. The disease has been found in many countries where Atlantic salmon are farmed.

3. ETIOLOGY

The causative agent is ISA virus (ISAV), a single-stranded RNA, enveloped orthomyxovirus (ISAV), proposed to be included in an Aquaorthomyxidae taxonomic family group. There are several distinct genotypes of ISAV, with European and North American strains comprising the main 2 subgroups. It is highly likely from genetic analyses that the pathogen has been in existence for many years.

Not all genotypes of ISAV have been found to result in disease or pathology, leading to the speculation that there may be a host-adapted, non-pathogenic variant (which has resisted successful culture to date), which is capable of mutating under favorable conditions to cause disease in certain populations of Atlantic salmon. Other theories include the possibility of concurrent existence of multiple subgroups that evolved to their current form many years ago, with periodic introductions to farmed salmon via exposure to wild reservoirs.

4. HOST RANGE

a. Domestic animals

Farmed Atlantic salmon (*Salmo salar)* are naturally susceptible. Other farmed salmonid species such as rainbow trout (*Oncorhynchus mykiss*) and brown trout (*Salmo trutta*) are experimentally or naturally susceptible to infection, but they do not develop clinical disease. ISA has been reported in Europe and North America; a somewhat ambiguous report of disease attributable to ISAV in farmed coho salmon (*Oncorhynchus kisutsch*) in Chile was also published, but this species is not generally considered to be susceptible.

b. Wild animals

Wild populations of Atlantic salmon (*Salmo salar*) throughout their natural range are naturally susceptible to infection, although the prevalence of disease in feral populations is not known. Sea-run brown trout (*Salmo trutta*) are naturally susceptible to infection, but are thought to function mainly as carriers of ISAV to susceptible populations of Atlantic salmon.

c. Humans

Humans are not susceptible to infection with ISAV.

5. EPIDEMIOLOGY

a. Transmission

The mechanism of infection in saltwater fish populations is thought to initiate mainly from infected or diseased wild fish that make contact with farmed salmon. ISA virus is generally transmitted horizontally, and generally in saltwater, although there may be a vertical component in fresh water through the infection of eggs by contaminated ovarian fluids (a truly *intra-ovum* mechanism of ISAV transmission has not been conclusively demonstrated). Virus is shed in feces and urine, and is present in skin mucus in diseased fish. Virus may also be liberated quickly from dead fish that are not promptly removed from production tanks or cages. Sea lice (arthropods such as *Lepeophtheirus* or *Caligus* spp.) have been demonstrated to carry infectious ISAV, which is thought to be transferred from one fish to another during lice infestations. Processing wastes (such as viscera, discarded heads or carcasses, and bloodwater) and fomites (such as cage and dip nets, processing equipment, boat hulls and other items) also serve as vectors of transmission. Certain husbandry and management practices such as transferring fish by wellboat also appear to contribute to viral transmission.

Stocking density may play a role in the rapidity of spread throughout a population of farmed Atlantic salmon. Distance to other infected or diseased farms, or to processing plants, has been demonstrated to present certain risks of pathogen exposure or disease spread. The speed of removal of infected and/or diseased cages at affected farms also likely plays an important role in the overall risks of spreading ISA.

ISAV may survive and remain infective up to several weeks in saltwater or freshwater at certain temperatures. The role of water movements in the transmission of the virus and spread of disease has been partially assessed but likely plays an important role in the epizootiology of ISA.

b. Incubation period

Incubation is thought to be environmental temperature-dependent (maximum *in vitro* viral replication occurs at 15°C), and ranges from 7 to 42 days, with an average of 2-4 weeks.

c. Morbidity

Morbidity ranges from 10 to 100%. An average has not been accurately determined, and may not be applicable, since many variables, including management strategies, affect morbidity. However, 50% morbidity in given populations is not uncommon.

d. Mortality

Mortality also can range from 0 to 100%. Many variables, including management strategies, affect overall mortality.

6. CLINICAL SIGNS

Clinical signs are not extremely predictive of ISA. A field necropsy of several moribund fish may provide external and internal parameters of clinical relevance. External signs include anorexia, darkening of the skin, exophthalmia (uni- or bi-lateral; with or without anterior chamber hemorrhages), petechiation of the skin, often ventrally, and generally listless behavior in the water column. Gills may or may not be pale. Blood may appear to be very thin, with hematocrits measured below 5%.

7. POST- MORTEM LESIONS

a. Gross

Internally, the liver may appear dark and congested (although this may be variable depending on the nutritional status of the fish as well as the extent of anemia present). The spleen is usually enlarged. Ascites (ranging from serous to sero-sanguinous, and typically a transudate) may be present and petechial or ecchymotic hemorrhages may be present on the liver, kidney, intestinal serosa and mucosa, visceral fat, and swim bladder. Alternatively, mortality may occur with few or none of the usual post-mortem findings present. Depending on the water temperature, fish decompose quickly after death, and some non-specific changes resulting from autolysis may present as post-mortem lesions consistent with ISA.

b. Key microscopic

A pattern of renal interstitial congestion, hemorrhage, and necrosis is considered highly indicative of ISA pathology, though not pathognomonic. There is often focal congestion and dilatation of hepatic sinusoids, with resulting erythrocytic infiltration into the space of Disse in early disease progression; later changes can result in a zonal appearance to hepatic parenchyma with confluent hepatic hemorrhage and necrosis of areas that are distributed among relatively unaffected sectors. Moderate sinusoidal congestion and erythrophagocytosis is often observed in the spleen. In addition to these microscopic lesions, osmotic balance and general metabolic processes may become disrupted as renal and/or hepatic function decreases due to necrosis of tissue.

8. IMMUNE RESPONSE

a. Natural infection

Circulating antibodies are produced in Atlantic salmon exposed to ISAV, but the immune response may not be adequate to stave off disease processes for a number of possible reasons. These include pathogen pressure, strain antigenicity, temperature, genetic makeup, nutritional status, or presence of concurrent infections. ISAV is similar to human influenza virus (another orthomyxovirus) in its capacity for genetic shift or drift, either of which may alter antigenicity. However, since mortality is sometimes less than 100% in infected and diseased populations, there are probably innate and acquired immune defense mechanisms capable of mitigating the effects of exposure and infection. It is not known if survivors of outbreaks have long-lasting immunity.

b. Immunization

A number of vaccines have been developed or are under investigational research for use in farmed Atlantic salmon, but efficacy has been difficult to demonstrate (due to generally low tolerance thresholds for infected farmed fish before eradication is ordered under the USDA-overseen ISA management scheme).

9. DIAGNOSIS

a. Field diagnosis

A field diagnosis for ISA is generally made through a combination of case history, potential exposure pathways, and observation of clinical signs and behavior. These factors are highly supported by prior knowledge of infection (as demonstrated through testing) in a given population, since many of the presenting signs are non-specific and could be attributed to other or concurrent diseases.

ISA-diseased fish are often listless, dark, exophthalmic, have petechial ventral skin hemorrhages, and remain near the top of the water column. Gills in affected fish may or may not be pale. A quick field necropsy on several fish (using freshly dead or moribund fish) is useful to provide additional criteria for a field diagnosis. Tissue samples should be taken for immediate submission if ISA is suspected. Mortality patterns also aid in a field diagnosis; mortality typically spikes upward over a period of days, and may reach 1-3% of a population per day if mitigations are not taken. However, there is substantial variation to this cycle, and mortality may be chronic and/or lower in magnitude.

b. Laboratory diagnosis

i. Samples

For ISAV detection by virus isolation, mid-kidney and spleen tissues should be sampled and preserved for transport in phosphate-buffered saline (PBS) or Hank's buffered salts solution (HBBS) placed on ice. Whole blood is also useful to obtain for both virus isolation and serology by ELISA, as the virus adheres well to red blood cells. Slides should be prepared using tissue imprints from small, lightly blotted mid-kidney sections for indirect fluorescent antibody testing. Small pieces of mid-kidney tissues should be collected in 2 ml Eppendorf tubes containing 1.5 ml of an RNA preservative such as RNA-Later® for PCR analysis. For histology, mid-kidney, spleen, liver and intestinal samples should be taken and placed in 10% formalin. For specific surveillance purposes (such as an ongoing epizootic investigation), gill lamellae may be collected in addition to the samples for virology assays listed above. Sea water may also be analyzed by molecular techniques.

ii. Laboratory procedures

No detection assays for ISAV have been validated. However, diagnostic accuracy trials for currently available ISAV detection assays suggest high specificity and moderate-to-high sensitivity. Virus isolation using the SHK, CHSE-214 or ASK cell lines is the "gold standard." PCR is useful for confirmation of virus identity or for rapid screening of populations. Indirect fluorescent antibody testing is also utilized using monoclonal ISAV antibodies. Serological tests (ELISA) are under development.

10. PREVENTION AND CONTROL

Prevention may be achieved through a variety of management-related practices such as developing detailed biosecurity protocols for individual farms; health pre-screening of prospective brood and production stock; single year-class production stocking; elimination of site-to-site marine transfers of fish; reducing stocking densities; minimizing stress; and, post-production cycle fallowing of sites with thorough disinfection of cages and nets between production cycles using steam, chlorine or iodophors. All fish culture and processing equipment should be routinely and frequently disinfected between uses. Additionally, Integrated Pest Management protocols may be standardized within designated zones to mitigate sea lice infestations. Land-based broodstock facilities or entirely land-based production systems are other alternatives to limit potential exposure through water sources. The use of dedicated wellboats to transport fish live for slaughter has been utilized in some countries.

In addition to prevention, control may be facilitated through rapid and reliable reporting of exposed or infected fish, continuous surveillance of farmed and wild fish populations, and (after exposure) through prompt action in harvesting infected but non-diseased fish cages as rapidly as is feasible to minimize potential viral transmission.

11. GUIDE TO THE LITERATURE

1. BOUCHARD, D., KELEHER, W., OPITZ, H.M., BLAKE, S., EDWARDS, K.D. and NICHOLSON, B.L. 1999. Isolation of infectious salmon anemia virus (ISAV) from Atlantic salmon in New Brunswick, Canada. *Diseases of Aquatic Organisms*. 35:131-137.
2. CIPRIANO, R. and MILLER, O., session coordinators. 2002. International Response to ISA: Prevention, Control and Eradication. USDA APHIS ISA Symposium, New Orleans, LA.
3. DANNEVIG, B.H., FALK, K. and NAMORK, E.1995. Isolation of the causal virus of infectious salmon anaemia (ISA) in a long-term cell line from Atlantic salmon head kidney. *J. Gen. Virol*. 76:1353-1359.
4. FALK, K., NAMORK, E., RIMSTAD, E., MJAALAND, S. and DANNEVIG, B.H. 1997. Characterization of infectious salmon anemia virus, an orthomyxo-like virus isolated from Atlantic salmon (*Salmo salar* L.). *J. Virol*. 71:9016-9023.
5. JONES, S.R.M., MACKINNON, A.M. and GROMAN, D.B. 1999. Virulence and pathogenicity of infectious salmon anemia virus isolated from farmed salmon in Atlantic Canada. *Journal of Aquatic Animal Health*. 11:400-405.
6. NYLUND, A., HOVLAND, T., HODNELAND, K., NILSEN, F. and LOVIK, P. 1994. Mechanisms for transmission of infectious salmon anaemia (ISA). *Dis. Aquat. Org*. 19:95-100
7. SPEILBERG, L., EVENSEN, O. and DANNEVIG, B. H. 1995. A sequential study of the light and electron microscopic liver lesions of infectious anemia in Atlantic salmon (*Salmo salar* L.). *Vet. Pathol*. 32:466-478

Peter Merrill, DVM, PhD, Aquaculture Specialist, USDA APHIS Import Export, Riverdale, MD. Peter.Merrill@aphis.usda.gov

29

JAPANESE ENCEPHALITIS

1. NAME
Japanese encephalitis, sometimes also called Japanese B encephalitis

2. DEFINITION
Japanese encephalitis (JE) is a mosquito-borne flavivirus that causes clinical disease in horses, humans, and pigs. Horses and humans display neurologic signs of disease; manifestations in pigs are primarily abortion in pregnant sows, infertility in boars, and neurologic diseases in piglets. The virus circulates normally between mosquitoes and wild birds. Pigs serve as amplifying hosts.

3. ETIOLOGY
The JE virus (JEV) is a member of the family Flaviviridae and, within that family, is in the genus *Flavivirus*. JEV is closely related to several other flaviviruses of public health importance, notably St. Louis encephalitis virus, West Nile virus, and Murray Valley encephalitis virus.

There is only a single serotype of Japanese encephalitis virus but several strains have been identified.

4. HOST RANGE
a. Domestic animals
Among the domestic animals, horses are the principal victims of Japanese encephalitis. Pigs serve as amplifier hosts, experiencing a high viremia that then serves to infect populations of mosquitoes. The main manifestations of the disease in pigs are reproductive: abortion and stillbirth in sows, and aspermia in boars. Occasionally, piglets may be infected postnatally and experience neurologic disease. Many other domestic animals, including cattle, goats, dogs, chickens, and ducks, may be infected but without clinical signs and without sufficient viremias to infect mosquitoes.

b. Wild animals
Ardeid birds (herons and egrets) are the natural reservoir for Japanese encephalitis virus. They are infected subclinically but have high viremias.

c. Humans

In endemic areas, Japanese encephalitis can be a major public health problem. Approximately 25% of human clinical cases are fatal and 50% result in neurologic sequelae, including psychiatric disturbances, ataxia, and catatonia.

5. EPIDEMIOLOGY

a. Transmission

The JEV is transmitted through the bite of an infected mosquito. The primary species for transmission are *Culex tritaeniorhynchus*, *C. fuscocephala*, *C. gelidus*, *C. vishnui* and *Anopheles annularis*. In temperate areas and subtropical areas, disease occurs most frequently in late summer and early fall after there has been time for sufficient amplification in mosquitoes (which first appear in the spring). There is vertical transmission in mosquitoes. In tropical areas, there is minimal seasonal variation, but occurrence of the disease may coincide with the end of the rainy season. It appears the *Culex* mosquitoes and birds are common factors in the epidemiology of JE, regardless of the region of occurrence, and that swine are involved where they are numerous in Asia. The mechanism of maintaining the virus over the winter in temperate areas has not been elucidated. Overwintering in mosquitoes is a possibility either in infected hibernating mosquitoes or by transovarial passage.

b. Incubation period

Experimentally, the incubation period for horses is between 4-14 days, with 10 days being the average. In experimentally infected piglets, incubation period was only 3 days, with severe clinical signs by 7 days postinfection.

c. Morbidity

As with many vector-borne diseases, morbidity is dependent on the extent of mosquito infection and exposure of the animal population to the infected mosquitoes.

d. Mortality

In horses, inapparent infections are more common than recognizable cases. Of those with clinical disease, case fatality rates range from 5-30%. Mortality in adult swine is negligible. Entire litters will die due to maternal and subsequently transplacental infection. In nonimmune piglets that become infected, mortality is close to 100%.

6. CLINICAL SIGNS

a. Horses

Initial signs are fever, impaired locomotion, stupor, and grinding of teeth. Blindness, coma, and death follow in more severe cases. A small percentage of cases exhibit hyperexcitability. Some cases may resemble wobbler syndrome.

b. Pigs

The principal manifestation of disease in swine is the expulsion of litters of stillborn or mummified fetuses, usually at term. Viable piglets may have tremors and convulsions before dying. Occasionally, piglets born clinically normal can be infected and suffer a short neurologic syndrome followed by death. Infected boars have decreased numbers and motility of sperm.

7. POST- MORTEM LESIONS

a. Gross

In horses, gross lesions are similar to those observed in animals dying from Eastern equine encephalomyelitis and Western equine encephalomyelitis virus infections and are not specific enough to establish an etiologic diagnosis. Litters from infected pigs contain fetuses that are mummified and dark in appearance. Hydrocephalus and cerebellar hypoplasia have been noted.

b. Key microscopic

Histologically, there is diffuse nonsuppurative encephalomyelitis with prominent perivascular cuffing. This may help to distinguish from Eastern and Western equine encephalitides but is not enough to render a diagnosis of JE.

8. IMMUNE RESPONSE

a. Natural infection

Immunity subsequent to natural infection is long-lasting.

b. Immunization

Vaccination of horses and swine in endemic areas is advised and consists of modified live or inactivated product made from either mouse brains or cell cultures infected with a strain of the virus.

9. DIAGNOSIS

a. Field diagnosis

A tentative diagnosis can be made in horses that manifest CNS disease accompanied by fever, particularly in an epizootic period or in an endemic area at a particular time of year. Similarly, in an endemic area, a presumptive diagnosis in swine is based on the birth of litters with a high percentage of stillborn or weak piglets. However, laboratory confirmation is critical.

b. Laboratory diagnosis

i. Samples

For virus isolation, cerebrospinal fluid and/or brain from affected animals is preferred. Serum for serologic confirmation is desirable.

ii. Laboratory procedures

Virus isolation, virus neutralization, hemagglutination inhibition, complement fixation may all be used to diagnose JE.

10. PREVENTION AND CONTROL

Options for control include elimination of the vectors, prevention of amplification of the infection cycle in birds and pigs, or immunization of horses, pigs, and people. Prevention of disease centers on 2 mechanisms – vaccination and vector control. Vaccination of horses is advised to prevent clinical disease. Decreasing mosquito populations will dampen transmission. Because swine serve as amplifying hosts for the normal mosquito-wild bird cycle, swine should be vaccinated to prevent outbreaks in endemic areas. The most promising approach to reducing livestock losses and simultaneously reducing the totality of infection in nature is widespread immunization of swine. Live attenuated vaccines are in use in Japan and Taiwan. It is anticipated that those animals retained for breeding will remain immune, and, because of immunity or natural seasonal lows in transmission, they will resist infection during pregnancy and therefore bear normal litters. Although controlling the disease in swine dampens the spread of infection in nature, there is a continued threat to horses and human beings from other sources.

11. GUIDE TO THE LITERATURE

1. ANONYMOUS. 2004. Japanese encephalitis. In: *Manual of Diagnostic Tests and Vaccines for Terrestrial Animals*. 5th edition, OIE, www.oie.int/eng/normes/mmanual/A_00092.htm.
2. ELLIS, P.M., DANIELS, P.W. and BANKS, D.J. 2000. Japanese encephalitis, in *Emerging Infectious Diseases*, P.J. Timoney, ed., *Veterinary Clinics of North America: Equine Practice*. 16(3):565-578.
3. ENDY, T.P. and NISALAK, A. 2002. Japanese encephalitis virus: ecology and epidemiology. *Curr Top Microbiol Immunol.* 267:11-48.

4. YAMADA, M. NAKAMURA, K., YOSHII, M., and KAKU, Y. 2004. Nonsuppurative encephalitis in piglets after experimental inoculation of Japanese encephalitis flavivirus isolated from pigs. *Vet. Pathol.* 41:62-67.

Corrie Brown, DVM, PhD, Department of Pathology, College of Veterinary Medicine, University of Georgia, Athens, GA 30602-7388, corbrown@uga.edu

30

JEMBRANA DISEASE

1. NAME
Jembrana Disease

2. DEFINITION
Jembrana disease (JD) is a severe acute disease of cattle, caused by a lentivirus, and characterized by inappetence, fever, lethargy, lymphadenopathy, oral erosions, and diarrhea. The disease was first recognized in 1964 when it affected Bali cattle, which are the domesticated form of the wild banteng (*Bos javanicus*). Three decades after it was first described, the causative agent was identified as a lentivirus.

3. ETIOLOGY
Early reports characterized the etiology as a rickettsial agent, but those theories have been disproven. The causative agent, Jembrana disease virus (JDV), has been shown to be a lentivirus, distinct from, but related to, bovine immunodeficiency virus (BIV). There is cross-reactivity between the major and immunodominant 26K capsid proteins of JDV and BIV. Jembrana disease virus has not yet been cultivated successfully in vitro.

4. HOST RANGE

a. Domestic animals
Severe clinical disease is seen only in Bali cattle (*Bos javanicus*). Other types of cattle (both *Bos taurus* and *Bos indicus*) as well as swamp buffalo (*Bubalus bubalus*) can be infected experimentally and mild disease results. In contrast, natural disease has not been reported in any species other than Bali cattle. Crossbred Bali cattle will not show disease but will be viremic for 3-6 months. Swamp buffalo can maintain a viremia for 9 months following infection.

b. Wild animals
Bali cattle are the domestic variety of the wild banteng (*Bos javanicus*). Other than disease occurring in this species, there are no reports of the disease in wild animals.

c. Humans
There is no evidence to suggest that humans can be infected with JDV.

5. EPIDEMIOLOGY

a. *Transmission*

The precise mechanism of transmission is not known. Proximity of infected with noninfected animals is a circumstantial requirement, so it is presumed that contact is the main means of transmission. The virus is present in saliva and milk, and can infect via a number of mucous membranes. Because viremias are so high during the clinical phase of the disease (10^8 ID_{50}/ml), blood-sucking insects could also serve as vectors of the infection.

b. *Incubation period*

The incubation period ranges from 5 to 12 days.

c. *Morbidity*

When the disease first emerged in 1964, morbidity was very high. Now that it has become endemic among Bali cattle, morbidity rates are much lower.

d. *Mortality*

In the current endemic situation, case-fatality rate in Bali cattle is about 20%.

6. CLINICAL SIGNS

a. *Bali cattle*

Animals experience clinical disease for as long as 12 days. There is fever, lethargy, and lymphadenopathy. Hematologically, there is severe leukopenia and anemia. The cattle can be severely ill, with death of up to 20% of clinically apparent cases.

b. *Other breeds of cattle*

Experimentally infected Friesian cattle had fever, lymphadenopathy, and leukopenia. Recovery rate is almost 100%.

7. POST- MORTEM LESIONS

a. *Gross*

There is enlargement of all lymph nodes as well as spleen. Serous to hemorrhagic exudates can be found on multiple surfaces. There are dramatic hematologic changes, including lymphopenia, eosinopenia, neutropenia, thrombocytopenia, and anemia (pancytopenia except for monocytes).

b. *Key microscopic*

Lymphoblastoid response is evident in many lymphoid organs, affecting the T-cell areas and sparing the B-cell zones. In addition, lymphocytic inflammation may be seen in multiple visceral organs, especially kidney and lung.

8. IMMUNE RESPONSE

a. Natural infection

Animals that survive the disease are resistant to a recurrence of the disease. Antibodies do not develop until several weeks postinfection but are still detectable at least a year after infection. It has been shown experimentally that during infection the animal's ability to mount an antibody response to any infectious agent is lowered.

b. Immunization

There are no commercially available vaccines. Experimentally, inoculating animals with a suspension of bovine-derived materials containing inactivated virus resulted in partial protection.

9. DIAGNOSIS

a. Field diagnosis

Bali cattle that are febrile with enlarged superficial lymph nodes should prompt a presumptive diagnosis, with samples taken for laboratory confirmation.

b. Laboratory diagnosis

i. Samples

Serum, lymph nodes, spleen, and a variety of tissues fixed in formalin should be collected.

ii. Laboratory procedures

An ELISA test is available. Diagnosis can also be made by PCR and histopathology.

10. PREVENTION AND CONTROL

Recovered Bali cattle may carry the virus for as long as 2 years after infection and can be important sources of new infections. Other breeds of cattle may also be persistently viremic after infection, but the period is shorter, on the order of 3-6 months. Because the disease in non-Bali types of cattle may be mild, infection could go undetected and spread to new areas. Control of cattle movements is essential. There is no vaccine.

11. GUIDE TO THE LITERATURE

1. SOEHARSONO, S., WILCOX, G.E., DHARMA, D.M., HARTANINGSIH, N., KERTAYADNYA, G. and BUDIANTONO, A. 1995. Species differences in the reaction of cattle to Jembrana disease virus infection. *Journal of Comparative Pathology,* 112(4):391-402.

2. SOEHARSONO, S., WILCOX, G.E., PUTRA, A.A., HARTANINGSIH, N., SULISTYANA, K. and TENAYA, M. 1995. The transmission of Jembrana disease, a lentivirus disease of *Bos javanicus* cattle. *Epidemiology and Infection*, 115:367-374.
3. WAREING, S., HARTANINGSIH, N., WILCOX, G.E. and PENHALE, W.J. 1999. Evidence for immunosuppression associated with Jembrana disease virus infection of cattle. *Veterinary Microbiology.* 68:179-185.
4. WILCOX, G.E. 1997. Jembrana Disease. *Australian Veterinary Journal.* 75:492-493.
5. WILCOX, G.E., CHADWICK, B.J. and KERTAYADNYA, G. 1995. Recent advances in the understanding of Jembrana disease. *Veterinary Microbiology.* 46:249-255.

Corrie Brown, DVM, PhD, Department of Pathology, College of Veterinary Medicine, University of Georgia, Athens, GA 30602-7388, corbrown@uga.edu

31

LOUPING-ILL

1. NAME

Louping-ill, ovine encephalomyelitis, infectious encephalomyelitis of sheep, trembling-ill

2. DEFINITION

Louping-ill (LI) is a tick-borne acute viral disease, primarily of sheep and red grouse, characterized by fever, neurologic signs, and death. Humans may also be infected.

3. ETIOLOGY

Louping-ill virus is a member of the tick-borne subgroup of the *Flavivirus* genus, family Flaviviridae. Other closely related members of this genus include yellow fever virus, tick-borne encephalitis virus, Kyasanur Forest disease virus and Japanese encephalitis virus. All of these viruses are small, enveloped, single-stranded RNA viruses.

4. HOST RANGE

a. Domestic animals

Many species can be affected by this virus, including sheep, cattle, horses, pigs, goats, dogs, and llamas. Sheep are the primary domestic species affected. Mice have served as an effective laboratory model. There is good indication that sheep may serve as a reservoir of the infection.

b. Wild animals

Red grouse, deer, hares, rabbits, rats, shrews, voles, and wood mice are infected naturally. Red grouse and hares have a viremia sufficiently high to allow for transmission. Because mortality in red grouse is so extensive, they are not considered very important in the maintenance of the virus in nature. Hares are implicated most strongly in maintenance of the virus in nature. The other wild mammals are not considered important in the epidemiology of the disease.

c. Humans

Humans can be infected through tick bite, entrance of the virus through skin wounds (abattoir workers especially), aerosol exposure in the laboratory, or ingestion of contaminated milk. Clinical disease in humans appears in a variety of forms: limiting 'flu-like illness, encephalitis, poliomyelitis-like syndrome, or hemorrhagic fever.

5. EPIDEMIOLOGY

a. Transmission

The virus is transmitted by the three-host sheep tick, *Ixodes ricinus*. This tick feeds on a variety of hosts, including sheep, deer, hares, and grouse. The adult feeds primarily on large animals only, such as sheep and deer. The larval and nymph stages will feed on a variety of small mammals and birds. Transmission of the disease is transstadial. Transovarial transmission has not been demonstrated. Peak activity of the tick occurs in the spring, and so the majority of cases are noted at this time. Although other ticks have been shown to be capable vectors of infection, specifically, *Rhipicephalus appendiculatus*, *Ixodes persulcatus*, and *Haemaphysalis anatolicum*, they are not thought to be important in the epidemiology of the disease. Studies have shown that the virus is present in the milk of infected ewes and does, and may be spread to lambs, kids, or humans through ingestion.

b. Incubation period

In sheep, the incubation period ranges from 6 to 18 days.

c. Morbidity

Morbidity in sheep depends on immune status and severity of virus challenge. In areas where there is widespread vaccination, morbidity may be seen only in lambs at the time of waning of colostral immunity, or in replacement breeding stock that are brought in.

d. Mortality

Mortality is highest in red grouse and sheep. In red grouse, mortality may be as high as 80%. In sheep, mortality rates depend on the breed of sheep and the prevalence of virus, but are rarely more than 15%.

6. CLINICAL SIGNS

For those infections that are clinical, initial signs are fever, depression, and anorexia. The fever tends to be biphasic, with the second spike of fever coinciding with the onset of neurologic signs. Evidence of brainstem or cerebellar dysfunction is seen, including ataxia, incoordination, and characteristic hopping, or "louping" gait. As the disease progresses, more cortical function is compromised, with hypersensitivity, head-pressing, convulsions, and coma. The clinical course lasts 7-12 days. Even those animals that recover may have persistent central nervous deficits of varying severity.

7. POST- MORTEM LESIONS

a. Gross

There are no characteristic gross lesions associated with LI infection. The most dramatic pathologic changes are in the CNS and they can only be appreciated histologically.

b. Key microscopic

The target of LI virus is neurons. Consequently, changes in the CNS consist of neuronal necrosis and neuronophagia. There is a nonsuppurative encephalomyelitis, with perivascular and meningeal infiltrates of lymphocytes, plasma cells and histiocytes.

8. IMMUNE RESPONSE

a. Natural infection

Natural infection may only precipitate partial immunity and recurring viremias, which will help to disseminate the virus through the feeding on these animals. There may be no clinical signs in these reservoir animals

These animals that serve as reservoirs of infection do not have any signs of disease.

b. Immunization

A formalin-inactivated commercial vaccine is available. Two doses are required for optimal protection. Vaccination of pregnant ewes during the last trimester is recommended for maximal passive immunity in lambs. If lambs are vaccinated as maternal immunity wanes, protection should be complete.

9. DIAGNOSIS

a. Field diagnosis

The presence of CNS disease in an endemic area allows for high suspicion but etiology must be confirmed by laboratory testing.

b. Laboratory diagnosis

i. Samples

Heparinized blood, brain, spinal cord, and serum should all be collected.

ii. Laboratory procedures

LI may be diagnosed by a variety of methods, including virus isolation and identification, RT-PCR, serology (ELISA, CFT, HI), histopathology, and immunohistochemistry.

10. PREVENTION AND CONTROL

The most effective control measure is vaccination. Treatment of animals with acaricide to diminish tick burden and therefore viral challenge is also helpful.

11. GUIDE TO THE LITERATURE

1. GILBERT, L., JONES, L.D., HUDSON, P.J., GOULD, E.A. and REID, H.W. 2000. Role of small mammals in the persistence of Louping-ill virus: field survey and tick co-feeding studies. *Medical and Veterinary Entomology*. 14:277-282.
2. LAURENSON, M.K., NORMAN, R., REID, H.W., POW, I., NEWBORN, D. and HUDSON, P.J. 2000. The role of lambs in louping-ill virus amplication. *Parasitology*. 120:97-104.
3. MARRIOTT, L., WILLOUGHBY, K., CHIANINI, F., DAGLEISH, M.P., SCHOLES, S., ROBINSON, A.C., GOULD, E.A., and NETTLETON, P.F. 2006. Detection of Louping ill virus in clinical specimens from mammals and birds using TaqMan RT-PCR. *Journal of Virological Methods*. 137:21-28.
4. SHEAHAN, B.J., MOORE, M. and ATKINS, G.J. 2002. The pathogenicity of louping ill virus for mice and lambs. *Journal of Comparative Pathology*. 126:137-146.
5. TIMONEY, P.F. 1998. Louping-ill, in *Foreign Animal Diseases*, eds. W Buisch, JL Hyde, C Mebus, USAHA, Richmond, Virginia. pp. 292-302.

Corrie Brown, DVM, PhD, Department of Pathology, College of Veterinary Medicine, University of Georgia, Athens, GA 30602-7388, corbrown@uga.edu

32

MALIGNANT CATARRHAL FEVER

1. NAME

Malignant catarrhal fever, bösartiges katarrhalfieber (Ger.), snotsiekte (Afr.), coryza gangréneux (Fr.)

2. DEFINITION

Malignant catarrhal fever (MCF) is a fatal viral disease of ruminant species, particularly cattle, bison, and deer. Clinical signs vary between species, yet within species they tend to be consistent. The most common presentation in domesticated cattle involves fever, rhinitis, depression, and bilateral kerato-conjunctivitis. Clinical signs in American bison and susceptible cervid species comprise depression, diarrhea, and a short clinical course terminating in death.

Two major epidemiologic types of MCF exist, and both are defined by the ruminant species that serves as natural reservoir host for infection. One, the African form, is generally referred to as wildebeest-associated MCF (WA-MCF). The other is sheep-associated MCF (SA-MCF).

3. ETIOLOGY

MCF is caused by one of several closely related ruminant gammaherpesviruses in the *Rhadinovirus* genus. The MCF group of ruminant rhadinoviruses currently comprises 10 known members. Of these, only 4 have clearly been shown to be pathogenic under natural conditions.The pathogenic viruses comprise alcelaphine herpesvirus 1 (AlHV-1), ovine herpesvirus 2 (OvHV-2), caprine herpesvirus 2 (CpHV-2), and a virus of unknown origin that causes MCF in white-tailed deer (MCFV-WTD). They exist as well-adapted, ubiquitous infections that generally cause no disease in their respective carrier species.

- AlHV-1 in wildebeest is the causative agent for wildebeest-associated MCF (WA-MCF), a serious disease problem in Africa and in zoological collections where clinically susceptible hosts are mingled with wildebeest.

- OvHV-2 causes sheep-associated MCF (SA-MCF), a significant problem for highly disease-susceptible species, particularly Bali cattle, farmed deer, bison, and many ungulate species in zoological collections worldwide.

- CpHV-2 is endemic in domestic and wild goats. To date this virus has been documented to cause disease in 2 cervid species, white-tailed and Sika deer. Clinical signs are characterized by alopecia, chronic weight loss and dermatitis. The potential impact of CpHV-2 on other species is unknown.

- MCFV-WTD causes a "classic" MCF syndrome in white-tailed deer in North America (see below). The carrier species has not been identified. Phylogenetic analysis based on the MCF viruses and corresponding hosts suggests that the host species of MCFV-WTD is a close relative of domesticated sheep or goats.

The remaining 6 members in the MCF virus group have not been reported to cause clinical disease. The recent recognition of an AlHV-2-like MCF virus in diseased Barbary red deer suggests that other members of the MCFV group may be pathogenic for species under some circumstances.

4. HOST RANGE

a. Domestic animals and wild animals

MCF viruses infect a range of host species within the Artiodactyl Families Bovidae, Cervidae, and Giraffidae. These hosts can be divided into 2 types: well-adapted asymptomatic carriers, and poorly adapted hosts in which both disease and subclinical infections occur. The well-adapted carrier hosts include species in the subfamilies Alcelaphinae, Caprinae, and Hippotraginae. The major carrier species are wildebeest, sheep and goats. The hosts shed virus into the environment and are capable of transmitting it to clinically susceptible species when contact is close, or when indirect means of transfer of virus, such as suitable fomites, occur. Poorly adapted hosts (or clinically susceptible hosts) are generally considered not to shed infectious virus, and are therefore dead-end hosts. Clinically susceptible species are usually within the families Cervidae and Giraffidae and subfamily Bovinae.

MCF has been described in 33 ungulate species. It has been reliably documented in, among others, domesticated cattle (*Bos taurus* and *B. indicus*), Indian gaur (*Bos gaurus*), Javan banteng or Balinese cattle (*Bos javanicus*), swamp or water buffalo (*Bubalus bubalis*), American bison (*Bison bison*), red deer (*Cervus elaphus*), Sika deer (*Cervus nippon*), white-tailed deer (*Odocoileus virginianus*), mule deer (*Odocoileus hemionus*), Axis deer (*Axis axis*), Père David deer (*Elaphurus davidiansus*), brown brocket deer (*Mazama gouazoubira*), Rusa deer (*Cervus timorensis*), Chinese water deer (*Hydropotes inermis*), and moose (*Alces alces*). MCF due to OvHV-2 occurs in domestic pigs, producing an acute fatal syndrome. Little is known about the epidemiology or pathogenesis in domestic pigs. The domestic rabbit (*Oryctolagus cuniculus*) can be infected experimentally with both AlHV-1 and OvHV-2 and develop MCF-like disease.

Increased susceptibility to disease →

Natural hosts	**Moderately susceptible**	**Highly susceptible**
Domestic sheep	*Bos taurus*	American & European bison
Blue and black wildebeest	*Bos indicus*	Bali cattle Water buffalo Many cervid species Gaur Bongo
Lesion profile		
None	Predominantly vasculitis/ lymphadenopathy	Predominantly hemorrhagic enteritis

b. Humans

No evidence indicates transmission of MCF viruses to human beings and MCF has never been documented in people.

5. EPIDEMIOLOGY

a. Transmission

Although both AlHV-1 and OvHV-2 cause indistinguishable disease in clinically susceptible hosts, the epidemiology of these 2 major MCF viruses within natural carrier hosts differs appreciably. Most wildebeest calves are infected perinatally by horizontal and occasional intrauterine transmission, and shed virus until 3-4 months of age. Wildebeest remain infected for life. Shedding of virus from adult wildebeest is rare, but can be induced by stress or steroid administration. Therefore, outbreaks of WA-MCF in Africa generally correspond to the wildebeest calving season.By contrast, a majority of lambs are not infected with OvHV-2 until after they are 2-3 months old under natural flock conditions. Intrauterine transmission in sheep is rare. Infected lambs begin to shed virus at about 6 months of age.Shedding episodes are short-lived but intense. After about 10 months of age, the frequency of shedding episodes declines. Adult sheep can shed virus, but at lower frequencies and in smaller amounts.Adolescents between 6-9 months of age are the highest risk group for transmission.

Natural transmission of MCF viruses from carriers to clinically susceptible hosts is predominantly through nasal and ocular secretions, by direct contact, or by poorly defined airborne routes. Mechanical vectors and contaminated water or feed materials may play a role in transmission.

Maximal transmission distances depend on several factors, including viral load and climatic conditions, which influence virus survival time. A recent SA-MCF outbreak on a bison ranch showed successful transmission over distances of up to 3 miles when bison were exposed to 20,000 adolescent sheep in a feedlot. Based on limited data, the epidemiology of CpHV-2 appears similar to that in sheep.

b. Incubation period

A relatively long incubation period is observed in both wildebeest-associated and sheep-associated MCF. From experimental studies, the average incubation time for cattle affected with AlHV-1 is about 20 days, with a range of 11-34 days. The average incubation time for cattle experimentally inoculated with large volumes of blood from OvHV-2 affected cattle is about 30 days, with a range of 11-73 days. Peak losses in bison with OvHV-2 occurred 40-70 days post-exposure under natural transmission conditions. Reported incubation periods of more than several months may represent recrudescence of previously established infections.

c. Morbidity

Morbidity in domestic cattle with WA-MCF in Africa is approximately 6-7%, although it occasionally approaches 50%. Morbidity in domestic cattle due to OvHV-2 is generally less than 1%. There are reports of higher morbidity rates in individual herds: 8.3% (Wyoming), 16.6% and 10% (California), 33% (Ireland), 37% (Colorado), 50% (Michigan), 50% (South Africa). These reports tend to enter the literature due to the unusually high rates of loss. A lower figure (1%) is more typical. In bison, morbidity approaches 50% when there is close contact between sheep and bison and/or exposure to large flocks. Heavy losses (94% and 90%) have occurred in individual herds. In some bison feedlots, up to 5% of all consignments die of MCF during the feeding period. In other susceptible species where epidemics of MCF occurred, morbidity rates of 65% (deer), 20% (Bali cattle), and 28% (water buffalo) are recorded.

d. Mortality

Mortality rates in cattle with clinical signs of MCF are typically 80-90%. Recovery is now well documented, but such cattle may develop recrudescent disease or remain unthrifty. Complete recovery is possible. Once clinical signs appear, mortality rates in American and European bison, susceptible cervid species, water buffalo and banteng approach 100%.

6. CLINICAL SIGNS

a. Cattle and water buffalo

Clinical signs begin with acute depression and fever (40- 42°C) during the first 1-3 days of illness, and oculonasal discharge. The clinical course in cattle of MCF due to AlHV-1 and OvHV-2 is similar, extending from 2-21 days (mean: 6.25 days). In water buffalo the mean duration of clinical signs is 7.4 days (range 1-30 days). In Rusa deer the clinical course is 4-34 days. The "head and eye" form is the most common presentation in cattle. Cattle develop bilateral mucoid nasal discharge that becomes mucopurulent, profuse, and, in some, hemorrhagic. Affected cattle may be dyspneic with open-mouthed breathing. Severe stomatitis and pain may cause cattle to drool, resembling the signs of FMD. There is severe bilateral keratitis, manifested as opacity of the cornea with epiphora and hypopyon. Such cattle are photophobic and partly blind. Neurological signs, such as circling, may occur.Diarrhea develops in some cattle. MCF is most commonly seen in yearlings and 2-year-olds. A Colorado survey reported that 72% of all cases of MCF occurred in cattle aged less than 1-2 years of age. In several large outbreaks there are reports of disease in calves as young as 6 weeks, but in general MCF is a disease of weaned animals.

b. American bison, Bali cattle, and many of the susceptible cervid species

The clinical course is shorter (1-3 days) than in cattle. Some are found dead. Handling of apparently mildly depressed bison in chutes may precipitate death or inhalation pneumonia. The primary manifestation is severe depression, separation from the herd, dehydration, and weight loss. Dysuria, hematuria, diarrhea and melena develop. Nasal discharge, crusting of the muzzle and corneal opacity is less pronounced in bison and deer than in cattle. In bison, epiphora is more difficult to detect from a distance due to the thick hair coat. The age range of MCF in bison and deer is broader than in cattle. Disease has been seen in unweaned calves, although bison and deer older than 6 months are most likely to develop disease.

c. Sheep

Experimentally, sheep exposed to high doses of OvHV-2 can develop clinical signs similar to MCF in domestic cattle. Some sporadic outbreaks of panvasculitis syndrome(s) in sheep (so called *polyartertis nodosa*) may be the spontaneous correlate of this experimental phenomenon. Pigs with MCF exhibit weakness, anorexia, fever, neurological symptoms, erosions of the snout, and cutaneous lesions.

7. POST- MORTEM LESIONS

a. Gross

Gross lesions of acute MCF are distinctive, predominantly affecting the lymphoid tissue, digestive and upper respiratory tracts, urogenital system, and eyes. Lesions in gut can be hard to identify, particularly in moderately autolytic carcasses. Cystitis is almost always invariable in acute MCF of cattle and bison. Severity varies from multifocal mucosal ecchymoses to severe hemorrhagic cystitis with intraluminal blood. Similar changes may be seen in the ureters when cut in cross-section. Other lesions are erosions and ulcers of oral mucosa (especially tongue, and hard and soft palate), fauces, esophagus, forestomachs, and abomasum. Identical lesions occur in the larynx, but rarely extend any distance down the trachea. Ulcers are typically small (2-5 mm), shallow and discreet. Vesicular lesions do not occur.

Bison and cervid species generally have diffuse to segmental typhlocolitis, which is the presumed basis for diarrhea/melena. This is best detected by attempting to wash off luminal contents; if material remains adherent in minimally autolytic carcasses, it is strongly suggestive of typhlocolitis. In some cervid species (e.g., Rusa deer), enteric lesions may be accompanied by serosal hemorrhage progressing to large hematomas. Enlargement of lymph nodes (2-5 times normal size) is a helpful clinical and gross diagnostic feature in cattle, but in deer and bison this feature may be minimal.

The cornea of affected cattle, deer and cervid species is diffusely or focally (limbal area) opaque; lesions are invariably bilateral. Animals with a longer clinical course (>7 days) have exudate in the anterior chamber. In some animals, keratitis progresses to perforating corneal ulcers. The muzzle of affected cattle and bison is diffusely crusted and ulcerated, so that layers of epithelium can be peeled with ease at necropsy. It is helpful to expose nasal passages for evidence of mucopurulent material.

Dermatitis is present in a minority of animals and detected by palpating thinly haired skin (thigh, axilla, teats, vulva, udder, coronet, or interdigital skin). A common finding in bison is terminal inhalation pneumonia. Terminally ill bison may be attacked by herd mates, and such cases have been misdiagnosed as death due to trauma. Some bison and water buffalo have pallor in heart and/or skeletal muscles, presumably of the "capture myopathy" variety. Animals with chronic MCF (bison, cattle, Sika deer) have distinctive obliterative arteriopathy changes that are grossly visible in multiple organs.

b. Key microscopic

The hallmark of MCF is lymphoproliferative inflammation with vasculitis. The "classical" lesions are:

- Disseminated vasculitis involving medium caliber arteries and veins. In bison and some cervid species, this may be associated with intravascular thrombosis indicative of disseminated intravascular coagulation (DIC).
- Widespread apoptosis of epithelium, progressing to intraepithelial necrosis followed by erosions/ulcers in digestive tract, upper respiratory tract, bladder, ureters, and skin.
- Lymphoproliferation in lymph nodes (cattle and many cervid species). This is sufficiently florid for some pathologists to regard the lesion as approaching or attaining neoplasia.

This vasculitis is readily detected microscopically in most organs of cattle dying of acute MCF. Vasculitis is less widespread in deer, water buffalo and bison, but is qualitatively identical. Few other infectious diseases of ruminants cause such disseminated arteritis-phlebitis. Other features of the disease, such as the consistent presence of non-suppurative encephalitis and panophthalmitis, corroborate a suspicion of MCF.

Although veterinary texts refer to MCF lesions as "pathognomonic," it is important for control purposes to establish which MCFV rhadinovirus is involved, particularly when exotic or species unfamiliar to the investigator are involved. The notion that MCF causes stereotypical pathognomonic lesions should be regarded skeptically. Once lesions suggestive of MCF are found, confirmation should be undertaken by molecular methods (polymerase chain reaction; PCR). Attempts to develop immunohistochemical methods to detect OvHV-2 in tissues of infected cattle and bison have failed uniformly to date.

8. IMMUNE RESPONSE

a. Natural infection

In sheep-associated MCF, most cattle (over 95%) are serologically positive using cELISA for the detection of antibody against the 15-A epitope at the time clinical signs develop. However, only about 70% of bison with clinical MCF are seropositive by cELISA. Reliance on cELISA alone for confirmation of MCF in this species is not recommended.

b. Immunization

There is no vaccine for wildebeest- or sheep-associated MCF. Attempts to develop a vaccine to protect against AlHV-1 have been unsuccessful. No efforts have been made to develop a vaccine against OvHV-2, due to the inability to propagate the virus *in vitro.*

9. DIAGNOSIS

a. Field diagnosis

A presumptive diagnosis of MCF in domestic cattle is made when typical head and eye lesions are observed, combined with a history of recent exposure to sheep.

Bison and deer dying of MCF often present with a history of 'sudden death' without premonitory signs, or depression with separation from the herd. Multifocal ulcers in the digestive tract and upper respiratory tract, particularly when hemorrhagic cystitis is found, suggest MCF. MCF should be high on any differential list when susceptible cervid species are found dead, acutely depressed and/or with diarrhea in zoological collections or game farms containing blue or black wildebeest.

b. Laboratory diagnosis

i. Samples

The best approach when MCF is on the differential list is to take a wide range of fresh and fixed samples from digestive, respiratory and urinary tracts, as well as from the brain, heart, lymph nodes, and serum (if available). Diagnostic pathologists make a presumptive diagnosis of MCF when severe necrotizing arteritis-phlebitis is identified in multiple organs. This is most readily detected in bladder, kidney, liver, heart, carotid rete (a rich vascular plexus surrounding the pituitary gland), and pampinform plexus. A minimum of 2 fresh tissues should be taken for PCR confirmation of MCF. Almost all types of tissues from a regular necropsy may be used for PCR confirmation, but lymphoid tissues, such as lymph nodes and spleen, are optimal. It is important to collect samples of tissues to exclude diseases that mimic MCF.

ii. Laboratory procedures

1) Virus identification methods

- Virus isolation (AlHV-1)
- PCR

2) Serological tests

- cELISA
- Indirect fluorescent assay
- Immunoperoxidase test
- Virus neutralization (AlHV-1)

10. PREVENTION AND CONTROL

No vaccine is available for sheep- or wildebeest-associated MCF. Preventing contact between carriers and clinically susceptible species is the primary method of control. Scrupulous attention should be taken to avoid sheep, wildebeest, and other carriers coming into contact with deer, bison, Bali cattle and water buffalo. For mixed species operations, such as petting zoos, game farms, and zoological collections, it is possible to use virus-free sheep or goats as an alternative control strategy to prevent clinically susceptible species from MCF losses. The production of OvHV-2 negative sheep is generally not practical for commercial operations.

The precise distance between carriers and clinically susceptible hosts that is needed to completely prevent MCF losses is difficult to determine. Many factors influence virus transmission, particularly host susceptibility and amount of virus that is shed. Establishing adequate yet practical separation distances is the most important factor for prevention.

Sheep-associated MCF in cattle is a low incidence problem in North America, Europe, Southeast Asia, Australia, New Zealand, and South Africa. It is most commonly seen in situations where cattle and sheep operations are co-located. The most effective way to reduce losses is to keep sheep and cattle operations separated by adequate distances (e.g. 1 or more miles apart). In many cases this is not feasible. Following the recognition of SA-MCF, some owners elect to sustain low-level losses and to manage cattle and sheep without precautions or separation.

In Southeast Asia, separation of small ruminants (sheep and goats) from Bali cattle and banteng can be undertaken in some situations. Due to the high case rates in these species, separation may be economically justifiable.

11. GUIDE TO THE LITERATURE

1. BRIDGEN, A. and REID, H.W. 1991. Derivation of a DNA clone corresponding to the viral agent of sheep-associated malignant catarrhal fever. *Res. Vet. Sci.* 50:38-44.

2. CRAWFORD, T.B., LI, H. and O'TOOLE, D. 1999. Diagnosis of sheep-associated malignant catarrhal fever by PCR on paraffin-embedded tissues. *J Vet Diagn Invest.* 11:111-116.
3. CRAWFORD, T.B., O'TOOLE, D. and LI, H. 1999. Malignant Catarrhal Fever. In: *Current Veterinary Therapy: Food Animal Practice*, 4th ed. J Howell, RA Smith, eds., Oklahoma: W.B. Saunders Company, pp. 306-309.
4. LI, H., SHEN, D.T., DAVIS, W.C., KNOWLES, D.P., GORHAM, J.R. and CRAWFORD, T.B. 1994. Competitive inhibition enzyme-linked immunosorbent assay for antibody in sheep and other ruminants to a conserved epitope of malignant catarrhal fever virus. *J Clin Microbiol*, 32:1674-1679.
5. LI, H., GAILBREATH, K., FLACH, E.J., TAUS, N.S., COOLEY, J., KELLER, J., RUSSELL, G.C., KNOWLES, D.P., HAIG, D.M., OAKS, J.L., TRAUL, D.L. and CRAWFORD, T.B. 2005. A novel subgroup of rhadinoviruses in ruminants. *J Gen Virol* 86:3021–3026.
6. O'TOOLE, D., LI, H., SOURK, C., MONTGOMERY, D.H. and CRAWFORD, T.B. 2002. Malignant catarrhal fever in a bison (*Bison bison*) feedlot 1994–2000. *J Vet Diagn Invest* 14:183–193.
7. PLOWRIGHT, W. 1990. Malignant catarrhal fever virus. In: *Virus Infections of Vertebrates*, series ed. M.C. Horzinik, vol. 3, Virus Infections of Ruminants, eds. Z. Dinter and B. Morein. Elsevier, pp.123-150.
8. WOBESER, G., MAJKA, J.A. and MILLS, J.H. 1973. A disease resembling malignant catarrhal fever in captive white-tailed deer in Saskatchewan. *Can Vet J* 14:106-9.

See Part IV for photos.

Donal O'Toole, DVM, PhD, Wyoming State Laboratory, Laramie, WY 82070, dot@uwyo.edu

- and -

Hong Li, DVM, PhD, Animal Disease Research Unit, USDA-ARS, Pullman, WA 99164-6630, hli@vetmed.wsu.edu

33

NAIROBI SHEEP DISEASE

1. NAME
Nairobi sheep disease

2. DEFINITION
Nairobi sheep disease (NSD) is a noncontagious, tick-borne, viral infection of sheep and goats in Eastern and Central Africa characterized by fever, hemorrhagic gastroenteritis, abortion, and high mortality in naïve populations. Ganjam virus is an Asian variant of NSD virus (NSDV), and causes a similar disease in India and Sri Lanka. Both viruses can cause mild field and laboratory infections in humans, and they are classified as Biosafety Level 3 agents.

3. ETIOLOGY
NSDV is a member of the genus *Nairovirus* and family Bunyaviridae. The genus *Nairovirus* has some of the most important tick-borne pathogens in animals and humans. The 34 members are predominantly tick-borne, and the genus consists of 7 serogroups. NSDV is closely related to Dugbe virus, which causes infection in cattle and humans, and to viruses within the Crimean-Congo hemorrhagic fever (CCHF) serogroup.

NSDV is spherical or slightly pleomorphic, enveloped, and has a 3-segmented, single-stranded, negative or ambi-sense RNA genome. The small (S) segment encodes the viral nucleocapsid protein, the medium (M) segment encodes the small surface glycoprotein projections, and the large (L) segment encodes the RNA polymerase for viral replication. Phylogenetic relationships among the Nairoviruses have been determined by sequencing conserved regions of the L and S segments and their respective encoded polymerase polypeptides. The virus is sensitive to lipid solvents and detergents and is rapidly inactivated at high and low pH. Its half-life at optimal pH (7.4-8.0) in 2% serum is only 6.8 days at 0°C and 1.5 hours at 37°C.

4. HOST RANGE
a. Domestic animals
Only sheep and goats are the natural vertebrate hosts of NSDV. Cattle, pigs, horses and poultry are refractory to infection. Goats are generally regarded as less susceptible than sheep, although mortality can also be high. There are also breed differences, and, contrary to expectation, indigenous breeds of sheep have

higher death rates than exotic breeds. For example, mortality in East African hair sheep and Persian fat-tailed sheep can reach 75% or greater while imported wool breeds like the Romney and Corriedale may only reach 30-40%.

In contrast to NSD, imported sheep breeds are more susceptible to Ganjam virus than local breeds or cross-breeds, and outbreaks even in imported sheep breeds are much less severe than in NSD.

b. Wild animals

A few fatal infections have been reported in blue duikers (*Cephalophus monticola*) in natural and zoo conditions. Also, the African field rat (*Arvicanthus abyssinicus nubilans*) became viremic after experimental inoculation in one study, however rodents are not a preferred feeding host for any stage of *R. appendiculatus,* the tick vector. The waterbuck (*Kobus ellipsiprymnus*) and other African wild ruminants are heavily infested with *R. appendiculatus* but have low prevalence of antibody. African buffalo are refractory. In summary, wild ruminants and wild rodents are not believed to play a significant role in maintenance of infection.

c. Humans

Human disease due to NSDV or Ganjam virus under conditions of natural exposure is rare. Laboratory infections by NSDV are also rare, even after needle-stick exposure or handling infected animals. However, several laboratory infections with Ganjam virus in India have been reported. Clinically the disease is self-limiting with fever, shivering, abdominal pain, back pain, headache, nausea and vomiting. Antibodies have been found in laboratory workers and in the general population in Uganda and India and in workers on a goat farm in Sri Lanka. For these reasons, NSDV and Ganjam virus are classified as Biosafety Level 3 agents.

5. EPIDEMIOLOGY

a. Transmission

NSD and Ganjam viruses are not contagious and are only transmitted by ticks. Although the virus is excreted in the urine and feces, transmission by contact or aerosol does not seem to occur. Experimentally, NSD can be transmitted by the inoculation of infectious blood, serum, or organ suspensions into susceptible animals. Large doses (50 ml) of virulent blood or serum given to sheep by mouth may also cause infection.

The major vector of NSD in East Africa is *R. appendiculatus*, but *R. pulchellus*, *R. simus* and *Amblyomma variegatum* can also transmit the virus but with less efficiency. In the haud (highland with *Acacia* bushes and grassland) of northern Somalia and Ethiopia, *R. pulchellus* is the principal vector. Transmission in the tick occurs transstadially in all susceptible ticks, and transovarially in *R. appendiculatus* and Somalian strains of *R. pulchellus* but not Kenyan strains. Ganjam virus in India and Sri Lanka is transmitted by *Haemaphysalis intermedia* ticks, although occasional virus isolations have been made from *H. wellingtoni* ticks and *Culex vishnui* mosquitoes in India.

NSD outbreaks occur when sheep or goats are moved from dry areas where the tick vector is absent into moist forest or grassland areas where infected ticks are abundant, or when prolonged rainfalls have allowed geographic extension of tick populations into areas populated by previously naïve sheep and goat herds. Most sheep and goats in enzootic areas have antibody to virus and are protected, and most lambs are protected by maternal immunity when first exposed to the tick and virus which acts to further boost immunity.

b. Incubation period

The incubation period in natural infections is generally 3-6 days after tick attachment. In experimental infection by syringe, incubation is generally 1-4 days, with a maximum of 6 days, depending on dose, route, virulence of the viral strain, and resistance of the individual animal.

c. Morbidity and Mortality

As with many tickborne diseases, morbidity will depend on tick challenge and pre-existing immunity. Mortality rate in field outbreaks of NSD can be 70-90% for indigenous breeds of sheep and 30-40% for exotic and cross-bred animals. Goats are generally regarded as less susceptible, however mortality in indigenous goat breeds can be 90%. Ganjam virus is reported to cause a similar illness as NSD, but in Africa outbreaks of this scale have not been reported.

6. CLINICAL SIGNS

Disease usually occurs 5-6 days after susceptible sheep or goats are transferred to pens infested with *R. appendiculatus*. The first clinical sign is fever above 41°C (106°F) which may last 1-7 days. Leukopenia and viremia usually accompany fever, with viremia ending about 24 hours before the drop in body temperature. Diarrhea usually appears 1-3 days after onset of fever. Diarrhea progressively worsens during the next few days, becoming watery and fetid often with mucus and blood. Animals become progressively depressed, anorexic, drop their heads, strain rectally, and sometimes groan with each expiration. Mucopurulent nasal discharge occurs in about half the cases, and conjunctivitis with lachrymal discharge may also occur. Pregnant ewes frequently abort.

The course of the disease in peracute and acute cases is usually 2-7 days, but may be as long as 11 days in less acute cases. Clinical signs in goats are similar but may be less severe. Sheep and goats with Ganjam virus are also less severely affected.

7. POST- MORTEM LESIONS

a. Gross

The most obvious lesions are those associated with hemorrhagic and catarrhal gastroenteritis. Externally the hindquarters are soiled with feces or a mixture of blood and feces, and internally there is hemorrhage and catarrhal inflammation. Hemorrhages are mainly in the mucosa of the abomasum (longitudinal folds), distal ileum, ileocecal valve, cecum, and colon. The most outstanding feature, and occasionally the only lesion at necropsy is congestion or hemorrhage in the cecum and colon which appear as longitudinal striations in the mucosa. Colon contents, if present, are liquid in consistency and blood tinged.

Hemorrhages may also occur in the serosa of the cecum and colon. In addition there may be hemorrhages in the submucosa of the gall bladder, subcapsular area of the kidneys, epicardium and endocardium of the heart, lower respiratory tract, and mucosa of the genital tract of ewes. The mucosa of the nasal cavity is often congested with catarrhal inflammation. The aborted fetus has numerous hemorrhages in its tissues and organs and fetal membranes may be edematous and hemorrhagic.

Generalized hyperplasia of lymphoid tissue is a prominent lesion. The spleen may be several times its normal size and engorged with blood. Peripheral lymph nodes are usually enlarged, especially those draining attached ticks, and mesenteric lymph nodes may be enlarged and edematous.

b. Key microscopic

The virus multiplies in lymphoid tissue, liver, lungs, spleen and other organs of the reticulo-endothelial system, and has a predilection for vascular endothelium. Microscopic lesions in the kidney are a constant feature and are important diagnostically - glomerulo-tubular nephritis, necrosis of the glomerular and tubular epithelium with hemorrhages, and hyaline and cellular casts within the renal tubules. Necrosis of cardiac myocytes is often severe, and coagulative necrosis of the gall bladder mucosa is usually present.

8. IMMUNE RESPONSE

a. Natural infection

Recovery from NSD leads to lifelong immunity. Because sheep and goats in enzootic areas are constantly exposed to ticks carrying virus, they maintain good immunity and have no clinical signs of illness. Lambs and kids in these enzootic stable areas are protected by colostral antibody until acquiring active immunity from bites of infected ticks.

b. Immunization

Two types of vaccine have been developed, a modified live virus (MLV) vaccine attenuated by serial mouse brain passage and an inactivated oil-adjuvant vaccine grown in cell culture. Both vaccines are experimental and have not been used extensively in thc field. A single dose of the MLV vaccine induces a rapid immunity, but must be given annually. It also induces viremia but reversion to virulence is not possible because the vaccine strain cannot be transmitted by *R. appendiculatus*. The inactivated vaccine requires 2 doses at a 1 month interval and has a limited duration of immunity. It was developed to protect sheep and goats entering enzootic areas.

9. DIAGNOSIS

a. Field diagnosis

An outbreak of NSD is nearly always associated with movement of susceptible animals into an enzootic area where *R. appendiculatus* is abundant. Recently introduced sheep and goats with signs of severe enteritis and nasal discharge, and with attached *R. appendiculatus* ticks, should raise suspicions, especially if locally grown sheep and goats are healthy. Introduced flocks can be severely compromised within a few weeks.

b. Laboratory diagnosis

i. Samples

Blood (both coagulated and non-coagulated) should be collected from live animals during the febrile stage. Spleen, lung and mesenteric lymph node are optimal tissues to collect from animals that died of acute disease (or were sacrificed during fever). For animals surviving infection, paired sera from acute and convalescent stages should be submitted. Freezing may decrease virus viability.

ii. Laboratory procedures

Laboratory confirmation is necessary for a definitive diagnosis. Inoculation of cell culture with suspensions of infected organs, whole blood or serum and subsequent staining in 24-48 hours with conjugated fluoroscein (Direct

Fluorescent Antibody Test, DFAT) or immunoperoxidase (IP) provides a reliable means of identifying NSDV and does not rely on cytopathic effects, which may not occur in cell culture during the first viral passage. Primary isolation in infant mouse brain and identification of antigen by DFAT or complement fixation (CF) tests is slightly more sensitive than cell culture but more laborious. Inoculation of sheep with tissue or whole blood suspensions is the most sensitive method of NSDV detection.

Viral antigen can be detected in tissue suspensions from animals dying acutely by agar gel immunodiffusion (AGID) tests and ELISA tests. AGID is simple and rapid and can be used in field laboratories without facilities for cell culture or immunofluorescence.

Viral antibodies can be detected by CF, virus neutralization (VN), AGID, indirect FAT and indirect hemagglutination (IHA) tests. CF antibody can be detected for only 6-9 months post infection, and VN antibodies are often difficult to demonstrate. Antibodies can first be detected within a few days after the body temperature returns to normal, and then persist for at least 15 months except for CF antibodies. Experimentally ELISAs have been developed but not yet validated.

10. PREVENTION AND CONTROL

In areas with *R. appendiculatus* tick populations, it is best to allow a stable enzootic state with the virus to occur, even though there may be occasional deaths from NSD. Long term tick control strategies are not warranted because they are expensive, environmentally destructive, and difficult to maintain leaving highly susceptible sheep and goat populations. 'Pour-on' synthetic pyrethroid acaricide treatments of small ruminant populations would be indicated in those areas bordering enzootic stable areas when high rainfall has allowed extension of the tick into naïve small ruminant populations. Remote sensing satellite techniques have made it possible to help predict extension of these areas for institution of vaccination or acaricide treatment programs. Animals in bordering areas and animals moving from free areas into enzootic stable areas should be vaccinated. Animals moving from affected to free areas should be treated for ticks.

Because infection is not transmitted by direct contact, quarantine procedures need not be as strict as for highly contagious diseases.

11. GUIDE TO THE LITERATURE

1. ANONYMOUS. The Universal Virus Database of the International Committee on Taxonomy of Viruses. Developed and maintained by Buchen-Osmond C. Accessed online January 6, 2006, Last updated June 15, 2004: www.ncbi.nlm.nih.gov/ICTVdb.
2. DAVIES, F.G. 1988. Nairobi sheep disease. In: *The Arboviruses: Ecology and Epidemiology*, vol. 4, T.P. Monath, ed., CRC Press, Boca Raton, FL, pp. 191-203.
3. DAVIES, F.G. and TERPSTRA, C. 2004. Nairobi sheep disease. In: *Infectious Diseases of Livestock*, 2nd ed., J.A.W. Coetzer, R.C. Tustin, eds., Cape Town (South Africa): Oxford University Press; 2004. pp. 1071-1076.
4. HONIG, J.E., OSBORNE, J.C. and NICHOL, S.T. 2004. The high genetic variation of viruses of the genus *Nairovirus* reflects the diversity of their predominant tick hosts. *Virology* 318:10-16.
5. MARCZINKE, B.I. and NICHOL, S.T. 2002. Nairobi sheep disease virus, an important tick-borne pathogen of sheep and goats in Africa, is also present in Asia. *Virology* 303:146-151.
6. PEIRIS, J.S.M. 2001. Nairobi sheep disease. In: *The Encyclopedia of Arthropod-transmitted Infections*, M.W. Service, ed., Wallingford (UK): CABI Publishing, Wallingford (UK), pp.364-368.

William R. White, BVSc, MPH, USDA-APHIS-VS-NVSL, Foreign Animal Disease Diagnostic Laboratory, Plum Island, PO Box 848, Greenport, NY 11944, William.R.White@aphis.usda.gov

34

NEWCASTLE DISEASE

1. NAME

Exotic Newcastle disease, avian pneumoencephalitis, Asiatic Newcastle disease, atypische geflugelpeste

2. DEFINITION

Newcastle disease (ND) is an acute viral disease of domestic poultry and many other bird species and a recognized worldwide problem. Occurrence of disease due to virulent strains of Newcastle disease virus (NDV) is referred to as Exotic Newcastle disease (END) in the U.S. and is a notifiable event, with subsequent trade restrictions. The clinical signs observed vary with the virulence of the infecting NDV strain, the avian species infected, and the predilection of the infecting strain for the respiratory, digestive, and/or nervous systems. The diagnosis of ND may require laboratory evaluation of virus isolates from infected birds rather than depending solely on the occurrence of severe clinical disease.

3. ETIOLOGY

Newcastle disease virus, also known as avian paramyxovirus-1 (APMV-1), belongs to the Genus *Avulavirus* in the Family Paramyxoviridae. As with other paramyxoviruses, 2 surface proteins are important to the identification and behavior of the virus. The first, hemagglutinin/neuraminidase (HN) is important in the attachment and release of the virus from the host cells in addition to its serologic identification. The other very important surface protein is the fusion (F) protein, which has a critical role in the pathogenesis of the disease. There are at least nine known serotypes of avian paramyxoviruses and of these, APMV-1 viruses are the most important.

Newcastle disease viruses were classically defined as members of 1 of 3 pathotypes: lentogenic, mesogenic, and velogenic, reflecting increasing levels of virulence. The most virulent (velogenic) isolates were further subdivided into neurotropic and viscerotropic types based on the clinical form of the disease they caused in chickens. Currently NDV strains are classified by methods adopted by the World Organization for Animal Health (OIE). This classification involves typing strains based on their virulence by 1 of 2 assays. The first test, the intracerebral pathogenicity index (ICPI) is a score reflecting the number of sick and dead chickens recorded daily after virus-containing fluid is inoculated into a group of 10 day-old chickens. Scores range from 0.0 – 2.0 and any strain with a score equal or greater than 0.7 is considered to be a virulent NDV strain and therefore the cause of a "notifiable" infection and the cause of ND (END). The second assay is based on the sequence analysis of the isolate's fusion protein cleavage site – any strain with multiple basic amino acids and phenylalanine at this locus is classified as a virulent NDV. Strains that do not meet these criteria are considered of low pathogenicity. In general, virulent NDV strains would correspond to those previously classified as mesogens and velogens as described above.

4. HOST RANGE

a. Domestic animals and wild animals

More than 250 species of birds can be infected with NDV. This virus has perhaps the largest host range of any viral disease, although all hosts recorded to date (with the exception of humans) are birds.

b. Humans

Human infection with NDV has been recorded several times, usually appearing as a transient conjunctivitis. Recovery is rapid, and the virus is no longer present in eye fluids after 4 to 7 days. Infection is most commonly seen in poultry workers, especially under circumstances of aerosol delivery of vaccine in the absence of sufficient ocular protection.

5. EPIDEMIOLOGY

a. Transmission

Transmission occurs through contact with infective material, including secretions and excretions from actively infected birds, as well as fomites. Once introduced into poultry, backyard birds, or other captive bird populations, the virus spreads between premises by the movement of inapparently infected birds, and/or on contaminated objects such as boots, sacks, egg trays, and crates. Inapparent carriers in the form of vaccinated but infected poultry or species susceptible to infection but unlikely to demonstrate signs of disease do occur with NDV. Therefore, transmission can also occur through the introduction to a flock of apparently healthy but infected birds.

b. Incubation period

The incubation period for ND after natural exposure varies from 2-15 days (average 5-6). Variables include virus dose, host species, age, immune status, and the occurrence of other infections and environmental conditions.

c. Morbidity

Morbidity with END will vary with the infected species but can be expected to be close to 100% in fully susceptible chickens or other Galliformes species.

d. Mortality

Mortality varies with the strain of the virus and also with the specific immune status of the bird. Exotic Newcastle disease (END) has a mortality rate of 70-100% in fully susceptible chickens or other Galliformes species.

6. CLINICAL SIGNS

a. Chickens

The introductions of virulent NDV infections of poultry have been eradicated by depopulation when they have occurred in the U.S. Consequently the endemic NDV infections of poultry are due to strains of low virulence that typically produce mild respiratory signs and production declines. Such mild signs including reduced feed and water consumption should not be overlooked because they are also typical of the disease associated with virulent NDV infections in a well-vaccinated flock.

In contrast to the mild clinical signs observed in well-vaccinated poultry, END is a devastating malady in unvaccinated chickens of any age. The first sign in laying chickens is usually a marked drop in egg production followed within 24-48 hours by high death losses. At the onset, 10-15% of a flock may be lost in 24 hours. After 7-10 days, deaths usually subside, and birds surviving 12-14 days generally do not die but may display permanent paralysis and other neurologic signs. The reproductive system may be permanently impaired, and thus egg production does not return to previous levels. In vaccinated chickens, or chicks protected by parental antibodies, the clinical signs are less severe and are proportional to the level of protective antibodies.

Infections with viscerotropic strains, those that produce hemorrhagic lesions typical of viscerotropic velogenic ND (VVND), are likely to cause conjunctival edema and reddening as the earliest prominent sign, appearing at 2-3 days. Birds are depressed and have ruffled feathers. Some diarrhea might be evident as well. Neurologic signs may be present by 5 dpi, about the time that massive deaths occur. However, perhaps the most noteworthy clinical sign in unvaccinated flocks is sudden death without prior indications of illness.

During infections with the neurotropic strains of NDV, clinical signs may not be so dramatic in the early stages. There is a brief period of depression, followed by neurologic signs, usually beginning about 5-7 days post-infection with mortality occurring at the same time. These neurologic signs include muscular tremors, paralysis of legs or wings, torticollis, and opisthotonos.

b. Turkeys

Turkeys are more refractory to infection with both viscerotropic and neurotropic viruses than are chickens. Morbidity is high but clinical disease is slightly less severe and mortality correspondingly lower.

c. Other birds

Clinical disease is not frequently seen in ducks and geese infected with virulent NDV, but it has been reported. Psittacine species vary in their susceptibility to clinical disease. Nervous signs and diarrhea are frequently seen in pigeons infected with virulent NDV. Nervous signs typically predominate as the clinical disease form in species other than Galliformes even when those infections are due to viscerotropic NDV strains.

7. POST- MORTEM LESIONS

a. Gross

END can take a wide range of forms. With the viscerotropic strains, the most characteristic features include an enlarged, friable and mottled spleen (necrosis), and hemorrhage in the cecal tonsils and other lymphoid patches in the intestines. Thymic and bursal hemorrhage are often noted but in older birds these organs are no longer so prominent. Hemorrhage in the caudal part of the pharynx and proximal trachea was a cardinal feature of the California 2003 END virus. Other possible lesions include pancreatic necrosis and pulmonary edema. With the neurotropic strains, even though neurologic signs may be dramatic, all tissues, including brain, may be grossly normal.

b. Key microscopic

Necrosis of splenic lymphoid tissue and lymphoid tissue at the cecal tonsil, with associated overlying necrosis of epithelium, occurs only with END, viscerotropic strains. When neurotropic strains infect, microscopic lesions may be present only in the brain. Cerebellum and brainstem are preferentially affected, with necrosis of Purkinje cells and multifocal glial nodules, usually in the vicinity of necrotic neurons in the cerebellum and medulla.

8. IMMUNE RESPONSE

a. Natural infection

Following natural infection with viruses that can replicate systemically, birds will develop neutralizing antibodies.

b. Immunization

Vaccination with low virulence live or inactivated oil emulsion vaccines, or both, can markedly reduce the losses from ND in poultry flocks. The frequency of vaccination and the types of vaccine used are directly related to the severity of the ND problem in any country. The available ND vaccines can induce an active immunity that will prevent disease from a challenge infection of vaccinates but will not protect them from becoming infected with the challenge virus. There is little doubt, however, that vaccination can make the flock more refractive to infection when exposed and reduces the quantity of virus shed by infected flocks. The reduction in virus shed can thereby reduce the potential of virus transmission to other birds.

9. DIAGNOSIS

a. Field diagnosis

A tentative diagnosis of END may be made on the basis of history, clinical signs, and gross lesions in those cases caused by virulent viscerotropic NDV strains, but because of similarities to other diseases such as fowl cholera and highly pathogenic avian influenza, confirmation requires virus isolation and identification.

b. Laboratory diagnosis

i. Samples

Oropharyngeal, tracheal, and cloacal swabs from sick or recently dead birds are excellent samples for submission. Optimal visceral tissues include: spleen, lung, trachea, brain, intestines (especially cecal tonsil), to be submitted fresh as well as in formalin. Blood samples can be collected for serology.

ii. Laboratory procedures

Virus isolation in embryonating eggs is the preferred method for virus recovery from field samples. The detection of hemagglutination in egg fluids and inhibiting that hemagglutinating activity with NDV specific antiserum is the standard method for identifying the virus. Pathogenicity of the isolate should be determined by completing an ICPI test in day-old specific-pathogen free chickens or determining the sequence analysis of the fusion protein cleavage site. Because NDV is a common infection of birds, the completion of the pathogenicity assessment is critical to the determination of whether the virus isolates is virulent and therefore the cause of a notifiable disease or is of low virulence and not notifiable. A reference laboratory will likely be utilized to complete the pathogenicity evaluation. During 2002-03 END outbreak in U.S. the development of the real-time reverse transcription PCR assay was instrumental in expediting the detection and presumptive pathogenicity evaluation of virus recovered with swab samples.

Serologic tests are most frequently completed by a hemagglutination-inhibition (HI) assay or by ELISA. The HI test measures antibody response to the virus attachment glycoprotein, the HN, therefore the antibody levels determined are a better predictor of protection against disease than antibody detected by ELISA, a test that typically utilizes whole virus as antigen and therefore detects antibody to all of the virus proteins.

10. PREVENTION AND CONTROL

Preventing introduction of the disease is the most reliable way to minimize economic losses with END. The most important component of preventing introduction includes avoiding contact with potentially infected birds or introduction of such birds into a flock, and housing management to prevent their entry, but it also includes for example controlling traffic, feed sources, manure and litter management, and fly and rodent control. Once introduced into a country or area, stamping out, vaccination, and enhanced biosecurity will greatly aid in eradication efforts. A limitation of current ND vaccines is the unavailability of a commercially marketed marker vaccine with a companion diagnostic test that would differentiate vaccinated birds from NDV infected birds.

11. GUIDE TO THE LITERATURE

1. ALEXANDER, D. J. 2003. Newcastle disease. In: *Diseases of Poultry*, 11th ed., Y.M. Saif, H.J. Barnes, J.R. Glisson, A.M. Fadly, L.R. McDougald, D.E. Swayne, eds., Iowa State University Press, Ames, IA, pp. 64-87.
2. BROWN, C., KING D.J. and SEAL, B.S. 1999. Pathogenesis of Newcastle Disease in Chickens Experimentally Infected With Viruses of Different Virulence. *Veterinary Pathology* 36:125-32.
3. BRUGH, M. and BEARD, C.W. 1984. Atypical disease produced in chickens by Newcastle disease virus isolated from exotic birds. *Avian Dis.* 28(2):482-488.
4. KINDE, H., HULLINGER, P.J., CHARLTON, B., McFARLAND, M., HIETALA, S.K., VELEZ, V., CASE, J.T., GARBER, L., WAINWRIGHT, S.H., MIKOLON, A.B., BREITMEYER, R.E. and ARDANS, A. A. 2005. The isolation of exotic Newcastle disease (END) virus from nonpoultry avian species associated with the epidemic of END in chickens in Southern California: 2002-2003. *Avian Diseases* 49:195-98.

5. KINDE, H., UTTERBACK, W., TAKESHITA, K. and McFARLAND, M. 2004. Survival of exotic Newcastle disease virus in commercial poultry environment following removal of infected chickens. *Avian Diseases* 48:669-74.
6. KING, D.J. and SEAL, B.S. 1998. Biological and Molecular Characterization of Newcastle Disease Virus (NDV) Field Isolates with Comparisons to Reference NDV Strains. *Avian Diseases* 42:507-16.
7. OIE (World Organization for Animal Health). 2004. Newcastle disease. In: *Manual of Diagnostic Tests and Vaccines for Terrestrial Animals*, Chapter 2.1.15., 5th edition, Volume 1, pp. 270-282. Available online at http://www.oie.int/.
8. PIACENTI, A.M., KING, D.J., SEAL, B.S., ZHANG, J. and BROWN, C.C. 2006. Pathogenesis of Newcastle disease in commercial and specific pathogen free turkeys experimentally infected with isolates of different virulence. *Veterinary Pathology*, 43:168-178.
9. WAKAMATSU, N., KING, D.J., KAPCZYNSKI, D.R., SEAL, B.S. and BROWN, C.C. 2006. Experimental pathogenesis for chickens, turkeys, and pigeons of exotic Newcastle disease virus from an outbreak in California during 2002-2003. *Veterinary Pathology* 43:925-933.
10. WILSON, T.M., GREGG, D.A., KING, D.J., NOAH, D.L., PERKING, L.E.L., SWAYNE, D.E., and INSKEEP, W. 2001. Agroterrorism, Biological Crimes, and Biowarfare Targeting Animal Agriculture – The Clinical, Pathological, Diagnostic and Epidemiological Features of Some Important Animal Diseases. *Clinics in Laboratory Medicine*. 21 (3):549-591.
11. WISE, M.G., SUAREZ, D.L., SEAL, B.S., PEDERSEN, J.C., SENNE, D.A., KING, D.J., KAPCZYNSKI, D.R. and SPACKMAN, E. 2004. Development of a Real-Time Reverse-Transcription PCR for Detection of Newcastle Disease Virus RNA in Clinical Samples. *Journal of Clinical Microbiology* 42:329-38.

See Part IV for photos.

Daniel J. King, DVM, PhD, Southeast Poultry Research Laboratory, USDA-ARS, Athens, GA 30605, jack.king@ars.usda.gov

35

NIPAH VIRUS

1. NAME

Nipah virus disease, barking pig syndrome, porcine respiratory and neurological syndrome, porcine respiratory and encephalitis syndrome

2. DEFINITION

Nipah virus is a newly emerged paramyxovirus that causes an acute febrile respiratory and/or neurologic disease in pigs, with subsequent transmission to and high case fatality rate in humans. The disease first appeared in Malaysia in 1998 and the origin was traced to pteropid fruit bats that are infected subclinically and shed virus in urine.

3. ETIOLOGY

Nipah virus (NiV), named for the town in Malaysia where the first human case was identified, is a member of the Family Paramyxoviridae. Originally classified as a morbillivirus, both NiV and Hendra virus have now been placed in a new genus, *Henipavirus*, within the Subfamily Paramyxovirinae in the Paramyxoviridae family. Because NiV is associated with numerous human deaths, it is considered a Biosafety Level 4 (BSL4) agent.

4. HOST RANGE

a. *Domestic animals*

Pigs are considered the chief amplifying host and main source for human infections. A wide range of domestic animals, namely dogs, cats, horses, and goats can be naturally infected with NiV. Guinea pigs and hamsters have been infected experimentally.

b. *Wild animals*

Fruit bats (Suborder Megachiroptera), particularly the flying foxes (genus *Pteropus*), are considered to be the natural reservoir host of NiV. In addition, some insectivorous bats (Suborder Microchiroptera) have shown evidence of infection.

c. *Humans*

Humans are susceptible to infection with NiV, with a resulting serious clinical disease. Mortality rate in humans is high, 40-75% recorded in different outbreaks.

5. EPIDEMIOLOGY

a. Transmission

Field and epidemiologic studies strongly suggest that pigs are infected through direct contact with pteropid bat fluids or tissues (saliva, urine, feces, dead carcasses, or placental tissues and fluids). Disease transmission among pigs and to humans is attributed to aerosol, as the virus replicates extensively in the airways and affected pigs display prominent coughing. In addition, direct contact with excretions or secretions (urine, saliva, pharyngeal fluids) may also be a means for transmission in groups of pigs and from pigs to humans.

b. Incubation period

The incubation period in pigs is 7-14 days.

c. Morbidity

Morbidity in swine in close confinement is remarkably high and is reported to reach almost 100% among pigs from 4 weeks to 6 months old.

d. Mortality

In swine, highest mortality rate is seen in piglets (40%), although neglect from sick sows may have contributed to deaths in this group. A lower mortality rate has been observed among 1- to 6-month-old pigs (1-5%).

6. CLINICAL SIGNS

a. Pigs

Clinical disease in suckling piglets presents as dyspnea, weakness, and CNS signs. Pigs from 1 to 6 months old can present with fever and respiratory difficulty, coughing, and hemoptysis in severe cases. Although respiratory signs are more prominent among this age group, some animals may also show neurologic signs such as myoclonus or spastic paresis. Lateral recumbency with paddling and seizures upon stimulation may be seen in more severe cases. In animals older than 6 months, sudden death, early abortion, or acute fever are common. In this age group, when neurologic signs occur, they might include head pressing, tetanus-like spasms, nystagmus, chewing gum fits, and inability to swallow. Concomitant respiratory signs may be present.

b. Dogs

Limited reports describe this as a canine distemper-like presentation with fever, respiratory distress, and conjunctivitis with mucopurulent nasal and ocular discharges.

c. Cats

Experimentally infected cats develop fever and show decreased mental status, increased respiratory rate, and dyspnea with open mouthed breathing.

d. Horses

A single case has been reported – there were nonspecific neurologic signs reported prior to death.

7. POST- MORTEM LESIONS

a. Gross

In swine dying of NiV, trachea and bronchi may be filled with clear to blood-tinged froth. The lungs are consolidated, emphysematous, with distended interlobular septa, and petechial to ecchymotic hemorrhages throughout. The brain is often congested and edematous while the kidneys may show cortical congestion.

b. Key microscopic

Most pathologic descriptions are from infected swine. There is a generalized mononuclear vasculitis with fibrinoid necrosis and often associated thrombosis. A bronchointerstitial pattern of pneumonia occurs. If the brain is affected, there are perivascular cuffs and gliosis. Syncytial cell formation and viral inclusions are often present in endothelium of affected blood vessels and in alveolar epithelium.

8. IMMUNE RESPONSE

a. Natural infection

NiV elicits a humoral immune response in animals and humans. The initial IgM peak and subsequent IgG rise are the bases for ELISAs and neutralization diagnostic tests.

b. Immunization

Experimentally, animals vaccinated with either recombinant G or F proteins or passively immunized with hyperimmune serum survive a lethal challenge. No commercially available vaccines exist.

9. DIAGNOSIS

a. Field diagnosis

Suspicion of NiV infection in pigs would be raised if there was extensive morbidity in swine populations in a region where pteropid bats are established. Death of young piglets, respiratory disease in older piglets, with some neurologic signs, would all be cause for concern. Because this virus is basically a pteropid bat virus that only makes periodic incursions into domestic animal (or human) populations, any suspicion should be carefully followed with appropriate and biosecure investigations.

b. Laboratory diagnosis

i. Samples

NiV is classified as a BSL4 agent and also as a select agent. Collection, transport, and submission of samples should all be done in close consultation with government authorities and by properly trained personnel to avoid accidental exposure or dissemination of the agent.

ii. Laboratory procedures

A variety of tests are available to diagnose NiV, including virus isolation, virus/serum neutralization test, ELISA, histopathology, immunohistochemistry, RT-PCR, and electron microscopy.

10. PREVENTION AND CONTROL

Keeping pteropid bats and their secretions separated from swine husbandry operations may be the best mechanism for preventing future outbreaks of this disease.

Because the disease has such a high morbidity rate in swine and the zoonotic disease is so severe, mass culling of affected animals is recommended. It is important to ensure that all workers in the eradication program are fully trained in personal protection.

There are no vaccines available for the prevention of NiV disease.

11. GUIDE TO THE LITERATURE

1. ANONYMOUS. *Manual of Diagnostic Tests and Vaccines for Terrestrial Animals*, 5th ed., July 2004, Chapter 2.10.10 World Organization for Animal Health (OIE), www.oie.int
2. ANONYMOUS. *Manual on the diagnosis of Nipah virus infection in animals*. Food and Agriculture Organization of the United Nations Regional Office for Asia and the Pacific & Animal Production and Health Commission for Asia and the Pacific. RAP Publication no. 2002/01; January 2002
3. FIELD, H., YOUNG, P., MOHD YOB J., MILLS, J., HALL, L. and MACKENZIE, J. 2001. The natural history of Hendra and Nipah viruses. *Microbes and Infection* 3:307-14.

4. GUILLAUME, V., CONTAMIN, H., LOTH, P., GEORGES-COURBOT, M.C., LEFEUVRE, A., MARIANNEAU, P., CHUA, K.B., LAM, S.K., BUCKLAND, R., DEUBEL, V. and WILD, T.F. 2004. Nipah virus: vaccination and passive protection studies in a hamster model. *Journal of Virology* 78(2): 834-40.
5. HSU, V.P., HOSSAIN, M.J., PARASHAR, U.D., ALI, M.M., KSIAZEK, T.G., KUZMIN, I., NIEZGODA, M., RUPPRECHT, C., BRESEE, J. and BREIMAN, R.F. 2004. Nipah virus encephalitis reemergence, Bangladesh. *Emerging Infectious Diseases* 10:2082-87.MIDDLETON, D.J., WESTBURY, H.A., MORRISSY, C.J., VAN DER HEIDE, B.M., RUSSELL, G.M., BRAUN, M.A. and HYATT, A.D. 2002. Experimental Nipah virus infections in pigs and cats. *Journal of Comparative Pathology* 126:124-36.
6. MOHD NOR, M.N., GAN, C.H., ONG, B.L. 2000. Nipah virus infection of pigs in peninsular Malaysia. Rev Sci Tech Off Int Epiz; 19(1): 160-65.
7. WEINGARTL, H., CZUB, S., COPPS, J., BERHANE, Y., MIDDLETON, D., MARSZAL, P., GREN, J., SMITH, G., GANSKE, S., MANNING, L. and CZUB, M. 2005. Invasion of the central nervous system in porcine host by Nipah virus. *Journal of Virology*; 79:7528-34.

Fernando J. Torres-Vélez, DVM, Department of Pathology, College of Veterinary Medicine, University of Georgia, Athens, GA, 30602-7388, ftorres@vet.uga.edu

36

PESTE DES PETITS RUMINANTS

1. NAME

Peste des petits ruminants, goat plague, pest of small ruminants, pneumonia-enteritis or stomatitis-pneumoenteritis complex or syndrome, pseudo-rinderpest, kata (Pidgin English for catarrh)

2. DEFINITION

Peste des petits ruminants (PPR) is an acute or subacute contagious viral disease of goats and sheep characterized by fever, conjunctivitis, erosive stomatitis, gastroenteritis, and pneumonia. Clinical and pathological aspects of PPR closely resemble rinderpest, hence the synonym pseudo-rinderpest. Goats are usually more susceptible and more severely affected than sheep.

3. ETIOLOGY

Peste des petits ruminants virus (PPRV) is a single-stranded RNA virus in the family Paramyxoviridae, genus *Morbillivirus.* Other members of the *Morbillivirus* genus include rinderpest virus (RPV), measles virus (MV), canine distemper virus (CDV), phocine distemper virus (PDV) of pinnipeds (seals and sea lions), and cetacean morbillivirus (dolphin morbillivirus, porpoise morbillivirus). Peste des petits ruminants virus, being an enveloped virus, is relatively fragile and easily inactivated by sunlight, heat, lipid solvents, acidity, and alkalinity.

4. HOST RANGE

a. *Domestic animals and wild animals*

Peste des petits ruminants is primarily a disease of goats and sheep. However, there are 2 reports of naturally occurring PPR in captive wild ungulates from 3 families: Gazellinae (dorcas gazelle), Caprinae (Nubian ibex and Laristan sheep), and Hippotraginae (gemsbok). Furthermore, the virus has been isolated from an Indian buffalo (*Bubalus bubalis*) with rinderpest-like disease and was suspected of involvement. Experimentally, the American white-tailed deer (*Odocoileus virginianus*) is fully susceptible. The role of wildlife in the epizootiology of PPR in Africa remains to be investigated.

Camels, cattle and pigs are susceptible to infection with PPRV, but they do not exhibit clinical signs. Such subclinical infections result in seroconversion, and cattle are protected from challenge with virulent RPV. Camels, cattle, and pigs do not, however, play a role in the epizootiology of PPR because they appear to be dead-end hosts.

b. Humans

There is no report of PPRV infection in humans.

5. EPIDEMIOLOGY

a. Transmission

Transmission of PPRV requires close contact because the virus is not very stable in the environment. Ocular, nasal, and oral secretions and feces are the sources of virus. Contact infection occurs mainly through inhalation of aerosols produced by sneezing and coughing. Fomites such as bedding may also contribute to the onset of an outbreak. Infected animals are capable of transmitting the virus during the incubation period. There is no known carrier state.

b. Incubation period

Peste des petits ruminants has an incubation period of 4-6 days, with a range of 3-10 days.

c. Morbidity

Clinical disease is more common and more severe in goats. In susceptible goat populations, a morbidity rate of 80-90% can be expected.

d. Mortality

A case fatality rate of 50-100% can be expected in susceptible goat populations. In an enzootic area, a low rate of infection exists continuously. When the susceptible population builds up, periodic outbreaks occur. These may be characterized by almost 100% mortality among affected goat and sheep populations.

6. CLINICAL SIGNS

The disease usually appears in the acute form, with an incubation period of 4-5 days followed by a sudden rise in body temperature to 104°-106°F (40°-41°C). The temperature usually remains high for about 5-8 days before slowly returning to normal preceding recovery or dropping below normal before death. Affected animals appear ill and restless and have a dull coat, dry muzzle, and depressed appetite. Accompanying these nonspecific signs are a series of changes that make up a highly characteristic syndrome. From the onset of fever, most animals have a serous nasal discharge, which progressively becomes mucopurulent. The discharge may remain slight or may progress, resulting in a profuse catarrhal exudate, which crusts over and occludes the nostrils. At this stage, animals have respiratory distress, and there is much sneezing in an attempt to clear the nose. Small areas of necrosis may be seen on the visible nasal mucous membranes. The conjunctiva usually becomes congested, and the medial canthus may have some crusting. There may also be profuse catarrhal conjunctivitis resulting in matting of the eyelids.

Necrotizing stomatitis is common. It starts as small, roughened, red, superficial necrotic foci on the gum below the incisor teeth. These areas may resolve within 48 hours or progressively increase to involve the dental pad, the hard palate, cheeks and their papillae, and the dorsum of the anterior part of the tongue. Necrosis may result in shallow irregular nonhemorrhagic erosions in the affected areas of the mouth and deep fissures on the tongue. Necrotic debris may collect at the oral commissures, and scabs may form along the mucocutaneous junction of the lips. There may be excessive salivation but not to the extent of drooling.

At the height of development of oral lesions, most animals manifest severe diarrhea that is often profuse but not hemorrhagic. As it progresses, there is severe dehydration, emaciation, and dyspnea followed by hypothermia, and death usually occurs after a course of 5-10 days. Bronchopneumonia, evidenced by coughing, is a common feature in the later stages of PPR. Pregnant animals may abort. Commonly, secondary bacterial infections occur or latent infections may be activated and complicate the clinical picture.

The prognosis of acute PPR is usually poor. The severity of the disease and outcome in the individual is correlated with the extent of mouth lesions. Prognosis is good in cases where the lesions resolve within 2-3 days. It is poor when extensive necrosis and secondary bacterial infections result in an unpleasant, fetid odor on the animal's breath. Respiratory involvement is also a poor prognostic sign. Young animals (4-8 months) have more severe disease, and morbidity and mortality are higher. Both field and laboratory observations indicate that PPR is less severe in sheep than in goats. Nevertheless, field outbreaks have been reported in the humid zones of West Africa in which no distinction could be made between the mortality rates in sheep and in goats. Poor nutritional status, stress of movement, and concurrent parasitic and bacterial infections enhance the severity of clinical signs.

7. POST- MORTEM LESIONS

a. Gross

The pathology caused by PPRV is dominated by necrotic and inflammatory lesions in the mouth and the gastrointestinal tract. There is also a definite, albeit inconstant, respiratory system component, hence the synonym stomatitis-pneumoenteritis complex. Emaciation, conjunctivitis, erosive stomatitis involving the inside of the lower lip and adjacent gum, cheeks near the commissures, and the free portion of the tongue are frequent lesions. In severe cases, lesions may also be found on the hard palate, pharynx, and upper third of the esophagus. The necrotic lesions do not evolve into ulcers because the basal layer of the squamous epithelium is not penetrated unless secondary bacterial infections occur.

The rumen, reticulum, and omasum rarely have lesions. Sometimes, there may be erosions on the pillars of the rumen. The abomasum is a common site of regularly outlined erosions and often oozes blood. Lesions in the small intestine are generally moderate, being limited to small streaks of hemorrhages and, sometimes, erosions in the first portion of the duodenum and the terminal ileum. Peyer's patches are the site of extensive necrosis, which may result in severe ulceration. The large intestine is usually more severely affected with congestion around the ileocecal valve, at the ceco-colic junction, and in the rectum. In the posterior part of the colon and the rectum, discontinuous streaks of congestion ("zebra stripes" or “tiger stripes”) form on the crests of the mucosal folds. These “stripes” are the result of straining and tenesmus due to the diarrhea.

In the respiratory system, small erosions and petechiae may be visible on the nasal mucosa, turbinates, larynx, and trachea. The lungs commonly exhibit pneumonia of a bronchointerstitial type, characterized by congestion, consolidation and enlarged regional lymph nodes. Suppurative bronchopneumonia, caused by secondary bacterial infections, is usually confined to the cranio-ventral areas with muco-purulent exudate in deeper airways and atelectasis of the affected areas. There may be sero-fibrinous pleuritis with inflammatory effusion within the pleural cavity. The spleen may be slightly enlarged and congested. Most lymph nodes throughout the body are enlarged, congested, and edematous. Erosive vulvovaginitis similar to the lesions in the oral mucocutaneous junction may be present.

b. Key microscopic

Microscopic lesions in the digestive tract include degeneration and necrosis of epithelial cells of the mucous membranes associated with mucosal and submucosal congestion and inflammatory infiltration at the border of the lesions. Multinucleated syncytial cells are evident in the squamous epithelium. In addition, eosinophilic intracytoplasmic and intranuclear inclusion bodies are found in epithelial cells. In Peyer's patches lymphoid cell depletion and occasionally syncytia are observed. In the upper respiratory system lesions are similar to those described for the upper digestive tract. In the lungs, bronchointerstitial pneumonia characterized by infiltration of lymphocytes, neutrophils and hyperplasia of type II pneumocytes are found. In addition, large multinucleated syncytia and intracytoplasmic and intranuclear eosinophilic inclusions in giant cells and alveolar macrophages are present. These lesions are frequently complicated with serofibrinous or suppurative pneumonia. Lymphocytolysis and occasional syncytia formation are observed in the lymphatic tissues.

8. IMMUNE RESPONSE

a. *Natural infection*

Animals that survive PPRV infection harbor circulating neutralizing antibodies against the virus for up to 4 years and are likely protected from re-infection. However, natural infection with virulent strains can induce severe immunosuppression resulting in leukopenia and lymphopenia, which may transiently diminish immune responses to other pathogens.

b. *Immunization*

A safe and efficacious modified live vaccine is available for PPR. The modified live rinderpest (RP) vaccine is also safe and efficacious against PPR. No method is available for distinguishing between vaccinated and naturally infected animals. For this reason, use of the non-homologous RP vaccine for PPR control is not recommended as RP occurs naturally in goats and sheep.

9. DIAGNOSIS

a. *Field diagnosis*

In the field, a presumptive diagnosis can be made on the basis of clinical, pathological, and epizootiological findings. However, laboratory confirmation is an absolute requirement, particularly in areas or countries where PPR has not previously been reported.

b. *Laboratory diagnosis*

i. Samples

Specimens to submit for laboratory diagnosis should include: blood in EDTA anticoagulant, clotted blood or serum (if possible, paired sera), mesenteric lymph nodes, spleen, lung, tonsils, and sections of the ileum and large intestine. Swabs of serous nasal and lachrymal discharges may also be useful. All samples should be shipped fresh (not frozen) on ice within 12 hours of collection.

ii. Laboratory procedures

A wide range of laboratory procedures are available, and the most frequently used are: agar gel immunodiffusion, virus isolation, monoclonal antibody-based antigen capture ELISA, and reverse-transcription PCR for detection of the virus; and, serum neutralization test and monoclonal antibody-based competitive ELISA for antibody detection.

10. PREVENTION AND CONTROL

There is no specific treatment for PPR. However, drugs that control bacterial and parasitic complications may decrease mortality. Eradication is recommended when PPR appears in new areas. Methods that have been successfully applied for RP eradication in many areas would be appropriate for PPR. These should include quarantine, slaughter with proper disposal of carcasses and contact fomites, decontamination, and restrictions on importation of sheep and goats from affected areas.

11. GUIDE TO THE LITERATURE

1. ABU ELZEIN, E.M.E., HOUSAWI, F.M.T., BASHAREEK, Y., GAMEEL, A.A., AL-AFALEQ, A.I. and ANDERSON, E. 2004. Severe PPR infection in gazelles kept under semi-free range conditions. *J. Vet. Med. B* 51:68-71.
2. BARRETT, T., PASTORET, P.-P. and TAYLOR, W.P. (editors). 2006. *Rinderpest and Peste des petits ruminants: virus plagues of large and small ruminants.* Elsevier Academic Press, New York. 341 pp.
3. BROWN, C.C., MARINER, J.C. and OLANDER, H.J. 1991. An immunohistochemical study of the pneumonia caused by peste des petits ruminants virus. *Vet. Pathol*. 28:166-170.
4. BUNDZA, A., AFSHAR, A., DUKES, T.W., MYERS, D.J., DULAC, G.C. and BECKER, S.A.W.E. 1988. Experimental peste des petits ruminants (goat plague) in goats and sheep. *Can. J. Vet. Res*. 52:46-52.
5. DIALLO, A. 2004. Peste des petits ruminants. In : *Manual of Diagnostic Tests and Vaccines for Terrestrial Animals*, 5th ed., OIE, Paris, France, pp. 153-162.
6. GIBBS, E.P.J., TAYLOR, W.P., LAWMAN, M.J.P. and BRYANT, J. 1979. Classification of peste des petits ruminants virus as the fourth member of the genus *Morbillivirus*. *Intervirol*. 11:268-274.
7. ROSSITER, P. B. 2004. Peste des Petits Ruminants. In: *Infectious diseases of livestock*, 2nd ed., JAW Coetzer, RC Tustin, eds., Oxford University Press, Oxford, UK, pp. 660-672.
8. SALIKI, J.T., HOUSE, J. A., MEBUS, C.A. and DUBOVI, E.J. 1994. Comparison of monoclonal antibody-based sandwich ELISA and virus isolation for detection of peste des petits ruminants virus in goat tissues and secretions. *J. Clin. Microbiol*. 32:1349-1356.

9. SCOTT, G.R. 1988. Rinderpest and peste des petits ruminants. In: *Virus Diseases of Food Animals,* Vol. II, E.P.J. Gibbs, ed., Academic Press, London, UK, pp. 401-432.
10. TAYLOR, W.P. 1984. The distribution and epidemiology of peste des petits ruminants. *Prev. Vet. Med.* 2:157-166.

See Part IV for photos.

Jeremiah T. Saliki, DVM, PhD, College of Veterinary Medicine, University of Georgia, Athens, GA 30602, jsaliki@vet.uga.edu.

- and -

Peter Wohlsein, Dr. Med. Vet., School of Veterinary Medicine, Hannover, Germany, Peter.Wohlsein@tiho-hannover.de

37

RABBIT HEMORRHAGIC DISEASE

1. RABBIT HEMORRHAGIC DISEASE

Rabbit hemorrhagic disease, rabbit calicivirus disease, viral hemorrhagic disease of rabbits

2. DEFINITION

Rabbit hemorrhagic disease (RHD) is a peracute to acute viral disease of the European rabbit (*Oryctolagus cuniculus*) with high morbidity and mortality, characterized by extensive hepatic necrosis and disseminated intravascular coagulation.

3. ETIOLOGY

The causative agent is a calicivirus, rabbit hemorrhagic disease virus (RHDV). It is thought that only 1 serotype of the virus exists. A closely related virus, European brown hare syndrome virus (EBHS) causes a similar disease in the hare. However the 2 viruses are antigenically distinct and have non-overlapping host ranges. RHD is resistant to physical and chemical agents.

4. HOST RANGE

a. Domestic animals

Only the European rabbit (*Orcytolagus cuniculus*) is affected by RHD. Because all domestic rabbits are of this genus and species, they are all susceptible. Although rabbits of all ages can be infected, animals younger than 40 days of age tend to be resistant.

b. Wild animals

Wild rabbits of the *Oryctolagus* genus are fully susceptible. Other lagomorphs tested, including volcano rabbit of Mexico (*Romerolagus diazzi*), the black-tail jackrabbit (*Lepus californicus*), and the cottontail (*Sylvilagus floridanus*) are not affected by RHD. No other animals have been reported as being infected with RHD.

c. Humans

People are not susceptible to RHD.

5. EPIDEMIOLOGY

a. *Transmission*

Spread is by direct contact with infected animals or, indirectly, by contact with objects contaminated with virus. The virus is hardy and survives well in the environment, so contamination by secretions and excretions can seed a local area. Rabbits are known to shed virus for at least 4 weeks after clinical recovery from this disease. Experimental transmission can be accomplished through inoculation by oral, nasal, subcutaneous, intramuscular, or intravenous routes. Because there is so much virus in blood and it survives freezing well, importation of infected rabbit meat, with subsequent fomite transmission to susceptible rabbits, has been a frequent means of transmission of the disease to a new area.

b. *Incubation period*

Period of incubation is short, varying between 1-3 days.

c. *Morbidity*

The morbidity rate is high and related to proximity of rabbits. In group situations, morbidity approaches 80%. In research colonies where animals might be well separated from one another, either through air flow or cage separations, morbidity may be more sporadic.

d. *Mortality*

The death rate with RHD is high, usually between 40-80%.

6. CLINICAL SIGNS

The most prominent sign is that young adult and adult rabbits die suddenly after 6-24 hours of fever with few clinical signs. Fever may be high (up to 105°F or 40.5°C) but often is not detected until rabbits show terminal clinical signs. Most rabbits appear depressed in the final hours and may have a variety of neurologic signs including excitement, incoordination, opisthotonos, and paddling. They sometimes emit a terminal squeal. A few rabbits may have a terminal serosanguineous, foamy, nasal discharge.

7. POST- MORTEM LESIONS

a. *Gross*

Because the virus replicates extensively in the liver, there is usually massive liver necrosis at postmortem. The liver will be mottled or tan, sometimes with hemorrhages evident, and it may be enlarged and soft. The virus also replicates in cells of the mononuclear phagocytic system and so there is often enlargement, necrosis, or hemorrhage in spleen and lymph nodes. Splenomegaly is common. Replication within the fixed intravascular macrophages is thought to be responsible for the disseminated intravascular coagulation often seen terminally. There may be hemorrhage and/or thrombosis in multiple organs.

b. Key microscopic

Periportal to massive necrosis in the liver is the most characteristic histopathologic feature of RHD. Viral inclusions are not reported. Microthrombi are often present in multiple organs.

8. IMMUNE RESPONSE

a. Natural infection

Immunity is solid following natural infection. However, because the virus is hardy in the environment and the disease becomes endemic in populations, it is probable that recovered animals are repeatedly exposed to the virus, so long-term protection may be due to recurring immune boosts.

b. Immunization

In endemic areas where control is desirable, a vaccine consisting of clarified liver suspension that has been inactivated and adjuvanted is used. This inactivated vaccine is administered initially twice, at a 2-week interval, and then annually. In some countries, vaccine is commercially available.

9. DIAGNOSIS

a. Field diagnosis

A presumptive diagnosis can be made in a rabbitry when there are multiple cases of sudden death following a short period of lethargy and fever, and characteristic hepatic necrosis and hemorrhages. A field diagnosis is more difficult when there are few rabbits on the premises or rabbits are relatively isolated, as in research colonies.

b. Laboratory diagnosis

i. Samples

Fresh liver and blood, as well as formalin-fixed samples of liver, spleen, and any other organs should be submitted.

ii. Laboratory procedures

The virus has not yet been propagated *in vitro*. There is a variety of tests available for detection of antigen or nucleic acid. These include: hemagglutination (with human type O red blood cells), ELISA, Western blotting, electron microscopy, and immunohistochemistry. Serologic tests include hemagglutination inhibition and competitive ELISA.

10. PREVENTION AND CONTROL

Prevention of introduction is the optimal control method. Countries free of this disease should restrict the importation of rabbits, frozen rabbit carcasses, raw rabbit pelts, and angora wool from countries where VHD is endemic. The disease could also be introduced inadvertently by purchasing breeding stock or raw angora wool from an endemic area.

If wild rabbits are not susceptible, control through stamping out is possible. However, once the disease is established in the wild population, as it is in the European rabbit, control measures are more focused on vaccination or isolation of the domestic population to prevent introduction from the wild.

11. GUIDE TO THE LITERATURE

1. ANONYMOUS. 2004. Rabbit Haemorrhagic Disease, Chapter 2.8.2, In: *Manual of Diagnostic Tests and Vaccines for Terrestrial Animals*, OIE, www.oie.int.
2. FERREIRA, P.G., COSTA-E-SILVA, A., OLIVEIRA, M.J., MONTEIRO, E., CUNHA, E. and AGUAS, A.P. 2006. Severe leukopenia and liver biochemistry changes in adult rabbits alter calicivirus infection. *Res. Vet. Sci.* 80:218-225.
3. GREGG, D.A., and HOUSE, C. 1989. Necrotic hepatitis of rabbits in Mexico: A parvovirus. *Vet. Rec.* 125:603-604.
4. RAMIRO-IBANEZ, F., MARTIN-ALONSO, J.M., PANCIA, P.G., PARRA, F. and ALONSO C. 1999. Macrophage tropism of rabbit hemorrhagic disease virus is associated with vascular pathology. *Virus Research*, 60:21-28.

See Part IV for photos.

Corrie Brown, DVM, PhD, Department of Pathology, College of Veterinary Medicine, University of Georgia, Athens, GA 30602-7388, corbrown@uga.edu

38

RIFT VALLEY FEVER

1. NAME

Rift Valley Fever, Enzootic Hepatitis, Slenkdalkoors (Afrik.)

2. DEFINITION

Rift Valley fever (RVF) is a peracute or acute arthropod-borne viral disease characterized by epidemic hepatitis of ruminants. The disease is most severe in sheep, goats, and cattle, causing high mortality rates in neonates and abortion in pregnant animals. Camels may abort after subclinical disease. Humans become infected from contact with tissues of infected animals and by mosquito bites. Disease in humans presents as a severe influenza-like illness, hemorrhagic fever, encephalitis and occasionally death.

3. ETIOLOGY

RVF virus is a member of the Bunyaviridae family and *Phlebovirus* genus. It is an enveloped icosahedral virus with a diameter of 90-110 nm. The virus contains tri-segmented, negative sense, single stranded RNA. The large (L), medium (M) and small (S) segments code for the viral polymerase, enveloped glycoproteins G1 and G2, and N nucleoprotein, respectively. All virus strains belong to 1 serotype. Virus strains are classified into 3 distinct lineages: Egyptian, West African, and Central-East African, which indicate distinct geographical origins. The virus is inactivated by lipid solvents, detergents and low pH.

4. HOST RANGE

a. Domestic and wild animals.

Rift Valley fever has a wide host range. Sheep, goats, cattle and camels are all susceptible hosts for RVF virus. The levels of disease susceptibility are dependent on age and breed. Host susceptibilities to RVF viral infection are further listed in Table 1 (below).

b. Humans

RVF presents in humans as an acute influenza-like illness. The onset is abrupt, with malaise, fever, chills, rigors, diarrhea, vomiting, retro-orbital pain, severe headache, and generalized aching and backache. During the outbreaks in Egypt 1977 and in Saudi Arabia 2000, 1% of patients developed unilateral or bilateral retinitis. A minority of patients develop ocular lesions, encephalitis, or severe hepatic disease with hemorrhagic manifestations in less than 1%. Case fatality rate in patients that develop hepatic disease is close to 50%.

Table 1. Susceptibility of vertebrates to Rift Valley virus infection

Extremely susceptible	Highly susceptible	Moderately susceptible	Less susceptible	Refractory
Lambs	Calves	Cattle	Camels	Birds
Kids	Sheep	Goats	Equines	Reptiles
Puppies		Buffaloes	Pigs	Amphibians
Kittens		Humans	Dogs	
Mice		South American monkeys	Cats	
Hamsters		Asian Monkeys	Guinea pigs	
			Rabbits	

5. EPIDEMIOLOGY

a. Transmission

Several arthropod species are involved in transmission of virus by biological and mechanical means. Biological vectors include hematophagous insects such as mosquito species of *Aedes*, *Culex*, *Anopheles*, *Eretmopodites* and *Monsonia.* Mechanical transmitters of the virus include biting flies such as *Culicoides* (midges), phlebotomids (e.g. sandflies), stomoxides (e.g. stable flies) and simulids (e.g. blackflies), and other biting insects. The disease in sub-Saharan Africa is endemic because of transovarial transmission in *Aedes Neomelaniconion.*

During epidemics, transmission also occurs by aerosol from the blood of viremic vertebrates and from contact with viscera of infected animals. Consumption of raw milk has been documented as a route of exposure. RVF virus does not spread from person to person.

b. Incubation period

Incubation period is 12-72 hours in newborn lambs, kids, and calves; 24-72 hours in adult sheep, goats, and cattle; and, 3-6 days in humans.

c. Morbidity

Morbidity is highly variable and dependent on susceptibility of hosts and presence of insect vectors. It can be very high and affect entire herds.

d. Mortality

In newborn lambs and kids mortality may reach 70-100%. Older lambs and kids, as well as adult sheep and goats, are less susceptible with mortality varying from 10-70%.

Mortality during epidemics is less than 10% for all ages of cattle and 20% for calves. Abortion rates are varied from 40-100%. Pregnant animals may abort as a result of infection of the fetus or as a consequence of the febrile reaction.

6. Clinical signs

In the south and east of Africa, outbreaks of RVF are associated with heavy rainfall, while in the drier north and west parts of Africa outbreaks are linked to irrigation projects. Disease epizootics are cyclical in nature and characterized by long inter-epizootic periods of 5-15 years in wetter areas and 15-30 years or more in drier areas.

Onset of the disease in newborn lambs and kids is marked by high fever which declines sharply before death. Lambs rarely survive more than 24-36 hours after the onset of first sign. Lamb and kids older than 2 weeks develop fever that lasts for 24-96 hours, anorexia, weakness, listlessness, and an increased respiratory rate. Some animals may develop melena or fetid diarrhea, regurgitation and a blood-tinged, mucopurulent nasal discharge. The disease in calves resembles that in lambs and sheep with a higher proportion of calves developing icterus. Death generally occurs 2-8 days after infection.

Infection is frequently subclinical in adult cattle but some animals develop acute disease characterized by fever for 24-96 hours, anorexia, lacrimation, salivation, nasal discharge, and bloody or fetid diarrhea. The course of the disease in cattle is 10-20 days.

Abortion storms occur in infected pregnant sheep, goats, cattle, and camels at any stage of gestation, and aborted fetuses are often autolyzed.

7. POST- MORTEM LESIONS

a. Gross

Hepatic necrosis is the most common lesion of RVF in affected animals, and severity is related to age. The most severe lesions occur in newborn lambs and aborted fetuses of sheep and cattle. Lesions in newborn lambs include an enlarged, friable, soft, and red to yellow-brown liver. Petechial to ecchymotic hemorrhages, patchy congestion, and small gray-white foci 1-2 mm in diameters are scattered throughout the liver parenchyma. Numerous petechiae and ecchymoses are seen in the mucosa of the abomasum with dark chocolate-brown content as a result of the presence of partially-digested blood. There may be edema and hemorrhages in the wall of the gall bladder and hepatic lymph nodes.

Calves and adult sheep and cattle display more localized liver lesions. Necrotic foci are visible with hemorrhages and edema in the abomasal folds, and sometimes copious amounts of free blood in the lumen of the intestine are observed. The spleen is slightly enlarged with hemorrhages in the capsule. Other changes include widespread subcutaneous and visceral hemorrhages, mild to moderate effusion of fluid in body cavities, and congestion and edema of the lungs. Icterus is more common in adult sheep.

In all species affected, the peripheral and visceral lymph nodes are enlarged, necrotic, and edematous with occasional petechiation.

b. Key microscopic

The most consistent histopathological finding is hepatic necrosis. A severe lytic necrosis with dense aggregates of cellular and nuclear debris and the presence of fibrin and macrophages with relatively few infiltrating neutrophils is seen in the livers of aborted fetuses and neonates. Eosinophilic oval or rod-shaped intranuclear inclusion bodies are also present in up to 50% of the cells in affected livers. In adult animals, the hepatic necrosis is less diffuse and mostly multifocal. Fibrin thrombi can be seen in multiple organs.

8. IMMUNE RESPONSE

a. Natural infection

Antibody response to RVF infection in domestic ruminants appears on day 4-5 post infection. Offspring of immune mothers may have passive maternal immunity for the first 3-4 months of life.

b. Immunization

The vaccine currently in use is a live attenuated Smithburn strain of RVF virus which has been used in South Africa since 1952. This vaccine induces life-long immunity in livestock. Vaccination of offspring of vaccinated sheep and goats confers protection for life when given in a single dose at 6 months of age when colostral immunity has waned. Vaccination of pregnant animals may precipitate abortion, fetal anomalies, or neonatal death.

Cattle vaccinated with the Smithburn vaccine have poor antibody responses. For efficient protection, cattle should be primed with the formalin-inactivated vaccine and given a booster dose 3-6 months after initial vaccination to ensure the transfer of colostral immunity to offspring. Annual boosters before the rains are required in cattle as immunity lasts only about 1 year.

Formalin inactivated vaccines used to immunize animals include the Entebbe (produced in South Africa) and ZH501 (produced in Egypt) vaccines. These vaccines have shown a delay in antibody response and require a boost every 6 months. A formalin-inactivated vaccine, TSU-GSD-200, produced by the U.S. Army Research Institute of Infectious Diseases has also been shown to be safe and produces long-term immunity in humans when 2 doses are administered. This vaccine is given only to at-risk individuals because it is expensive and difficult to produce and is in short supply.

Additional vaccines under development include the mutagen attenuated MP12 vaccine and the clone 13 vaccine. The MP12 vaccine is developed by multiple mutations in the L, M and S segments in the presence of 5-fluorouracil. The vaccine is immunogenic and nonpathogenic to neonatal lambs when challenged with the wild type virus. Nevertheless, it produces a low viremia in cattle and is abortogenic when given to sheep at the first trimester. The clone 13 vaccine has a large deletion in the S segment which makes it unlikely to regain virulence. It is highly immunogenic, and has been shown to produce life-long immunity in sheep. It could be a marker vaccine candidate for livestock and humans for use in the face of an outbreak in non-endemic countries. Safety in humans and pregnant animals has not been determined.

9. DIAGNOSIS

a. Field diagnosis

RVF should be considered in the differential diagnosis whenever the following observations are made in a disease outbreak: acute onset of abortions at all stages of pregnancy in ruminants; acute febrile disease with high mortality rates in young ruminants; liver lesions present in all cases; influenza-like disease in man – particularly in individuals associated with livestock; high incidence coincides with high mosquito populations, rains, and/or flooding.

b. Laboratory diagnosis

i. Samples

Samples to be taken from live animals are heparinized blood, EDTA blood and clotted blood. Samples to be taken at necropsy are liver, spleen, brain, kidney, lymph nodes and heart blood. A complete set of tissues in 10% buffered formalin should also be submitted for histopathological examination. Paired serum samples taken 2 weeks apart are useful to demonstrate a rise in antibody titer.

ii. Laboratory procedures

The virus can be isolated in various types of cell cultures (vero, BHK, primary kidney or testis cells of calves and lambs), in suckling and adult mice, hamsters, embryonated chicken eggs or 2-day old lambs. A rapid diagnosis can be achieved by demonstration of virus particles in liver homogenates by negative staining electron microscopy; by detection of viral RNA using reverse transcriptase-polymerase chain reaction (RT-PCR); by immunoflourescent staining of impression smears of liver, spleen, brain, or infected cell culture; or, by the demonstration of virus in serum taken during the febrile stage using enzyme-linked immunosorbent assay (ELISA) or immunodiffusion.

Antibodies can be demonstrated in the serum of infected animals by virus neutralization, ELISA for IgM and IgG detection, and hemagglutination inhibition. Additional serological tests, used less often, include immunoflourescence, complement fixation and immunodiffusion.

10. PREVENTION AND CONTROL

The sporadic nature of RVF outbreaks and the long inter-epizootic periods in endemic countries present a major challenge to prevention and control of the disease. Thus there is an urgent need to develop an early warning system for RVF to guide improved disease control and international trade in livestock. Such an early warning system would predict when specific climatic events and environmental changes are more likely to be associated with increased mosquito vector populations and RVF virus activity in specific geographical areas. This would ideally provide sufficient lead time to vaccinate susceptible animals and to control the vectors to minimize transmission to humans.

Control measures may include the following:

1. Movement control with respect to trade and export.
2. Vector control with larvicides in vector breeding sites rather than aerial spraying of adults.
3. Vaccination of livestock, which is important for 2 reasons: livestock amplify the virus, and the young/newborn are most at risk of dying – but colostral immunity confers protection against virulent RVF virus challenge.
4. Use of appropriate personal protective equipment by high risk individuals when handling infectious materials.
5. Antiviral drugs such as ribavirin and interferon alpha have also been shown to be protective in non-human primates infected with RVF virus.

11. GUIDE TO THE LITERATURE

1. ANONYMOUS. 2004. Rift Valley fever. Chapter 2.1.8. *Manual of Diagnostic Tests and Vaccines for Terrestrial Animals,* OIE, www.oie.int.
2. BRAY. M. and HUGGINS J. 1998. Antiviral therapy of hemorrhagic fevers and arbovirus infections. *Antiviral Therapy*. 3:53-79.
3. COETZER, J.A.W. 1977. The pathology of Rift Valley fever. I. Lesions occurring in natural cases in new-born lambs. *Onderstepoort J. Vet. Res.* 44:205-212.
4. GERDES, G.H. 2004. Rift Valley fever. *Rev. Sci, tech. Off. Int. Epiz.* 23:613-623.
5. GERDES, G.H. 2002. Rift Valley fever. *Veterinary Clinics Food Animal Practice.* 18:549-555.
6. PAWESKA, J.T., BURT, F.J., ANTHONY, F., SMITH, S.J., GROBBELAAR, A.A., CROFT, J.E., KSIAZEK, T.G. and SWANEPOEL, R. 2003. IgG-sandwich and IgM-capture enzyme-linked immunosorbent assay for the detection antibody to Rift Valley fever virus in domestic ruminants. *J. Virological Methods*. 113:103-112.
7. PITTMAN, P.R., LIU, C.T., CANNON, T.L., MAKUCH, R.S., MANGIAFICO, J.A., GIBBS, P.H. and PETERS, C.J. 2002. Immunogenicity of an inactivated Rift Valley fever vaccine in humans: a 12-year experience. *Vaccine* 18:181-189.
8. SWANEPOEL, R. and COETZER, J. A.W. 2002. Rift Valley fever. In: *Infectious Diseases of Livestock*, 2nd ed., J.A.W. Coetzer, R. Tustin, eds., Oxford Press, UK, pp. 1037-1070.
9. VIALAT, P., MULLER, R., HANG VU, T., PREHAUD, C. and BOULOY, M. 1997. Mapping of the mutations present in the genome of the Rift Valley fever virus attenuated MP12 Strain and their putative role in attenuation. *Virus Research* 52:43-50.

Samia Metwally, DVM, PhD, Foreign Animal Disease Diagnostic Laboratory,
USDA-APHIS-VS-NVSL, Plum Island, Greenport, NY 11944,
samia.a.metwally@aphis.usda.gov

39

RINDERPEST

1. NAME
Rinderpest, cattle plague

2. DEFINITION
Rinderpest (RP) is a contagious viral disease of cattle, buffalo, and some species of wildlife. It is characterized by fever, oral erosions, diarrhea, lymphoid necrosis, and high mortality in susceptible populations.

3. ETIOLOGY
Rinderpest virus (RPV) is a single-stranded RNA virus in the family Paramyxoviridae, genus *Morbillivirus*. Other members of the *Morbillivirus* genus include peste des petits ruminants virus (PPRV), measles virus (MV) of humans, canine distemper virus (CDV), phocine distemper virus (PDV) of pinnipeds (seals and sea lions), and cetacean morbillivirus (dolphin morbillivirus, porpoise morbillivirus). Rinderpest virus, being an enveloped virus, is relatively fragile and easily inactivated by sunlight, heat, lipid solvents, acidity and alkalinity.

4. HOST RANGE
a. Domestic and wild animals
Most wild and domestic cloven-footed animals can be infected by RPV. Cattle and African buffalo are highly susceptible. Among wild ungulates, African buffalo, wildebeest, kudu, eland, giraffe, and warthog are highly susceptible, while Thompson's gazelle and hippopotamus are fairly susceptible. Goats and sheep are less susceptible and are either infected subclinically or exhibit milder clinical signs relative to cattle. Subclinically infected or recovered goats and sheep develop protective immunity against both RPV and PPRV. Pigs can also be naturally infected and some breeds (Swayback pigs in Thailand and the Malay Peninsula) may exhibit clinical signs and die.

b. Humans
There is no report of RPV infection in humans.

5. EPIDEMIOLOGY
a. Transmission
Secretions and excretions, particularly nasal-ocular discharges and feces, contain large quantities of virus 1-2 days before clinical signs to 8-9 days after onset of

clinical signs. Spread of RP is by direct and indirect (contaminated ground, waterers, equipment, clothing) contact with infected animals. Aerosol transmission is not a significant means of transmission and occurs only in a confined area over a short distance. A major reason why RP spread so easily in Africa is linked to the nomadic husbandry system in which cattle follow the grass and thus move great distances. Moreover, during the dry season, many herds use the same well or watering area, and there is ample opportunity for cross-infection. It is believed that a good fence will control RP. There is no vertical transmission, arthropod vector, or carrier state. Although many wildlife species can be affected by rinderpest, there is no evidence that wildlife serve as a reservoir for the virus. Indeed, it is widely believed that in the absence of RPV in cattle, the virus will disappear in wildlife.

b. Incubation period

Following natural exposure, the incubation period is 4-5 days, with a range of 3-15 days. The incubation period varies with the strain of virus, dosage, and route of exposure. Infected animals are capable of transmitting the virus during the incubation period.

c. Morbidity

Rinderpest affects cattle of all ages with morbidity approaching 100% in susceptible animals when infected with a highly virulent strain.

d. Mortality

Mortality rates can also approach 100% in susceptible animals infected with a highly virulent strain.

6. CLINICAL SIGNS

Depending on the strain of virus, immune status of the animal affected, and concurrent infections, RP can appear as a peracute, acute, or mild infection. In the peracute form, usually seen in highly susceptible and young animals without colostral immunity, the only signs of illness are a fever of 104°-107°F (40°-41.7°C), congested mucous membranes, and death within 2-3 days after the onset of fever. The acute or classic form is characterized by the following sequential signs: fever of 104°-106°F (40°-41.1°C), serous to mucopurulent nasal-ocular discharges, depression, anorexia, constipation, oral erosions resulting in abundant and frothy salivation, watery and/or hemorrhagic diarrhea, dehydration, emaciation, prostration, and death 6-12 days after onset of illness. Leukopenia is a common finding.

7. POST- MORTEM LESIONS

a. Gross

The most remarkable lesions occur in the digestive tract. Some strains of RPV cause oral lesions and others do not. Oral lesions start as small grey foci that may coalesce. The grey (necrotic) epithelium then sloughs off and leaves a red erosion. In the mouth lesions occur on the gums, lips, hard and soft palate, cheeks, and base of the tongue. Early lesions are grey, necrotic, pinhead-sized areas that later coalesce and erode and leave red areas. In the rest of the anterior portion of the digestive tract, brownish necrotic or eroded areas occur in the esophagus, erosions and hemorrhages occur occasionally in the omasum, and congestion and edema are observed in the abomasum. Lesions are rare in the rumen and reticulum.

In the small intestine, necrosis or erosions of the intestinal epithelium adjacent to Peyer's patches occur in the jejunum and ileum (ingesta adhering to the intestinal mucosa by fibrinous exudation indicates areas of necrotic epithelium). In the large intestine, the walls of the cecum and colon may be edematous, and there may be blood in the lumen and blood clots on the mucosa. Lesions are usually more severe in the upper colon (edema of the wall, erosions in the mucosa, and congestion). The lesions may be accentuated at the cecocolic junction. Further down the colon, the colonic ridges may be congested; this is referred to as "tiger striping." Tiger striping can occur in other diarrheas and probably results from tenesmus. The severity of intestinal lesions varies among strains of RPV.

Except for lymphoid tissues, the involvement of other organs is usually minimal. Lymph nodes are generally swollen and edematous, and the spleen is slightly enlarged. There may be petechial to ecchymotic hemorrhages in the gall bladder. The lung may exhibit emphysema, congestion, and areas of secondary bronchopneumonia.

b. Key microscopic

The microscopic lesions of RP are a direct result of virus-induced cytopathogenicity. Generally, the severity of lesions is directly related to the virulence of the virus strain involved. Both mild and highly virulent strains of RPV display epithelial and lymphoid tropism. Lesions in the mucous membranes of the upper digestive tract are characterized by ballooning degeneration of epithelial cells in the spinous layer, necrosis, formation of multinucleated syncytia, and eosinophilic intracytoplasmic inclusion bodies. The necrotic cells slough and erosions develop with mild perilesional infiltrations of mononuclear cells and neutrophils. The erosions may heal or ulcerate in cases with secondary bacterial infections. Depending on the virulence of the RPV strain involved, lymphoid tissues display variable degrees of lymphocytic depletion.

8. IMMUNE RESPONSE

a. Natural infection

There is only one serotype of RPV. Recovered or properly vaccinated animals are immune for life.

b. Immunization

The most commonly used vaccine for RP is the cell-culture-attenuated vaccine which was developed by Walter Plowright in Kenya in the 1960's. This is a safe vaccine for many species and produces life-long immunity in cattle (animals challenge-inoculated 7 years after vaccination were protected). In endemic areas where cattle have been vaccinated, colostral immunity will interfere with the vaccination of calves up to 11-12 months of age. Because the duration of colostral immunity is variable, the recommendation is to vaccinate calves annually for 3 years.

Experimental recombinant vaccinia-vectored vaccines containing the fusion and hemagglutinin genes of RPV have protected against challenge inoculation of virulent virus. Animals immunized with recombinant vaccines can be differentiated serologically from animals having antibody induced by live virus. These vaccines would be useful at this time when RP eradication is in sight by enabling countries to maintain herd immunity to RP without using the modified live vaccine.

9. DIAGNOSIS

a. Field diagnosis

Rinderpest should be considered in all ages of cattle whenever there is a rapidly spreading acute febrile disease accompanied by the preceding clinical signs and lesions of RP. The “all ages” stipulation is important because this will be one of the major differences between bovine virus diarrhea-mucosal disease, which predominantly affects animals between 4-24 months of age.

b. Laboratory diagnosis

i. Samples

Because the viral titer drops when the fever falls and diarrhea starts, specimens should preferably be collected from animals with a high fever and oral lesions. The following samples should be collected from live animals: blood in EDTA or heparin, clotted blood for serum, swabs containing lacrimal fluid, necrotic tissue from the oral cavity, and aspiration biopsies of superficial lymph nodes. At necropsy, the following samples should be collected: spleen, mesenteric lymph nodes, and tonsil. For the best specimens, a febrile animal should be slaughtered and specimens collected. If this cannot be done, then collect specimens from euthanized moribund animals. All samples should be transported to the laboratory on wet ice - not frozen. A complete set of tissues, including sections of all lesions, should also be collected in 10% formalin.

ii. Laboratory procedures

A wide range of laboratory procedures are available. The most frequently used of these are: agar gel immunodiffusion, virus isolation, monoclonal antibody-based antigen capture ELISA, and reverse-transcription PCR for detection of the virus. Serum neutralization test and monoclonal antibody-based competitive ELISA are available for antibody detection. To confirm the initial diagnosis in a free area, the virus has to be isolated and identified.

10. PREVENTION AND CONTROL

In the last 2 decades, rinderpest was present in Sub-Saharan Africa, the Indian subcontinent, and the Near East. By the end of 2005, most countries were self-declared free of the disease. However, OIE certification of disease-free status has not been made for many countries and information on disease status is not available from some countries due to civil strife. Nevertheless, judging from the success of the Global Rinderpest Eradication Program, the goal of global eradication of rinderpest appears to be achievable in the foreseeable future.

Countries and areas free of RP should prohibit unrestricted movement of RP-susceptible animals and uncooked meat products from areas infected by RP or practicing RP vaccination. Because recovered animals are not carriers, and there are good serological techniques, zoological ruminants and swine can be imported with proper quarantine and testing. If an outbreak occurs, the area should be quarantined, infected and exposed. Animals should be slaughtered and buried or burned, and ring vaccination considered.

High-risk countries (those trading with, or geographically close to, infected countries) can protect themselves by having all susceptible animals vaccinated before they enter the country or by vaccinating the national herd, or both. If an outbreak occurs, the area should be quarantined and ring vaccinated.

Endemic countries should vaccinate the national herd. Owing to the uncertainty of vaccine potency in areas without an adequate cold chain, the recommendation is to vaccinate annually for at least 4 years, followed by annual vaccination of calves. The foci of infection should be quarantined and stamped out. Wildlife, sheep, and goats should be monitored serologically. Serological monitoring of sheep and goats could be complicated by the practice of using RP vaccine to protect against peste des petits ruminants.

11. GUIDE TO THE LITERATURE

1. BARRETT, T., PASTORET, P.P. and TAYLOR, W.P. (eds). 2006. *Rinderpest and Peste des petits ruminants: virus plagues of large and small ruminants.* Elsevier Academic Press, New York. 341 pp.
2. BROWN, C.C. and TORRES, A. 1994. Distribution of antigen in cattle infected with rinderpest virus. *Vet. Pathol.* 31:194-200.
3. PLOWRIGHT, W. 1984. The duration of immunity in cattle following inoculation of rinderpest cell culture vaccine. *J. Hyg.* Camb. 92:285-296.
4. ROSSITER, P. B. 2004. Rinderpest. In: *Infectious Diseases of Livestock*, 2nd ed., JAW Coetzer, RC Tustin, eds., Oxford University Press, Oxford, United Kingdom. pp. 629-659.
5. TAYLOR, W.P. and ROEDER, P. 2004. *Manual of Diagnostic Tests and Vaccines for Terrestrial Animals*, 5th ed., OIE, Paris, France. pp. 142-152.
6. WAMWAYI, H. M. and FLEMING, M. and BARRETT, T. 1995. Characterization of African isolates of rinderpest virus. *Vet. Microbiol.* 44:151-163.
7. WOHLSEIN, P., TRAUTWEIN, G., HARDER, T.C., LIESS, B. and BARRETT, T. 1993. Viral antigen distribution in organs of cattle experimentally infected with rinderpest virus. *Vet. Pathol.* 30:544-554.
8. WOHLSEIN, P., WAMWAYI, H.M., TRAUTWEIN, G., POHLENZ, J., LIESS, B. and BARRETT, T. 1995. Pathomorphological and immunohistological findings in cattle experimentally infected with rinderpest virus isolates of different pathogenenicity. *Vet Microbiol.* 44:141-149.
9. YILMA, T. HSU, D., JONES, L., OWENS, S., GRUBMAN, M., MEBUS, C., YAMANAKA, M. and DALE B. 1988. Protection of cattle against rinderpest with infectious vaccina virus recombinant expressing the HA or F gene. *Science.* 242:1058-1061.

See Part IV for photos.

Jeremiah T. Saliki, DVM, PhD, College of Veterinary Medicine, University of Georgia, Athens, GA 30602, jsaliki@vet.uga.edu

- and -

Peter Wohlsein, Dr. Med. Vet., School of Veterinary Medicine, Hannover, Germany, Peter.Wohlsein@tiho-hannover.de

40

SCREWWORM MYIASIS

1. NAME

Screwworm myiasis, gusanos, mosca verde, gusano barrenador, gusano tornillo, gusaneras

2. DEFINITION

Myiasis is the infestation of live vertebrate animals with dipterous larvae that, for at least a period, feed on the host's dead or living tissues, liquid body substances, or ingested food. Such larvae are classified as either obligatory or facultative parasites, depending on their reliance on host viability. Screwworms are obligatory parasites because they feed only on live hosts. These larvae penetrate deeply into a wound on a warm-blooded animal and feed on living tissues and body fluids. Facultative parasite larvae also may be present in wounds, even simultaneously with screwworm larvae, but they feed on only dead tissue and decaying matter.

3. ETIOLOGY

Screwworm myiasis is caused by 2 species of fly larvae (also known as maggots) in the dipteran family Calliphoridae, subfamily Chrysomyinae: *Chrysomya bezziana* (Villeneuve), the Old World screwworm, and *Cochliomyia hominivorax* (Coquerel), the New World screwworm.

4. HOST RANGE

a. Domestic animals and wild animals

Any living warm-blooded animal is susceptible to screwworm myiasis, but infestations in poultry or fowl are rare.

b. Humans

Humans are susceptible to screwworm myiasis, and many cases are documented.

5. EPIDEMIOLOGY

a. Transmission

Myiasis is not transmitted from host to host in the conventional sense. In nature, screwworms infesting one animal cannot move directly to and infest another animal. Instead, new infestations arise only indirectly, when maggots of one generation continue their life cycles to become adult flies that disperse and lay eggs on new hosts.

Female screwworm fly dispersal depends on ecological conditions, food supply, and availability of hosts with suitable wounds. Individual female flies tend to range less than 3 km in moist tropical environments with a high density of potential hosts. In less favorable environments with lower densities of animals, single screwworm flies have traveled as far as 290 km, but usually more like 20-25 km. In arid areas, screwworm flies often travel along water courses. In mountainous areas, they may travel the course of valleys, where the climate is warmer, moisture is more abundant, and animal density is greater. Vehicles, especially those transporting animals, may assist the movement of screwworm flies over long distances, and prevailing winds may be a factor as well.

Life Cycle of screwworm fly:

The female screwworm fly seeks a superficial wound on a warm-blooded animal and lays eggs along the edge of the wound in a shingle-like pattern. Female New World screwworm flies usually deposit 100-350 eggs (mean = 200) in a single egg mass, and one fly may oviposit 6-8 (or more) times in her life, but usually only twice. After the first 2 egg masses, surviving females make non-ovipositional visits at animal wounds to imbibe protein-rich fluids that feed the maturation of any additional egg masses. An Old World screwworm egg mass typically contains about 100-250 eggs. Larvae eclose from the eggs in 8-12 hours, enter the wound, and begin feeding. Within 5-7 days, the feeding larvae are mature. At this stage of development (third instar), larvae exit the wound and drop to the ground.

Mature maggots are negatively phototactic (i.e., they move away from light) and burrow a short distance into the soil, where they pupate. Each pupa is enclosed in a seed-like case, or puparium, formed by contraction and hardening of the last larval skin. Transformation into the fly occurs during the pupal stage and may take as little as 7 days at 28° C or as long as 60 days at temperatures of 10°-15°C.

Adult flies emerge from their puparia, spread their wings, and harden for about 2 hours. Then they seek roosting places and sustenance, such as water and nectar. Survival of adult flies is dependent on temperature, humidity, food resources, host availability, and other ecological factors. Ambient air temperatures of 25°-30° C with relative humidities of 30-70 % are ideal for screwworm fly activity and survival. Male screwworm flies may survive for up to 14 days; females usually live for about 10 days, but about 10% may survive for 30 days or more. After 3-5 days of feeding and resting, the flies are ready to mate. Male screwworm flies will mate several times; females usually mate only once. Shortly after mating, the female flies seek out hosts for oviposition, and the cycle begins anew.

b. Incubation period

See Life cycle, above

c. Morbidity

In some areas of the Western Hemisphere, where screwworm populations are high and climatic and ecological conditions are ideal, livestock owners report that screwworms infest the navel wounds of every newborn animal if it is not treated soon after birth.

d. Mortality

Screwworm myiasis seriously affected white-tailed deer populations on the King Ranch in South Texas, USA, during the 1950s, when annual mortality rates in fawns varied from approximately 20-80%. Infestation rates in superficial wounds on wildlife varied directly with the size of local fly populations, which in turn, varied with the number of untreated domestic animal wounds.

Screwworm infestations that are treated and those that result from single egg masses often are not lethal to an animal, although death is a possibility, especially in very small animals. Secondary infections are also common.

Untreated screwworm infestations often attract additional ovipositions at the wound site from several female flies or even from the same fly. Left untreated, these multiple infestations often result in death of the host within 7-10 days, depending on the size and condition of the animal, the anatomical site of the infestation, and whether there are other complications, such as infection or toxicity. Animal deaths due to Old World screwworm myiasis seem less common than those from the New World screwworm.

6. CLINICAL SIGNS

Wounds vulnerable to screwworm infestation include those caused by tick bites, vampire bats, castration, dehorning, branding, barbed wire, mouth sores, shedding of velvet or antlers in deer, and a host of other causes. Navels of newborn mammals are common sites for screwworm infestation. Early stage maggots feeding in a wound are very difficult to see; only slight movement may be observed. As the larvae feed and grow, they destroy tissue, thus continually making the wound wider and deeper. By the third day of infestation, as many as 100-200 tightly packed, vertically oriented larvae easily may be observed embedded deep in the wound. Screwworms tend to burrow deeper when disturbed and generally do not crawl on the surface as do other maggots.

After 5-7 days of infestation with larvae from a single screwworm egg mass, a wound may grow to 3 cm or more in diameter and 5-20 cm deep. By this time, additional screwworm flies are likely to have deposited eggs, resulting in a multiple infestation. As many as 3,000 larvae may occur in a single wound. A serosanguinous discharge often exudes from infested wounds, and a distinct odor may be detected. In some cases, infested openings in the skin may be deceptively small, with extensive pockets of screwworm larvae beneath. Screwworm infestations in anal, preputial, vaginal, and nasal orifices may be easily overlooked, even in the later stages.

Animals with screwworm infestation usually display discomfort, may go off feed, and may produce less milk. Typically, animals with screwworm myiasis separate themselves from the rest of the flock or herd and seek dark or shady areas in which to lie down. Goats frequently hide in caves. Fawns with infested navels often stand in streams with water up to their abdomens. Brahman-type cows often lick the screwworm-infested navel wounds of calves, a process that cleans most maggots from the wound and reduces losses in this breed.

Animals with screwworm myiasis may die in 7-14 days if wounds are not treated to kill the larvae, especially in cases of multiple infestation. Death probably results from toxicity, secondary infections, or both. Smaller animals usually die of screwworm myiasis in a shorter time than larger animals, and anatomical site of the infestation also may affect time to death.

7. POST- MORTEM LESIONS

Necropsy usually is not helpful in diagnosing a screwworm infestation. Screwworms do not feed on dead tissue, so post-mortem examinations are unlikely to find them, unless death was very recent. After death, however, other blow fly larvae can quickly obscure screwworm wounds and infestations, confounding diagnosis.

8. IMMUNE RESPONSE

There is no available vaccine. However, preliminary work with the Old World screwworm suggests that at least partially efficacious immunoprotection may be possible.

9. DIAGNOSIS

a. Field diagnosis

Screwworm myiasis should be suspected when the described clinical manifestations are evident. New World screwworms appear first as 1mm long creamy white eggs, usually deposited in shingle-like fashion on the border of a superficial wound. Small screwworm maggots eclose from the eggs in 8-12 hours and soon grow to a length of 2 mm. Egg masses of Old World screwworms are indistinguishable, except the individual eggs are slightly larger. Eggs in the masses deposited by other species of blow flies are not well organized in a distinct pattern, but they may be on the margin of or in the hair close to a wound. Microscopic examination is required to distinguish individual eggs and small maggots of facultative blow fly parasites like *Cochliomyia macellaria, Phormia regina,* or *Phaenicia sericata* from those of screwworms. A few parasitic flesh fly species (Sarcophagidae) sometimes deposit their live larvae into a wound or in soiled wool or hair on vertebrate hosts. Maggots of these species are facultative parasites and also may be seen in wounds, usually near the surface, feeding on necrotic tissues or organic matter.

Maggots can be removed from a wound with tweezers. Second- and third-instar screwworms are cylindrical, pointed at one end and blunt at the other, and have a series of complete rings of minute, dark brown spines encircling their bodies. The common name, screwworm, derives from the resemblance of these maggots in size, shape, and contour to a wood screw. Field diagnosis is difficult, even for trained individuals, because most facultatively parasitic blow fly maggots display these same basic characteristics to some degree. The most distinctive difference, present only in larger screwworm maggots, is a pair of internal, dark-pigmented, longitudinal stripes (tracheal trunks) posteriodorsally. A magnifying glass or microscope is necessary to see the distinguishing characteristics of the various screwworm stages, and a diagnosis in the field should always be considered presumptive.

Female screwworm flies are very secretive in the field, but they occasionally may be observed visiting a wound to feed or oviposit. They are about 2 times the size of the common house fly. New World screwworm flies have a dark metallic blue to blue-green abdomen and thorax, an orange face with dark red eyes, and a pair of semi-clear wings. They have 3 longitudinal dark stripes on the back of the thorax, with the center stripe incomplete. Old World screwworm flies are similar, but their bodies are metallic green or bluish-green, and they have dark transverse bands on their thoraces, rather than longitudinal stripes. Adults of other facultatively parasitic New World blow flies are generally similar in appearance to *C. hominivorax*, although they vary in color and most lack thoracic stripes.

Most similar is *C. macellaria*, which has the 3 thoracic stripes, but all of them are equally long.

b. Laboratory diagnosis

i. Samples

Before treatment, a sample of larvae should be removed from the wound using forceps for submission to the laboratory. Eggs should be carefully removed from the edge of a wound using a scalpel. For laboratory diagnosis, specimens of eggs, larvae, or flies should be place in 70% alcohol and sent to a recognized diagnostic laboratory (note: do not use formalin as a preservative). Because screwworm maggots penetrate deep into a wound, and other facultative larvae may exist more superficially in the same wound, specimens of larvae for laboratory diagnosis are best collected from the deepest part of the wound. In the United States, send specimens to the U.S. Department of Agriculture's National Veterinary Services Laboratories, 1800 Dayton Avenue, Ames, IA 50010. Experienced professional personnel will identify the specimens and report the results.

ii. Laboratory procedures

Visual identification of parasite

10. PREVENTION AND CONTROL

Preventing the introduction of screwworms into ecologically susceptible areas that are currently pest-free is an important aspect of control. Blocking such introductions is accomplished through voluntary and regulatory actions. Immediately before transport from screwworm-endemic areas, animals (including pets) should be thoroughly inspected for the presence of a superficial wound subject to screwworm infestation. All wounds should be treated with an approved insecticide, followed by a precautionary general spraying or dipping of the animals before they are moved. Any animal with wounds suspected to contain screwworms should not be moved until the wounds have been properly treated and have healed.

Where screwworms are present, animals must be inspected at least every 3–4 days to discover and treat cases of myiasis. Open wounds on animals not infested with screwworm maggots should be treated to prevent infestation. In areas where screwworm myiasis is a seasonal problem, animal breeding can be regulated so births occur during times when myiasis incidence is lowest. Similarly, management practices that create wounds, such as branding, castration, dehorning, docking, or other operations, can be programmed for the season with lowest screwworm incidence.

Treating wounds and spraying or dipping animals with an approved insecticide will provide protection against screwworm infestation for 7-10 days. Should a screwworm egg mass be deposited on the edge of a wound on an animal so treated, the neonate larvae will encounter residual insecticide as they crawl into the wound and will be killed. Such care usually gives wounds sufficient time to heal and become unattractive oviposition sites. In the past, organophosphates were the products of choice for this purpose, but more recently, several avermectins and insect growth regulators have shown even better results. If wounds are already infested with second- or third-instar screwworms when an animal is sprayed or dipped, the treatment usually does not kill all larvae present. Therefore, this form of treatment should be used only as a preventative measure and not as a cure.

Eradication of the New World screwworm over vast geographic areas has been accomplished through application of the sterile insect technique (SIT). Overwhelming numbers of mass-reared screwworms 5.5 days into the pupal stage are exposed to 5,000 to 7,000 rads of gamma radiation, thus rendering the soon-emerging flies sexually sterile without any other adverse effects. Fly factories with immense capacity produce great quantities of sterile insects each day for mass aerial releases in targeted screwworm-infested areas. Once released, the sterile male screwworm flies mate with native females. Because females mate only once, any eggs they subsequently lay are infertile and do not hatch. With sufficiently overwhelming numbers of sterile males and unproductive matings, the reproductive cycles of targeted screwworm populations are disrupted, and the populations die out.

11. GUIDE TO THE LITERATURE

1. GRAHAM, O.H. (ed.) 1985. Symposium on eradication of screwworm from the United States and Mexico. Misc. Pub. *Entomol.* Soc. Am. 62:1-68.
2. GUIMARÃES, J.H. and PAPAVERO, N. 1999. Myiasis in Man and Animals in the Neotropical Region; Bibliographic Database. Sao Paulo, Brazil: Pleiade/FAPESP, 308 p.
3. KNIPLING, E.F. 1960. The eradication of screwworm fly. *Sci. Am.* 203(4):54-61.
4. KRAFSUR, E.S. 1998. Sterile insect technique for suppressing and eradicating insect population: 55 years and counting. *J. Agric. Entomol.* 15:303-317.
5. LINDQUIST, D.A., ABUSOWA, M. and HALL, M.J.R. 1992. The New World screwworm fly in Libya: a review of its introduction and eradication. *Med. Vet. Entomol.* 6:2-8.

6. McGRAW, L. 2001. Squeezing out screwworm. *Agric. Res.* 49(4):18-21.
7. MAYER, D.G. and ATZENI, M.G. 1993. Estimation of dispersal distances for *Cochliomyia hominivorax* (Diptera: Calliphoridae). *Environ. Entomol.* 22:368-374.
8. SUKARSIH, PARTOUTOMO, S., SATRIA, E., WIJFFELS, G., RIDING, G., EISEMANN, C. and WILLADSEN, P. 2000. Vaccination against the Old World screwworm fly (*Chrysomya bezziana*). *Paras. Immunol.* 22:545-552.
9. ZUMPT, F. 1965. *Myiasis in Man and Animals in the Old World; A Textbook for Physicians, Veterinarians and Zoologists.* London: Butterworth and Co., 267 p.

See Part IV for photos.

Jack Schlater, DVM, and Jim Mertins, USDA-APHIS-VS-National Veterinary Services Laboratories, Ames, IA 50010, Jack.L.Schlater@aphis.usda.gov, James.W.Mertins@aphis.usda.gov

41

SPRING VIREMIA OF CARP

1. NAME

Spring viremia of carp, formerly Infectious Dropsy of Carp, or Infectious Ascites

2. DEFINITION

Spring viremia of carp is an OIE-notifiable viral disease of certain coolwater and warmwater finfish species, often of high morbidity and mortality, and which may have serious economic and ecological consequences.

3. ETIOLOGY

Spring viremia of carp is caused by infection with a rhabdovirus (Spring viremia of carp virus, or SVCV, also known as *Rhabdovirus carpio*). There are several known strains generally classified as "European" and "Asian," of apparently varying antigenicity.

4. HOST RANGE

a. Domestic animals

A number of commercially important farmed fish species are susceptible to SVC disease, including goldfish (*Carassius auratus)* and common carp (*Cyprinus carpio*, including *C. carpio carpio*, also known as koi carp). Crucian carp (*Carassius carassius*), occasionally farmed as an ornamental fish, is also a susceptible species. Several other species of cultured fish (Zebra danios [*Brachydanio rerio*]), and guppies (*Lebistes reticulates*) are experimentally infectable with SVCV. An SVC-like virus has also been isolated from shrimp (Litopennaeus vannamei and Penaeus stylirostris), an aquatic animal that has both farmed and wild populations worldwide.

b. Wild animals

Other carp species such as grass carp (*Ctenopharyngodon idella*), bighead carp (*Aristichthys nobilis*), and silver carp (*Hypophthalmichthys molitrix*) are susceptible to SVC. Sheatfish (*Silurus glanis*) and tench (*Tinca tinca*) are also susceptible species. A number of other wild cyprinid and non-cyprinid species (which may also be cultured) such as golden shiners (*Notemigonus crysoleucas*), pumpkinseeds (*Lepomis gibbosus*), roach (*Rutilus rutilus*), and pike (*Esox lucius*) are considered experimentally infectable.

c. Humans

Humans are not susceptible to infection with SVCV.

5. EPIDEMIOLOGY

a. Transmission

The spread of SVC is generally through horizontal transmission of the virus shed by infected and/or diseased fish. A vertical component may exist through contamination of eggs with ovarian fluids that may contain virus, but transmission has not been adequately demonstrated to be truly intra-ovum in nature. A number of vectors have been implicated, including piscivorous birds and invertebrates such as fish lice (*Argulus* spp.) and leeches (*Pisciola* spp). Fomites may also serve as physical vectors.

There are distinct temperature-related and other variables in the epizootiology of SVC development, as the immune function of a number of fish species is impacted by water temperature. There is significant difference of disease expression among species; however, if fish are generally exposed to and infected by the virus in conditions between 50° and 60°F (10°-15°C), they may subsequently develop disease when environmental temperatures rise to between approx. 62° and 66°F (16° and 18°C). Some species of fish may become infected up to 76°F (23°C) but disease typically does not develop above 66°F (18°C) as immune function increases. Such fish remain infected but not clinically diseased, and (along with survivors of epizootics) may become carriers that can break or recrudesce when temperatures decrease to a more facilitative temperature regime.

b. Incubation period

Incubation is generally a function of environmental temperatures, and ranges typically from 5 to 17 days in many species of infected fish in facilitative conditions.

c. Morbidity

Morbidity may range from 10-100%, with an average of about 50%.

d. Mortality

Mortality is generally in the 30-70% range, depending on species, but may approach 100% in some epizootics.

6. CLINICAL SIGNS

Clinical signs of SVC include exophthalmia, edema, inflammation and hemorrhage of the swim bladder, ascites, and petechial hemorrhages of gills and skin. Affected fish may be lethargic, and display loss of equilibirium or uncoordinated swimming behavior. These signs are all non-specific and may be confused with a variety of other alternative or simultaneous infectious processes in susceptible species.

7. POST-MORTEM LESIONS

a. Gross

Post mortem lesions are non-specific and include ascites and congested internal organs, often with petechial hemorrhages. The skin, fins, and swim bladder may have visible hemorrhages ranging from petechial to ecchymotic.

b. Key microscopic

Many of the histopathological lesions associated with SVC are also non-specific, including: peri/panvasculitis of liver blood vessels; adipose degeneration and necrosis of the hepatic parenchyma; hyperemia of the spleen; and, epithelial desquamation and inflammation of the intestine with subsequent villous atrophy. Kidney tubules become clogged with casts, with vacuolation and hyaline degeneration of tubular cells. The lamina epithelia of the swim bladder may become a discontinuous multilayer, with submucosal hemorrhage and dilated vasculature. Myodegeneration and pericardial inflammation may be observed in heart tissue.

8. IMMUNE RESPONSE

a. Natural infection

Exposed fish probably mount an immune response consisting of circulating SVC antibodies directed against the surface glycoproteins of the virus, but this process is presumed to be temperature- and age-dependent, and is not well understood at the present time. Surviving fish may become lifelong carriers that can recrudesce or re-infect susceptible populations under temperature-facilitative conditions, or under other stress conditions.

b. Immunization

No effective vaccines currently exist for SVC.

9. DIAGNOSIS

a. Field diagnosis

There are no definite criteria for establishing a field diagnosis of SVC, since there are no pathognomonic signs associated with the expression of disease. However, if SVC-susceptible species of appropriate lifestage exhibit the clinical signs listed above, and do so within a temperature-facilitative context, SVC should always be suspected.

b. Laboratory diagnosis

i. Samples

Mid-kidney, spleen, liver, and brain tissues are useful to submit for virus isolation. Heparinized blood may be collected for immediate serology, or if samples are to be assayed later by serological assays, serum may be collected and frozen at -20°C.

ii. Laboratory procedures

Virus isolation, ELISA, PCR, and sequencing of strains to determine genotype are all used to diagnose SVCV.

10. PREVENTION AND CONTROL

Prevention may be enhanced treatment of influent water supplies by UV and/or sand filtration systems as well as through health status pre-screening of any fish stocks imported into aquaculture facilities. Good biosecurity protocols should be observed to minimize transmission of the virus through fomites or personnel. Stress should be kept to a minimum, and age-susceptible fish might be raised at temperatures above 18°C to delay or prevent the onset of disease.

SVC virus may be neutralized through several means, including heat (>60°C for 10 min.), ultraviolet exposure (254 nm for about 10 min), pH (<3 or >12), or exposure to various disinfecting agents such as lipid solvents, chlorine, potentiated peroxides or formalin. There are no vaccines currently available for the prevention of infection with SVCV or the development of SVC, and there are no effective treatments for the disease.

In the case of outbreaks in farmed fish, the benefits of retaining surviving fish that may become carriers capable of re-infecting populations must be weighed against the option of eradicating all exposed stocks and cleaning/disinfecting all culture equipment. Thorough disinfection is recommended. Although it is an enveloped virus, SVCV can remain infective in water for up to 4 weeks, and in mud up to 6 weeks, between 4° and 10°C.

11. GUIDE TO THE LITERATURE

1. AHNE, W., BJORKLUND H.V., ESSBAUER, S., FIJAN, N., KURATH, G. and WINTON, J.R. 2002. Spring Viremia of Carp (SVC). *Diseases of Aquatic Organisms.* 52:261-272.
2. ANONYMOUS. Office International des Epizooties, 2003. *Manual of Diagnostic Procedures*; Chptr. 2.1.4 4th ed., OIE, Paris, France.

3. DIKKEBOOM, A., RADI, C., TOOHEY-KURTH, K., MARQUENSKI, S., ENGEL, M., GOODWIN, A., WAY, K., STONE, D. and LONGSHAW, C. 2004. First Report of Spring Viremia of Carp Virus (SVCV) in Wild Common Carp in North America. *Journal of Aquatic Animal Health,* 16:169-178.
4. GOODWIN, A.E. 2002. First Report of Spring Viremia of Carp Virus (SVCV) in North America. *Journal of Aquatic Animal Health.* 14:161-164.

Peter Merrill, DVM, PhD, Aquaculture Specialist, USDA–APHIS Import Export, Riverdale, MD, Peter.Merrill@aphis.usda.gov

42

SWINE VESICULAR DISEASE

1. NAME
Swine Vesicular Disease, Enfermedad Vesicular del Cerdo (Sp), Maladie Vésiculose du Porc (Fr)

2. DEFINITION
Swine Vesicular Disease (SVD) is a contagious viral disease of pigs caused by an enterovirus. This is a vesicular disease characterized by fever, vesicular lesions, and subsequent erosions of the epithelium of the mouth, snout, feet, or teats.

3. ETIOLOGY
SVD virus (SVDV) belongs to the genus *Enterovirus* within the family Picornaviridae. SVDV is antigenically related to the human Coxsackie Virus B5. There is only one serotype with strain variations between different isolates.

As an enterovirus, the SVDV is quite resistant to a wide range of pH including low pH conditions (<2.0 to >12.0). Consequently, the SVDV remains viable in meat after *rigor mortis* (pH is <6.0) as well as in most other tissue and visceral organs after death. SVDV may remain in dry-cured hams for 180 days, dried sausages for over 1 year, and in processed intestinal casings for over 2 years. Depending on the temperature, ultraviolet radiation (sunlight), humidity, and pH conditions, SVDV can persists in the environment for more than 1 month.

SVDV can be inactivated by alkaline solutions (1% sodium hydroxide) combined with detergents. Other disinfectants such as oxidizing agents and iodophores are suitable if combined with a detergent in the absence of heavy organic matter.

4. HOST RANGE

a. Domestic animals
As the name implies, this is a disease primarily of domestic swine.

b. Wild animals
Although perhaps possible, there is no evidence of SVD infections in wild *Suidae* sp.

c. Humans

There are recorded infections in humans, primarily as a result of accidental laboratory exposures. In those cases, the clinical signs were similar to those of Coxsackie B5 infections, including influenza-like illness of varying severity. Despite the exposure of farmers, veterinarians, and abattoir workers to SVDV during large outbreaks of this disease in Europe and in Asia, there have been no reports of human illness under those conditions.

5. EPIDEMIOLOGY

a. Transmission

Infection can occur through damaged skin, by ingestion, or by inhalation. The most common means of transmission of SVD has been by the movement of infected pigs into a susceptible population. Infected SVDV pigs can then infect others by fighting and biting each other and by shedding SVDV into the environment. The disease has also been introduced into new herds by the feeding of infected swill containing meat scraps from infected swine. Exposure of pigs to contaminated environments (trucks, sale barns, etc.) has also resulted in SVD infection of susceptible pigs. There are reports of contaminated trucks remaining infectious for up to 4 months.

b. Incubation period

Clinical signs of SVD develop in 2-3 days after eating contaminated feed and in 2-7 days after contact with infected pigs.

c. Morbidity

The severity of SVD depends on many factors including the strain of virus and the conditions under which the animals are kept. Morbidity tends to be quite low and the course of the disease much less severe than in the case of FMD in pigs. Many pigs suffer sub-clinical infections. The disease is believed to persist in sub-clinical infections in many parts of the world.

d. Mortality

There is essentially no mortality due to SVD in pigs.

6. CLINICAL SIGNS

As a vesicular disease, SVD produces clinical signs similar to those caused by FMD, (see FMD chapter) or by other vesicular diseases (VS and/or VES). In addition, there are descriptions of an unsteady gait with shivering and chorea-type leg movements due to encephalitic involvement.

7. POST- MORTEM LESIONS

Lesions in pigs due to SVD are indistinguishable from those caused by FMD (see FMD chapter) or by other Vesicular Diseases. In contrast to the mortality seen in piglets during FMD outbreaks (with the appearance of "tiger heart" lesions), there is no mortality in pigs due to SVD.

8. IMMUNE RESPONSE

a. Natural infection

There is some immunity acquired after infection, but similar to most enterovirus infections, the duration of this immunity tends to be short.

b. Immunization

Although there have been some effective experimental vaccines, there are no commercial vaccines available.

9. DIAGNOSIS

a. Field diagnosis

SVD should be part of any full investigation of a vesicular disease condition in pigs, even in areas of the world (like the Americas) where this disease has never been reported.

b. Laboratory diagnosis

i. Samples

As in the case of any vesicular disease, the ideal samples for SVD diagnosis are: vesicular fluid, vesicular epithelium, scrapings or deep swabs of erosions, nasal swabs (for RRT-PCR), and acute and convalescent serum samples.

ii. Laboratory procedures

Laboratory procedures for the diagnosis of SVD or other vesicular diseases include tests for the detection of virus (virus isolation), detection of viral antigens (antigen-capture ELISA, complement fixation), detection of antibodies (virus neutralization, agar-gel immunodifussion, antibody detection ELISA), or the new technology of nucleic acid detection (real-time reverse-transcriptase polymerase chain reaction – RRT-PCR).

Note: Since pigs are the only species susceptible to all vesicular diseases, there is a need to test any pig with vesicular lesions for at least FMD, VS and SVD (vesicular exanthema of swine –VES – is mostly a historical disease and tests should only be performed if the history may indicate contact of swine with marine mammals or fish). This is particularly important in countries where SVD is endemic. The lesson of large FMD outbreaks in Taiwan in 1997, wrongly assumed to be due to SVD due to its clinical presence only in pigs, should be a reminder of the importance of a full differential diagnosis in all cases of vesicular diseases in swine (see FMD chapter).

10. PREVENTION AND CONTROL

The most effective way to prevent the introduction of SVD into previously free areas is by implementing proper import and movement controls for animals, as well as for animal products including the collection and destruction of foreign garbage at ports of entry. Controls of feeding of meat-containing garbage are also important mitigations in the prevention and control of this disease. Given the absence of an effective vaccine, SVD controls depend on testing herds and slaughtering those testing positive to SVDV infection.

11. GUIDE TO THE LITERATURE

1. BROWN, F., TALBOT P. and BURROWS R. 1973. Antigenic Differences between Isolates of Swine Vesicular Disease Virus and their Relationship to Coxsackie B5 Virus. *Nature.* 245:315 – 316.
2. GRAVES, J. H. 1973. Serological Relationship of Swine Vesicular Disease Virus and Coxsackie B5 Virus. *Nature.* 245:314 – 315.
3. MEBUS, C., et al. 1997. Survival of several porcine viruses in different Spanish dry-cured meat products. *Food Chem* Vol. 59 (4):555-559.
4. NARDELLI, L., LODETTI, E., GUALANDI, G.L., BURROWS, R., GOODRIDGE, D., BROWN, F. and CARTWRIGHT, B. 1968. A foot and mouth disease syndrome in pigs caused by an enterovirus. *Nature.* 219:1275 – 1276.
5. WOOLDRIDGE, M., HARTNETT, E., COX, A. and SEAMAN, M. 2006. Quantitative risk assessment case study: smuggled meats as disease vectors *Rev. sci. tech.* Off. int. Epiz. 25 (1):105-117.
6. ZHANG, G., HAYDON, D.T., KNOWLES, N.J., and McCAULEY, J.W. 1999. Molecular evolution of swine vesicular disease virus. *Journal of General Virology.* 80:639-651.

Alfonso Torres, DVM, MS, PhD, Associate Dean for Public Policy, College of Veterinary Medicine, Cornell University, Ithaca, NY, 14852, at97@cornell.edu

43

TROPICAL THEILERIOSIS

1. NAME
Mediterranean Coast fever

2. DEFINITION
Tropical theileriosis is a tickborne disease, caused by *Theileria annulata*, and characterized by anemia, jaundice, ill thrift, dyspnea, and hemorrhagic diarrhea.

3. ETIOLOGY
The causative agent is a protozoan parasite, *Theileria annulata*. Sporozoites in the saliva of the *Hyalomma* tick vector are inoculated during feeding and these sporozoites invade macrophages, and B-lymphocytes to a lesser extent. In these cells they mature to become macroschizonts, which eventually differentiate into microschizonts, leading to merozoites that invade erythrocytes. Within the erythrocytes, the organisms are usually referred to as piroplasms. Ticks feeding on infected animals will ingest the infected erthyrocyte and there is further development to a sporozoite within the salivary gland of the tick.

4. HOST RANGE

a. Domestic and wild animals
Cattle are the principal victims of tropical theileriosis. The taurine breeds of cattle, introduced into endemic areas, have a much more severe form of the disease than do indigenous zebu cattle. The parasite will also infect buffalo and camels.

b. Humans
There is no evidence that humans are susceptible to infection with *Theileria annulata*.

5. EPIDEMIOLOGY

a. Transmission
The disease is transmitted by ticks of the *Hyalomma* spp.

b. Incubation period
As is true of many tickborne pathogens, incubation period is 10-25 days.

c. Morbidity

The number of animals sick is influenced by a variety of factors, including tick infectivity, tick challenge, pre-existing immunity in the host herd.

d. Mortality

In susceptible cattle introduced into endemic areas, mortality is high, between 40-90%. Mortality among local breeds is much lower.

6. CLINICAL SIGNS

In the acute disease, death occurs 15-25 days after infection. Clinical signs might include pale mucous membranes (anemia) or jaundice, as the piroplasms will precipitate destruction of red blood cells. During the stage when there is great production of macroschizonts within macrophages, there could be enlarged lymph nodes, and a generalized loss of condition and muscle wasting due to massive release of cytokines from infected cells. Hemorrhagic diarrhea may be present in the terminal stages.

7. POST- MORTEM LESIONS

a. Gross

There are no specific lesions associated with tropical theileriosis. Shortly after infection, the lymph node draining the site of the tick bite will be enlarged. At the time of severe clinical disease or death, anemia, jaundice, enlarged lymph nodes, muscle wasting, pulmonary edema, and hemorrhagic enterocolitis may all be present.

b. Key microscopic

Unlike the better known theilerial disease, East Coast fever (caused by *Theileria parva*), which is characterized by a marked lymphoproliferative response due to massive infection of lymphocytes, tropical theileriosis (caused by *Theileria annulata*) is primarily a macrophage infection. It is thought that extensive infection of macrophages stimulates a huge outpouring of cytokines, predominantly TNFα, which accounts for many of the lesions seen. Macroschizonts may be seen in infected macrophage type cells within various organs.

8. IMMUNE RESPONSE

a. Natural infection

Following infection, animals are not susceptible to development of disease. Cross-protection between strains is considered good. Protective immunity is cell-mediated.

b. Immunization

Animals can be protected by through an intentional low dosing of sporozoites. Alternatively, vaccination using an attenuated product of schizont-containing cells is protective.

9. DIAGNOSIS

a. Field diagnosis

In an endemic area, anemia, poor condition, and muscle wasting, and the presence of *Amblyomma* ticks, should put tropical theileriosis on the list of rule-outs. Laboratory confirmation is essential.

b. Laboratory diagnosis

i. Samples

Blood smears, impression smears of lymph nodes, fixed or frozen specimens of lymphoid tissue.

ii. Laboratory procedures

Giemsa-stained samples to visualize the macroschizonts within impression smears or tissues are helpful. Polymerase chain reaction is sensitive. An immunofluorescent antibody (IFA) test is used but antibodies may not be detectable for prolonged periods after infection.

10. PREVENTION AND CONTROL

Endemic stability between tick challenge and host immunity is operative in many areas but this can be easily disturbed through a number of factors, including sudden increase or decrease in tick burdens, failure to vaccinate, or introduction of naïve stock. The greatest problems occur when nonnative cattle, especially of the taurine breeds, are introduced into an area of endemicity. Protecting these animals through vaccination is practicable.

Anti-protozoal drugs, such as buparvaquone, parvaquone, and halofuginone, can all be used in the face of clinical infection and will diminish clinical signs, but do not sterilize the infection.

11. GUIDE TO THE LITERATURE

1. AHMED, J.S. and MEHLHORN, H. 1999. Review: the cellular basis of the immunity to and immunopathogenesis of tropical theileriosis. *Parasitol Res* 85:539-549.

2. BENIWAL, R.K., NICHANI, A.K., SHARMA, R.D., RAKHA, N.K., SURI, D. and SARUP, S. 1997. Responses in animals vaccinated with the *Theileria annulata* (Hisar) cell culture vaccine. 1997. *Trop Anim Health Prod* 29(4 Suppl):109S-113S.
3. GLASS, E.J. 2001. The balance between protective immunity and pathogenesis in tropical theileriosis: what we need to know to design effective vaccines for the future. *Res Vet Sci* 70:71-75.
4. MARTIN-SANCHEZ, J., VISERAS, J., ADROHER, F.J. and GARCIA-FERNANDEZ, P. 1999. Nested polymerase chain reaction for detection of *Theileria annulata* and comparison with conventional diagnostic techniques: its use in epidemiology studies. *Parasitol Res* 85:243-245.
5. PRESTON, P.M., HALL, F.R., GLASS, E.J., CAMPBELL, J.D.M., DARGHOUTH, M.A., AHMED, J.S., SHIELS, B.R., SPOONER, R.L., JOHGEJAN, F. and BROWN, C.G.D. 1999. Innate and adaptive immune responses co-operate to protect cattle against *Theileria annulata*. *Parasitology Today* 15:268-274.

Corrie Brown, DVM, PhD, Department of Pathology, College of Veterinary Medicine, University of Georgia, Athens, GA 30602-7388, corbrown@uga.edu

44

TRYPANOSOMIASIS (AFRICAN)

1. NAME
African Animal Trypanosomosis (AAT), African Animal Trypanosomiasis, nagana

2. DEFINITION
African animal trypanosomosis is a chronic, debilitating disease of cattle and other livestock, usually with anemia as the primary disorder, and caused by 1 of 3 species of protozoa–*Trypanosoma congolense*, *T. brucei brucei*, and *T. vivax*. All are transmitted biologically through the bite of an infected tsetse fly; additionally *T. vivax* can be mechanically transmitted by other biting insects.

3. ETIOLOGY
The 3 trypanosomes responsible for the disease complex known as African Animal Trypanosomosis are categorized into 3 different subgenera within the family Trypanosomatidae, genus *Trypanosoma*. *Trypanosoma congolense* is in the subgenus *Nannomonas*; *T. vivax* is within the subgenus *Duttonella* and *T. brucei brucei* in the subgenus *Trypanozoon*. As a fly bites, metacyclic forms of the trypanosome are injected into the skin; these then differentiate into bloodstream forms, and cause a systemic infection. All 3 are extracellular pathogens. *T. vivax* and *T. congolense* tend to remain in the bloodstream, whereas *T. brucei brucei* is often found extravascularly as it invades tissues. Although the morphology of the 3 varies somewhat, all are elongated, with undulating membranes, a flagellum, prominent nucleus, and kinetoplast.

4. HOST RANGE
a. *Domestic animals*
The number of domestic species that can be infected with the agents of AAT is considerable. Cattle, sheep, goats, pigs, horses, camels, dogs, cats, and monkeys are all susceptible to natural infection. In the laboratory, rats, mice, guinea pigs, and rabbits can all be infected.

b. *Wild animals*
At least 30 species of wild animals can harbor the agents of AAT. Wild ruminants, especially, are noted for their ability to act as reservoirs.

c. Humans

Tsetse transmitted trypanosomosis, or human sleeping sickness is a serious and widespread problem in many parts of Africa, and is caused by the trypanosomes *T. brucei gambiense* and *T. brucei rhodesiense*. There is often confusion regarding cross transference of trypanosomes between animals and humans. Both of these agents have wild animals as their reservoir. *T. brucei rhodesiense* can also cause disease in cattle, and cattle may sometimes serve as one of the reservoirs for human infection. *T. brucei gambiense* will also infect pigs and is a zoonotic disease as well.

5. EPIDEMIOLOGY

a. Transmission

The disease spreads from animal to animal through the bite of an infected tsetse fly (primarily 3 species, *Glossina morsitans*, *G. palpalis*, and *G. fusca*). *Trypanosoma vivax*, in addition, can be spread by biting flies, especially tabanids.

b. Incubation period

The time from the bite of the tsetse fly to clinical disease is quite variable, from 4 days up to 5 weeks.

c. Morbidity

When the tsetse challenge is high, morbidity can also be expected to be very high.

d. Mortality

Unless treated, most animals infected with any 1 of these 3 species of trypanosomes, will eventually die of the disease.

6. CLINICAL SIGNS

AAT is most important in cattle, but clinical disease can be seen in sheep, goats, horses, and pigs. In all, clinical signs have some similarities, with intermittent fever, emaciation, anemia, loss of body condition, and reduced productivity. The earliest sign is a chancre at the site of the fly bite, but usually this period passes unnoticed. The most prominent clinical feature is anemia, as the parasite decreases production of red blood cells. This is the most common reason that animals are unable to function normally. In addition, there is an elevated, but ineffective, production of antibodies. As covered in more detail below (see: Natural infection), these trypanosomes can change their surface coat easily, so that each time the antibody response can effectively thwart the level of parasitemia, there is a change to a new surface protein. As a result, the immune system of the animal is in a constant state of stimulation, and there may be clinical signs related to hyperproduction of antibodies and the deposition of antigen-antibody complexes. These can include lymphadenopathy, or there may be vascular or renal compromise.

7. POST-MORTEM LESIONS

a. *Gross*

Lesions are usually related to the development of anemia and also to the prolonged nature of the antigen-antibody response. Consequently there may be pallor of mucous membranes, serous atrophy of fat, perhaps pulmonary edema, subcutaneous edema, ascites, hydrothorax (due to hypoxia and hypoproteinemia), and enlarged lymph nodes and spleen. Alternatively, lymphoid tissue may be atrophic in the terminal phases of the disease, as the animal becomes too debilitated to mount any immune response. With infection with *T. brucei brucei*, which has a tendency to invade tissues in addition to causing anemia, there may also be inflammation and/or degeneration of multiple tissues.

b. *Key microscopic*

The histologic changes are not specific for trypanosomosis. Bone marrow may reveal a nonregenerative anemia, and in the kidney there might be a chronic membranous glomerulonephritis.

8. IMMUNE RESPONSE

a. *Natural infection*

The trypanosomes possess a unique method for evading the immune response through sequential switching of the variable surface glycoprotein (VSG). This VSG covers the surface of the parasite, and as soon as sufficient numbers are present in the blood to generate an antibody response to the VSG, there is a genetic switch to a new VSG, rendering the antibodies produced useless to combat the new variant. It is estimated that trypanosomes have about 1,000 of these VSGs available, so conquering the infection through immune activation is not possible. In addition, AAT causes an immunosuppression to other disease agents, either through preoccupation of the immune response to the trypanosomes or through general debilitation, or both.

b. *Immunization*

There is no vaccine available for these trypanosome diseases.

9. DIAGNOSIS

a. Field diagnosis

Severe anemia in an endemic area should prompt collection of specimens for laboratory diagnosis.

b. Laboratory diagnosis

i. Samples

Serum, blood, thick and thin blood smears, and lymph node aspirates should all be collected.

ii. Laboratory procedures

Trypanosomiasis may be diagnosed by microscopic examination of wet and stained thick or thin blood films, impression smears from lymph node aspirates, PCR, ELISA, or IFA.

10. PREVENTION AND CONTROL

Trypanosomiasis is a major constraint to ruminant livestock production in many areas of Africa. It has often been stated that the livestock-carrying capacity of much of the African continent could be tremendously expanded if AAT could be eliminated. The chronic and debilitating nature of the disease causes economic losses to be underestimated. Animals are unthrifty for prolonged periods and may linger without any kind of productivity for months to years prior to succumbing to the infection.

There are several trypanocidal drugs available for use in infected animals. Many have experienced problems with drug resistance in the parasites. Currently, isometamidium chloride (trade names: Samorin, Trypamidium, M&B 4180A) is the most widely used chemoprophylactic drug, and diaminazine aceturate (trade name Berenil) is used most commonly for chemotherapeutics.

Breaking the cycle of transmission requires decreasing the number of tsetse flies or the parasite load within the tsetse flies. Brush clearing was a method used years ago to destroy the habitat of the tsetse flies. Because the reservoir is in wild hoofed stock, killing of these animals has also been used in the past. Insecticide treatment of tsetse habitats is also a method that can be utilized. All of these methods have negative environmental impacts. More recently, pheromone-loaded tsetse traps have been used to attract and kill the tsetse flies. To decrease the prevalence of AAT in endemic areas, there is a general consensus that tsetse eradication is the most effective solution. This will require a large-scale, well-coordinated effort.

A great deal of research has been expended to explore the development of trypanotolerant animals through selective breeding. Cattle breeds that have been in the endemic area for centuries, such as the West African Shorthorn and the N'Dama, also from West Africa, have an innate resistance to the development of disease. They are infected by the tsetse flies but do not exhibit clinical disease. These cattle are often also quite small in stature and not optimally productive. Cross-breeding experiments aimed at generating resistant strains of cattle that are also productive have been promising.

11. GUIDE TO THE LITERATURE

1. ANONYMOUS. 2004. Trypanosomosis. Chapter 2.3.15, In: *Manual of Diagnostic Tests and Vaccines for Terrestrial Animals*, Office of International Epizootics, OIE. www.oie.int.
2. NAESSENS, J. 2006. Bovine trypanotolerance: A natural ability to prevent severe anaemia and haemophagocytic syndrome? *International Journal for Parasitology,* 36:521-528.
3. OMAMO, S.W. and D'IETEREN, G.D.M. 2003. Managing animal trypanosomosis in Africa: issues and options. *Rev. sci. tech. Off. Int. Epiz.*, 22(3):989-1002.
4. MARE, C.J. 1998. African animal trypanosomiasis. In: *Foreign Animal Diseases,* 6th ed., WW Buisch, JL Hyde, CA Mebus, eds., USAHA, Richmond, Virgina, pp. 29-40.

See Part IV for photos.

Corrie Brown, DVM, PhD, Department of Pathology, College of Veterinary Medicine, University of Georgia, Athens, GA 30602-7388, corbrown@uga.edu

45

VENEZUELAN EQUINE ENCEPHALOMYELITIS (VEE)

1. NAME
Peste loca, Venezuelan encephalitis

2. DEFINITION
Venezuelan equine encephalomyelitis (VEE) is a zoonotic, mosquito-borne, viral disease affecting both Equidae and humans. In equines, infection may produce either an acute, fulminating disease that terminates in death or recovery without development of encephalitic signs, or the more classical disease with progressive clinical encephalitis. In human beings, a flu-like syndrome predominates with an accompanying high fever and frontal headache. A wide variety of hosts and vectors may be infected.

3. ETIOLOGY
The etiologic agent of VEE is an alphavirus in the family Togaviridae (formerly the group A arboviruses). The virions are 60-75 nm in diameter and have an essential lipid membrane.

Only minor antigenic variations exist among different VEE virus isolates. Six subtypes (I, II, III, IV, V, and VI) have been identified within the VEE complex. Within subtype I, only 2 (A/B and C) of the 5 variants (A/B through F) have been associated with epizootic activity in equines. The other variants (I-D through I-F) and subtypes (II through VI) generally have been associated with nonequine, sylvatic, or enzootic activity, although limited focal outbreaks of equine disease have been reported occasionally from variant I-E virus strains. Infection with one variant or vaccination with attenuated virus generally results in production of neutralizing antibodies and cross-protection of varying duration to infection with the other subtypes and variants.

4. HOST RANGE
a. Domestic animals
As the name implies, horses are the main species affected by the virus. Several other domestic animals, including cattle, swine, and dogs have demonstrated serologic and virologic evidence of infection, but generally without clinical signs during epizootics of VEE. However, with the possible exception of human beings, no evidence exists to date to incriminate any animal species other than equines as prime amplifiers of VEE epizootics. A wide variety of laboratory

animals are susceptible in varying degrees to both enzootic and epizootic variants of VEE virus.

b. Wild animals

Rodents are important in the maintenance of the virus in nature.

c. Humans

Human infections resulting from bites of infected mosquitoes occur tangentially to equine infections. Transmission also can occur by exposure to aerosolized infectious material and has been demonstrated after numerous laboratory accidents. In human beings, a 'flu-like' syndrome predominates accompanied by high fever and frontal headache. Human deaths may occur in the young or the aged.

5. EPIDEMIOLOGY

a. Transmission

Sylvatic or Enzootic Cycle

Variants I-D through I-F and subtypes II through VI of VEE virus are invariably associated with a sylvatic or enzootic cycle in which rodent-mosquito transmission occurs; human beings and horses are involved only incidentally in this cycle. Although sylvatic variants and subtypes are pathogenic for human beings and have caused occasional epidemics with a few deaths, these variants and subtypes are normally nonpathogenic for horses. However, it is clear that under certain ideal but undefined conditions sylvatic I-E variant viruses are pathogenic to horses. Generally, the sylvatic virus tends to cycle in rodents in areas where a highly efficient vector such as *Culex* (*Melanoconion*) spp. is found.

Epizootic Cycle

No known epizootic virus variant (I-AB or I-C) has ever been shown to cycle enzootically in rodents. Recent molecular genetic studies of VEE virus isolates indicate very close phylogenetic relationships between epizootic variant I-AB and I-C isolates and sylvatic variant I-D isolates. The results support the hypothesis that epizootic VEE virus variants emerge from sylvatic variant I-D viruses. During epizootics of VEE, many species of mosquitoes and possibly other hematophagous insects are involved in the explosive movement of the outbreak. Horses are the most important amplifiers of VEE virus during epizootics owing to the extremely high viremias that they develop and the large numbers of hematophagous insects that can feed on an animal of such size. Human infections occur tangentially to equine infections, but in spite of moderately high viremia levels, human beings probably do not contribute significantly to the maintenance and movement of an epizootic wave. The maintenance cycle of equine virulent (epizootic) VEE virus during the interepizootic period and the origin of epizootic VEE virus variants are unknown. Recent studies have indicated the persistence of epizootic variant I-C virus strains in a cryptic postepidemic transmission cycle since 1995. Efficient vectors of epizootic VEE include mosquitoes of the genera *Aedes*, *Anopheles*, *Culex*, *Deinocerites*, *Mansonia*, and *Psorophora*.

b. Incubation period

The incubation period from the inoculation of the virus until the febrile response generally is 0.5-2 days but may be as long as 5 days, depending on the virus strain or quantity of virus in the inoculum. Typically, detectable viremia occurs concurrently with the onset of fever and persists for 2-4 days. Onset of encephalitic signs occurs 4.5-5 days after infection at a time when circulating virus is disappearing, neutralizing antibody is first detectable, and body temperature is returning to a normal range.

c. Morbidity

Estimated case morbidity rates vary from 50-100% in some areas to 10-40% in other areas. Infection rates with or without clinical signs may be as high as 90%. Occasional limited outbreaks of clinical encephalomyelitis in horses from infection with sylvatic variant I-E viruses have been documented and VEE attenuated virus, strain TC-83, has been used effectively to stop recent outbreaks.

d. Mortality

Epizootic VEE due to virus variants I-A/B and I-C may be highly fatal. Mortality rates vary from 50-90%. In most cases, infection with sylvatic or enzootic VEE virus variants I-D, I-E, or I-F, or subtypes II, III, IV, V, and VI is considered to be nonlethal for equines.

6. CLINICAL SIGNS

In equines, VEE virus infection may be expressed as (a) subclinical, with no overt signs; (b) moderate, and characterized primarily by anorexia, high fever, and depression; (c) severe, but nonfatal, and characterized by anorexia, high fever, stupor, weakness, staggering, blindness, and, occasionally, permanent neurologic sequelae; or, (d) fatal, with the same clinical signs. Not all fatal cases of VEE in equines are accompanied by definite neurologic signs. In general, 2 forms of the disease exist: (a) the fulminating form in which signs of generalized, acute, febrile disease predominate, and (b) the encephalitic form in which the more impressive signs of central nervous system (CNS) involvement usually dominate.

An incubation period of 0.5-5 days precedes a rise in body temperature to 39°-41°C (103°-105°F) which is accompanied by a hard, rapid pulse and depression. The onset of VEE virus infection is insidious, with fever, inappetence, and mild excitability being among the earliest clinical signs of disease. Frequently, a rapid progression ensues with depression, weakness, and ataxia followed by overt signs of encephalitis such as muscle spasms, chewing movements, incoordination, and convulsions. Early encephalitic signs include loss of both cutaneous neck reflexes and visual responsiveness; diarrhea and colic also may develop. Some animals may stand quietly in their surroundings whereas others may wander aimlessly or press their heads against solid objects. A braced stance or circling may occur late in the disease. A characteristic paddling motion of the limbs may be observed with lateral recumbency.

The course of the disease may be interrupted at any point by recovery or prostration and death. The course of the disease may be rapid with death ensuing within hours after the observation of the first clinical manifestations of encephalitis (during epizootics, reports of sudden death are not uncommon), or more protracted with dehydration and extreme loss of weight occurring before an encephalitic death or recovery.

7. POST-MORTEM LESIONS

a. *Gross*

The macroscopic appearance of the CNS of horses inoculated with VEE virus varies from no visible lesion to extensive necrosis and hemorrhages. Lesions reported in other tissues have been too variable to be of any diagnostic significance.

b. *Key microscopic*

Histopathologic lesions consistent with a diagnosis of VEE are a diffuse necrotizing meningoencephalitis that ranges from a slight perivascular mixed cellular reaction to marked vascular necrosis with hemorrhages, gliosis, and frank neuronal necrosis. Lesions are usually most severe in the cerebral cortex and become progressively less severe toward the cauda equina. The degree and severity of the CNS lesions vary with the progression and duration of the clinical signs. Necrotic lesions may involve the adrenal cortex, liver, myocardium, and the walls of small and medium blood vessels.

8. IMMUNE RESPONSE

a. Natural infection

After natural infection with one strain, there are sufficient neutralizing antibodies to cross-protect against infection with another subtype or variant. But periods of protection may vary.

b. Immunization

Both attenuated and inactivated vaccines are available. After immunization with one strain, there are usually enough antibodies to protect against another strain, but the period of protection may vary.

9. DIAGNOSIS

a. Field diagnosis

A field diagnosis of VEE can rarely be made unless an epizootic of encephalitic disease is in progress and a prior etiologic diagnosis of VEE has been made. Seasonality of the disease and association with large populations of mosquitoes would suggest a diagnosis of arboviral encephalomyelitis. The initial signs of VEE may go undetected. When signs of encephalitis predominate, the disease in equines is indistinguishable from other arboviral equine encephalomyelitides, such as eastern equine encephalomyelitis (EEE) and western equine encephalomyclitis (WEE). In contrast to EEE or WEE, hcrd morbidity and mortality with VEE are high.

b. Laboratory diagnosis

i. Samples

Specimens for diagnostics include heparinized blood, serum (paired [acute and convalescent] sera, if the animal survives), and half the brain and piece of pancreas unfixed, and a complete set of tissues in 10% formalin if the animal dies.

ii. Laboratory procedures

A specific diagnosis can be made only by laboratory procedures, including virus isolation or demonstration of a specific rise in hemagglutination-inhibiting or neutralizing antibody with paired (acute and convalescent) sera. Frequently, animals die before a convalescent serum sample can be obtained. Experimental studies and field experiences have shown that viremia terminates before signs of clinical encephalitis are exhibited by VEE virus-infected equines. In this case, the highest probability of successful viral isolation is obtained by taking sera from other horses with marked elevations of body temperature that are in the vicinity of the encephalitic horse. Virus also may be isolated from brain, pancreas, or whole blood of dead or dying horses, but with a lower frequency of success. Virus is isolated by the intracranial inoculation of suckling mice or in various cell culture systems.

10. PREVENTION AND CONTROL

During epizootics, restriction of horse movement between the epizootic zone and noninfected areas is important to control the spread of VEE. Because of the high levels of VEE viremia in equines (>$10^{5.5}$ infectious virions/ml of blood), introduction of infected animals into noninfected areas readily establishes new foci of infection. However, control of movement of the equine population is not sufficient to curb the spread of VEE.

Mosquito control measures such as aerial spraying with ultralow volumes of insecticides have been instituted during epizootics. Vector control in the absence of other control measures can do little more than slow the spread of VEE and decrease its severity in the human population. Physical disruption or insecticide treatment of the aqueous larval habitats also can reduce adult mosquito populations.

For adequate epizootic control, the preceding measures must be accompanied by a large-scale equine immunization program. An attenuated VEE virus vaccine has been used in many areas of the Americas, both to combat the disease during an epizootic and to administer preventive vaccination in nonepizootic zones where a high risk of infection is present. Although preexisting neutralizing antibodies to EEE and WEE viruses may interfere with the neutralizing antibody response to VEE virus vaccination, the interference is not sufficient to affect immunity. Simultaneous vaccination with VEE attenuated virus and EEE-WEE inactivated virus produces a VEE neutralizing antibody response equivalent to that produced with the administration of attenuated VEE vaccine alone. An inactivated trivalent EEE-VEE-WEE virus vaccine has been shown to immunize equine recipients effectively.

11. GUIDE TO THE LITERATURE

1. GREENE, I.P., PAESSLER, S., AUSTGEN, L., ANISHCHENKO, M., BRAULT, A.C., BOWEN, R.A., et al. 2005. Envelope glycoprotein mutations mediate equine amplification and virulence of epizootic Venezuelan equine encephalitis virus. *J. Virol.* 79:9128-9133.
2. NAVARRO, J-C., MEDINA, G., VASQUEZ, C., COFFEY, L.L., WANG, E., SUAREZ, A., et al. 2005. Postepizootic persistence of Venezuelan equine encephalitis virus, Venezuela. *Emerg. Infect. Dis.* 11:1907-1915.

3. POWERS, A.M., OBERSTE, M.S., BRAULT, A.C., RICO-HESSE, R., SCHMURA, S.M., SMITH, J.F., et al. 1997. Repeated emergence of epidemic/epizootic Venezuelan equine encephalitis from a single genotype of enzootic subtype I-D virus. *J. Virol.* 71: 6697-6705.
4. WALTON, T.E. 1981. Venezuelan, Eastern and Western encephalomyelitis. In: *Virus Diseases of Food Animals* Vol. II: Disease Monographs. EPJ Gibbs, ed., San Francisco: Academic Press, pp. 587-625.
5. WALTON, T.E., and GRAYSON, M.A. 1989. Venezuelan Equine Encephalomyelitis. In *The Arboviruses: Epidemiology and Ecology*. Vol. 4, TP Monath, ed., Boca Raton, FL: CRC Press, Inc., pp. 203-231.

Thomas E. Walton, DVM, PhD, 5365 N Scottsdale Rd., Eloy, AZ 85231, vetmedfed@comcast.net

46

VESICULAR EXANTHEMA OF SWINE

1. NAME:

Vesicular Exanthema of Swine, Exantema Vesicular del Cerdo (Sp)

2. DEFINITION

Vesicular Exanthema of Swine (VES) is a disease caused by the infection of pigs with a number of marine-dwelling caliciviruses. As a vesicular disease, VES is characterized by fever, vesicular lesions, and subsequent erosions of the epithelium of the mouth, snout, feet, or teats.

3. ETIOLOGY

VES virus (VESV) belongs to the genus *Vesivirus* within the family Caliciviridae. There were some 13 strains of VESV identified during the period of 1932-59 when this disease was eradicated from the U.S. In 1972 a group of serologically indistinguishable caliciviruses, identified as the San Miguel Sea Lion Viruses (or SMS viruses), were isolated from marine mammals and from opaleye fish (*Girella nigricans*). This SMS group of viruses includes 16 different serogroups. Seven additional serologically distinct caliciviruses have been identified in cetaceans, cattle, primates, skunks, walrus, reptiles and humans, raising the number of distinct VESV-related viruses to at least 35.

VESV is stable in meat products even when decomposed at 7°C over a 2-4 week period. It was observed that heavily contaminated farms remained infectious for several months.

4. HOST RANGE

a. Domestic animals

This is a disease primarily of domestic swine, which most likely originated from the exposure of swine to infected marine mammals or fish, and transmitted from pig to pig through contaminated meat products and contact.

b. Wild animals

The primary reservoirs of these VESV-related caliciviruses are the marine mammals of the Pacific coast of the U.S. They include the California sea lion (*Zalophus californianus*), Stellar sea lions (*Eumetopias jubatus*), northern fur seals (*Callorhinus ursinus*), northern elephant seals (*Mirounga agustrirostris*), captive Atlantic bottle-nosed dolphins (*Tursiops truncates*), Pacific dolphins

(*Tursiops gillii*), Pacific walruses (*Odobenus rosmarus*), and opaleye fish (*Girella nigricans*). Some VES or SMS related caliciviruses have also been identified in snakes, toads, skunks, cattle, and 5 species of primates.

c. Humans

While there have been reports of laboratory infections with these VESV-related caliciviruses, this disease is not considered to be zoonotic.

5. EPIDEMIOLOGY

a. Transmission

It is thought that VES originated from the consumption by pigs of contaminated tissue from infected marine mammals. Once a vesicular disease condition was established, the VESV spread horizontally from pig to pig due to the shedding of large amounts of virus after vesicles ruptures. It was thought at the time that pigs could become infected by ingestion of contaminated feeds or by percutaneous infection through skin abrasions in a contaminated environment. The disease spread across the U.S. through the distribution of contaminated meat scraps used to feed pigs all along the railroad lines. This fact led to the promulgation of Federal laws in the U.S. dealing with the collection and decontamination of garbage as well as meat-containing garbage feeding of pigs. Recurrence of VES in some farms in California during the 1950s suggested the possibility of a carrier state. However, no detailed studies were conducted to elucidate this possibility.

b. Incubation period

Clinical signs of VES develop within 18-72 hours of natural exposure.

c. Morbidity

High morbidity rates were observed during the VES outbreaks, nearing 100%.

d. Mortality

Mortality was very low due to VES in pigs.

6. CLINICAL SIGNS

As a vesicular disease, VES produces clinical signs similar to those caused by foot-and-mouth disease (FMD, see FMD chapter) or by other vesicular diseases (vesicular stomatitis (VS) and/or swine vesicular disease (SVD)).

7. POST- MORTEM LESIONS

Lesions in pigs due to VES are indistinguishable from those caused by FMD (see FMD chapter) or by other vesicular diseases.

8. IMMUNE RESPONSE

a. *Natural infection*

Serological responses to VESV were observed in infected pigs. They tended to be serotype-specific, and no vaccines were available at the time of VES outbreaks.

b. *Immunization*

No commercial vaccines are available or desirable.

9. DIAGNOSIS

a. *Field diagnosis*

VES should be part of any full investigation of a Vesicular Disease condition in pigs only if there is a history of potential contact of pigs with marine mammals or fish, or in cases of vesicular disease in pigs not associated with FMD, VS, or SVD viruses.

b. *Laboratory diagnosis*

i. Samples

As in the case of any vesicular disease, the ideal samples for VES diagnosis are: vesicular fluid, vesicular epithelium, scrapings or deep swabs of erosions.

ii. Laboratory procedures

Laboratory procedures for the diagnosis of VES or other vesicular diseases include tests for the detection of virus (virus isolation), detection of viral antigens (antigen-capture ELISA, Complement Fixation), or detection of antibodies (virus neutralization, antibody detection ELISA).

Note: Given the historical aspects of VES, this infection should be part of the differential diagnosis of any Vesicular Disease in pigs where the history may indicate contact of swine with marine mammals of fish, or where a vesicular disease of pigs has tested negative to FMD, VS, or SVD viruses.

10. PREVENTION AND CONTROL

The most effective way to prevent the introduction of VES into the swine population is by preventing the feeding of pigs with meats derived from dead marine mammals or fish. VES was eventually controlled by an aggressive depopulation of affected farms, movement controls of pigs and strict controls on the feeding of meat-containing garbage to pigs.

VES has only been recognized in the U.S. The disease was first recognized in California in 1932 and remained there until 1952 when it spread to 31 other states. A state of emergency was declared in that year implementing control and eradication procedures that led to the final eradication in 1956. After the discovery in 1972 of the SMS viruses, these agents were detected in free-living and captive marine mammals along the Pacific coastline of the U.S. and on islands in the Eastern Pacific.

Due to its historical presence only in the U.S., and of its absence for half a century, VES is no longer included on the list of internationally reportable disease by the World Organization for Animal Health (OIE).

11. GUIDE TO THE LITERATURE

1. BURROUGHS, N., DOEL, T. and BROWN, F. 1978. Relationship of San Miguel sea lion virus to other members of the calicivirus group. *Intervirol.* 10:51-59.
2. CHU, R.M., MOORE, D.M. and CONROY J.D. 1979. Experimental swine vesicular disease, pathology and immunofluorescence studies. *Can J Comp Med.* 43:29–38.
3. GELBERG, H.B. and LEWIS, R.M. 1982. The pathogenesis of VESV and SMSV in swine. *Vet. Path.* 19:424-443.
4. NARDELLI, L., LODETTI, E., GUALANDI, G.L., BURROWS, R., GOODRIDGE, D., BROWN, F. and CARTWIGHT, B.. 1968. A foot-and-mouth disease syndrome in pigs caused by an enterovirus. *Nature*, 219:1275-1276.
5. NEILL, J.D., MEYER, R.F., and SEAL, B.S. 1995. Genetic relatedness of the caliciviruses: San Miguel sea lion and vesicular exanthema of swine viruses constitute a single genotype within the Caliciviridae. *J. Virol.* 69:4484-4488.
6. SMITH, A.W., AKERS, T.G., MADIN, S.H. and VEDROS, N.A.. 1973. San Miguel sea lion virus isolation, preliminary characterization and relationship to vesicular exanthema of swine virus. *Nature*, 244:108-10.
7. SMITH, A.W. and AKERS, T.G. 1976. Vesicular exanthema of swine. *J Am Vet Med Assoc.* 169:700-703.
8. SMITH, A.W. 1998. Calicivirus Emergence from Ocean Reservoirs: Zoonotic and Interspecies Movements. *Emerging Infectious Diseases*. 4:13-20.
9. ZHANG, G., HAYDON, D.T., KNOWLES, N.J. and McCAULEY, J.W. 1999. Molecular evolution of swine vesicular disease virus. *Journal of General Virology.* 80:639-651.

Alfonso Torres, DVM, MS, PhD, Associate Dean for Public Policy, College of Veterinary Medicine, Cornell University, Ithaca, NY, 14852, at97@cornell.edu

47

VESICULAR STOMATITIS

1. NAME

Vesicular stomatitis (VS)

2. DEFINITION

Vesicular stomatitis is an insect-transmitted acute disease, primarily of horses, cattle, and pigs, with less frequent infections of sheep and goats, and characterized by the formation of vesicles, on the snout, mouth, udder, and feet.

3. ETIOLOGY

The causative agent is vesicular stomatitis virus (VSV), a member of the genus *Vesiculovirus* in the family Rhabdoviridae. As with other rhabdoviruses, the virion is enveloped and consists of a bullet-shaped ribonucleoprotein containing a single-stranded negative sense strand of RNA, of approximately 11kb. Four vesicular stomatitis viruses have been found that cause outbreaks of vesicular disease in domestic animals. These viruses have been serologically classified as New Jersey (VSNJV), Indiana 1(VSIV), Indiana 2 (VSIV2) (e.g. Cocal virus; COCV), and Indiana 3 (VSIV3) (e.g. Alagoas virus; VSAV). VSNJV and VSIV are endemic from northern South America and throughout Central America into Southern Mexico, and they occur sporadically in northern Mexico and the U.S. In contrast, VSIV-2 and VSIV-3 cause vesicular disease only in South America (Brazil, Argentina, and Trinidad).

Other vesiculoviruses have been found in various geographic regions infecting insects, or, in some cases, humans. These include Piry (Brazil), Chandipura (India), Isfahan (Iran), and Calchaqui (Argentina). Although some of these viruses cause vesicles when experimentally inoculated in cattle, none of them are known to naturally cause vesicular disease outbreaks in livestock.

4. HOST RANGE

a. Domestic animals

Of the domestic animals, horses, donkeys, and mules, as well as cattle and swine are affected most severely. South American camelids are susceptible. Sheep and goats tend to be quite resistant to clinical disease, and infection of these species is rare. In the laboratory, a variety of rodents, both domestic and wild, can be infected with the virus and suffer clinical disease that is not vesicular in nature but rather is a systemic disease with central nervous system involvement.

Interestingly, VSV is not found in the blood of domestic animals (no viremia) during experimental or natural infection.

b. Wild animals

Evidence of infection has been found in a wide range of wild animals, including white-tailed deer (*Odocoileus virginianus*), many species of bats, howler monkeys (*Alloata palliatta*), field mice (*Peromyscus* spp. and *Reithrodontomys* spp.), cotton rats (*Sigmodon* spp.), and feral swine (*Sus scrofa*). Wild birds have also been found to have antibodies to VSV. The role of these species in the maintenance of the natural cycle of VSV remains unclear as none of these species is capable of sustained viremia necessary to serve as reservoirs.

c. Humans

Humans can be infected by contact or by aerosol. Most infections are seen in laboratory workers, veterinarians, and animal caretakers. The incubation period in humans is 24-48 hours. There is an influenza-like illness with fever, headache, muscular aches, and vesicles forming in the oral cavity. The period of illness in humans may last 4-7 days. The Piry, Isfahan, and Chandipura vesiculovirus strains are much more virulent for humans than are the New Jersey and Indiana strains. Human to human transmission has not been reported.

5. EPIDEMIOLOGY

a. Virus ecology and natural cycle

In endemic areas, VSV seems to exist in stable natural cycles with the same virus genotype persisting in specific ecological zones over long periods of time. Although the natural maintenance cycle is poorly understood, insects such as sand flies (*Lutzomyia* spp) carry the virus in these endemic foci and transmit the infection to susceptible hosts (e.g. feral swine) without causing clinical signs. The presence of neutralizing antibodies is evidence of widespread infection among wild animals living in endemic foci.

Ecological conditions that favor VSV endemic cycle include forested tropical and subtropical areas with humid and sub-humid conditions and defined rainy and dry seasons. Domestic animals living near these areas become infected during the season when insect populations increase (generally the rainy season).

In the U.S., VS occurs endemically in wildlife in selected regions of the southeast with no reported cases in domestic animals in the last 40 years. In contrast, in the southwest, sporadic VS outbreaks occur at 7-10 year intervals. Genetic evidence and geo-spatial analyses strongly suggest that the viral strains causing these outbreaks originate in endemic areas in southern Mexico. The factors determining these outbreak cycles and the mechanisms of incursion into the U.S. remain unknown.

b. Transmission

There are a number of methods of transmission, including insect vectors, contact, and fomites. Sand flies (*Lutzomyia spp*), black flies (*Simuliidae*), biting midges, or culicoides (Culicoides spp) are thought to play an important role in moving the disease during outbreaks. Replication of the virus and transovarial transmission occurs in the above mentioned insect groups. However, experimental data suggest that transovarial tansmission is not efficient enough for long-term maintenance of the virus cycle. Therefore it has been suggested that a yet to be identified natural reservoir might exist. Horizontal fly-to-fly transmission has been experimentally demonstrated with black flies co-feeding on various mammalian hosts including horses, swine and cattle. This might play an important role during VSV outbreaks and might help explain insect transmission in the absence of viremic hosts.

Vesicular stomatitis virus has also been isolated from other insects including mosquitoes (*Culex* spp. and *Aedes* spp.), and even domestic flies (*Musca domestica*). The importance of insect vectors is reflected in the fact that most outbreaks disappear after the first frost. Virus present on mucosal surfaces where vesicles have ruptured is an important source of virus for contact transmission. Milking machines may spread the virus from cow to cow if not properly decontaminated after being on an infected cow. There is no virus shed in urine, feces, or milk.

c. Incubation period

After experimental infection, clinical signs usually appear within 24-72 hours. Under natural conditions, the incubation period is probably slightly longer.

d. Morbidity

The morbidity depends largely on whether or not the area is endemic for VS, on the numbers of infected insect vectors, and, once established in a herd, the closeness of animal contact. In nonendemic areas, when the disease occurs, it is usually in outbreak form, with morbidity approaching 40-60%. Recent serological studies of VSV outbreaks in non-endemic areas of the U.S. showed a significant number of subclinical infections. In endemic areas, the morbidity is likely to be much lower since most of the natural VSV infections are subclinical. Serological studies in endemic areas show antibody prevalence nearing 100% in adult dairy cattle.

e. *Mortality*

The disease rarely causes death on its own. However, in the case of horses that develop laminitis, animals may be euthanized. Similarly, in cattle that go on to develop mastitis, culling is common.

6. CLINICAL SIGNS

In experimental infections in the tongue, fever is a common clinical sign. In natural infections fever is not always present. Blanching areas that develop into vesicles may develop in the mouth (gum, lips, togue), snout, coronary bands of the feet, or in the teats of lactating cattle, sows, and mares. Excess salivation, gaiting, lameness, and signs of pain during milking are common. Lesions on the coronary band in horses and cattle can lead to laminitis and even hoof loss. In swine, loss of the claws is not uncommon. Teat lesions in dairy cattle predispose to bacterial mastitis. Severe tongue lesions can lead to malnourishment and dehydration. With proper care, animals usually recover fully from clinical disease, but loss of hooves and severe mastitis might lead to culling.

7. POST-MORTEM LESIONS

a. *Gross*

Vesicles, which begin as small blanched areas that progress to fluid-filled bulges, rupture to reveal raw, eroded, or ulcerated areas which can be appreciated clinically or grossly. Vesicles in this disease are no different from the other vesicular diseases (FMD, SVD) in that they rupture quickly so the vesicular stage is often missed, but a raw ulcerated area remains longer. These erosions and ulcers are most frequently seen on the lips, muzzle, and tongue in all species. In cattle, they are especially common on the teats and in the interdigital spaces of the feet. In swine, lesions are more likely to occur on the snout as well as on the coronary band and interdigital spaces. There are no significant lesions recorded in any internal organs. Recent studies have shown that in experimentally infected cattle, horses, and swine, virus distribution is restricted to the site of inoculation in the skin and to the lymph nodes draining those areas. Viremia, if it occurs, is very rare.

b. *Key microscopic*

Unfortunately, the microscopic lesions associated with the vesicles are no different than those of any other vesicular diseases. Histologically, early after infection, microvesicles form in the stratum spinosum of the epithelia that later coalesce into larger vesicles. There is progressive degeneration and then necrosis as vesicles become larger and filled with fluid (48-72 hpi). A mixed population of inflammatory cells infiltrates the lesion site and some distinct VSV-antigen containing cells with dendritic cell-like morphology are observed in the lower layers of the dermis. Virus can also be detected in lymph nodes draining the lesions sites, usually associated with cells of dendritic morphology.

8. IMMUNE RESPONSE

a. *Natural infection*

Subsequent to subclinical infection in endemic areas, animals develop an immune response that is detectable by relatively modest neutralizing antibody titers. Although neutralizing antibodies may persist for several years, reinfection can occur, with lesions forming at the site of insect inoculation. It is probable that the neutralizing antibodies cannot prevent the formation of this first lesion. Low levels of neutralizing activity have often been reported in naïve animals in non-endemic areas. This non-specific neutralizing activity is poorly understood and does not prevent infection.

b. *Immunization*

Several experimental vaccines have been tested against VSV. A killed bivalent, oil-adjuvanted VSNJ and VSIV-1 vaccine has been successfully used in parts of South America to control VSV occurrence in endemic regions. These vaccines induce high titers of neutralizing antibodies and prevent disease after experimental challenge. Experimental subunit vaccines and attenuated viruses have been shown to protect against clinical disease in experimental challenge. In the U.S. there are no commercially available vaccines for vesicular stomatitis.

9. DIAGNOSIS

a. *Field diagnosis*

Presence of a vesicular disease as noted in the field should always raise suspicion of foot-and-mouth disease and laboratory confirmation is essential. Although there is the belief that FMD causes more severe disease and affects larger number of animals in a herd, this is not always the case. In susceptible populations, attack rates of 80-100% have been observed for VSV. Likewise some FMDV strains can cause mild clinical disease in a small number of less-susceptible animals (e.g. strain O-UK/2001 in sheep). For these reasons, it is of outmost importance to submit samples for diagnostic laboratory confirmation. One clear difference between VSV and FMDV is that the latter does not cause disease in horses, whereas VSV does. So, if a vesicular disease includes a large number of horses, the suspicion of VS increases.

b. Laboratory diagnosis

i. Samples

Vesicular fluid or epithelium from ruptured vesicles both have ample amounts of virus and are good specimens for detecting the agent. Swabs of lesions are also suitable samples. Even dried-up lesions can be useful samples as virus remains active in these lesions, and even if the virus is no longer infectious, viral antigens or RNA can be detected. For serology, collect blood for serum. Serum can be extremely valuable as recent infections are evidenced by the presence of VSV-specific IgM.

ii. Laboratory procedures

Tests for the presence of virus include virus isolation, antigen-capture ELISA, and RT-PCR (both conventional and real-time).

Tests for the presence of antibody include virus neutralization, IgM capture ELISA, competitive ELISA and CF test.

10. PREVENTION AND CONTROL

Prevention of the disease can be effected through insect control, which is usually quite difficult for the species of animals involved. During an outbreak, bringing animals indoors will help to decrease transmission. Also, movement of animals should be limited during an outbreak, perhaps through quarantines of affected premises. In dairy operations, infected animals should be milked last and milking machines should be thoroughly decontaminated.

Because of its clinical similarity with FMD it is important to rapidly diagnose outbreaks of VSV. Control and eradication might prove difficult due to the existence of natural cycles involving insect and wild animal species.

11. GUIDE TO THE LITERATURE

1. MEAD, D.G., RAMBERG, F.B., BESSELSEN, D.G. and MARE, C.J. 2000. Transmission of vesicular stomatitis virus from infected to noninfected black flies co-feeding on nonviremic deer mice. *Science,* 287:485-7.
2. MEAD, D.G., GRAY, E.W., NOBLET, R., MURPHY, M.D., HOWERTH, E.W., and STALLKNECHT, D.E. 2004. Biological transmission of vesicular stomatitis virus (New Jersey serotype) by *Simulium vittatum* (Diptera: Simuliidae) to domestic swine (Sus scrofa). *J.Med.Entomol*, 41(1):78-82.
3. MEBUS, C.A. 1998. Vesicular stomatitis. In: *Foreign Animal Diseases,* 6th ed., WW Buisch, JL Hyde, CA Mebus, eds., Richmond (VA): United States Animal Health Association, pp. 419-423.

4. RAINWATER-LOVETT, K., PAUSZEK, S.J., KELLEY, W.N. and RODRIGUEZ, L.L. 2007. Molecular epidemiology of vesicular stomatitis New Jersey virus from the 2004-2005 U.S. outbreak indicates a common origin with Mexican strains. *J.Gen.Virol.*, 88(Pt 7):2042-51.
5. RODRIGUEZ, L.L. and NICHOL, S.T. 1999. Vesicular stomatitis viruses. In: *Encyclopedia of Virology* 2nd ed. RG Webster, A Granoff, eds. Academic Press; London, pp. 1910-1919.
6. SCHERER, C.F.C., O'DONNELL, V., GOLDE, W.T., GREGG, D., ESTES, D.M. and RODRIGUEZ, L.L. 2007. Vesicular stomatitis New Jersey virus (VSNJV) infects keratinocytes and is restricted to lesion sites and local lymph nodes in the bovine, a natural host. *Veterinary Research*, 38:375-390.
7. SCHMITT, B. 2002. Vesicular stomatitis. *Vet Clin Food Anim.*, 18:453-459.

See Part IV for photo.

Luis Rodriguez, DVM, PhD, Plum Island Animal Disease Center, USDA-ARS, luis.rodriguez@ars.usda.gov

48

WESSELSBRON DISEASE

1. NAME
Wesselsbron disease

2. DEFINITION
Wesselsbron disease is an acute arthropod-borne viral disease of sheep, cattle, and goats, characterized primarily by abortion and neonatal deaths.

3. ETIOLOGY
Wesselsbron virus is an arthropod-borne virus and is a member of the subgroup Flavivirus in the Family Togaviridae. The virus has hemagglutinating properties.

4. HOST RANGE
a. Domestic animals
A wide range of animals is susceptible to infection with Wesselsbron virus, including sheep, goats, cattle, horses, pigs, camels, and farmed ostrich. Clinical disease is rarely seen except in pregnant sheep, goats, and occasionally cattle.

b. Wild animals
Many game species are susceptible to infection with Wesselsbron virus but disease is rare. Rodents can be infected with Wesselsbron virus.

c. Humans
Subclinical infection is more common than disease. Those who become ill experience a flu-like illness with muscle pain that may persist for a week or more. Infections have been recorded in laboratory workers exposed to aerosols of virus.

5. EPIDEMIOLOGY
a. Transmission
The virus is transmitted by a variety of aedine mosquitoes. Outbreaks of the disease typically occur at irregular intervals, in regions just at the periphery of endemic zones, and after periods of heavy rains. Animals in these drier areas are typically not exposed to mosquitoes carrying the infection and so are fully susceptible when heavy rains allow more mosquitoes to become established.

b. *Incubation period*
Subsequent to infection, a fever and clinical disease will develop usually in 3-6 days.

c. Morbidity
Serologic evidence indicates that infection is far more common than disease. Adult non-pregnant animals usually show no clinical signs. Morbidity among newborn lambs and kids may be very high, especially if mosquito levels are also high.

d. *Mortality*
In susceptible areas, there may be 20-30% mortality among newborn lambs and kids. Mortality in adults is rare.

6. CLINICAL SIGNS
Illness is noted only in pregnant animals and newborns. Neonatal mortality in lambs and kids is high – animals are febrile, weak, and may have mucoid diarrhea and a rough hair coat. Occasionally edema of the head is present as well. Icterus is noted frequently. Abortion is most common in sheep. Congenital defects of the CNS are often noted, as is arthrogryposis.

7. POST- MORTEM LESIONS
a. *Gross*
The livers of aborted fetuses and newborns dying of the disease may be large and orange-brown, with multifocal pinpoint white areas (multifocal hepatic necrosis). Icterus is often present. Grossly, the CNS may have defects, such as hydranencephaly or microcephaly. Additionally, arthrogryposis is seen in some aborted animals and newborns.

b. *Key microscopic*
The liver is the most rewarding organ to examine as the virus is decidedly hepatotropic. Key features include multifocal hepatic necrosis, with mixed mononuclear cell infiltrates and rare intranuclear eosinophilic inclusions. It should be noted that the liver damage is much less severe than that seen in the livers of animals infected with Rift Valley fever.

8. IMMUNE RESPONSE
a. *Natural infection*
A humoral response begins by 4 days postinfection. Antibodies are protective.

b. Immunization

A modified live vaccine is available but is associated with significant reproductive morbidity, including fetal malformations, dystocia, and hydrops amnii.

9. DIAGNOSIS

a. Field diagnosis

The presence of abortions and neonatal mortality associated with liver disease in an endemic area after a period of rainfall would all raise the suspicion of Wesselsbron. However, Rift Valley fever may look very similar in presentation.

b. Laboratory diagnosis

i. Samples:

Liver, blood, and brain are the most useful tissues to collect for diagnosis.

ii. Laboratory procedures:

Serology tests may be performed – HI, ELISA, and CFT, as well as virus isolation in suckling mouse brain inoculation. All tests need to differentiate Wesselsbron from other related flaviviruses.

10. PREVENTION AND CONTROL

Vaccination has to be done carefully, avoiding pregnant animals. In the laboratory, prevention of human infection requires the use of personal protective equipment and good laboratory technique.

11. GUIDE TO THE LITERATURE

1. COETZER, J.A.W. and THEODORIDIS, A. 1982. Clinical and pathological studies in adult sheep and goats experimentally infected with Wesselsbron disease virus. *Onderstepoort J Vet Res*, 49:19-22.
2. MUSHI, E.Z., BINTA, M.G. and RABOROKGWE, M. 1998. Wesselsbron disease virus associated with abortions in goats in Botswana. *J Vet Diagn Invest*, 10:191.
3. THEODORIDIS, A. and COETZER, J.A.W. 1980. Wesselsbron disease: virological and serological studies in experimentally infected sheep and goats. *Onderstepoort J Vet Res*, 47:221-229.
4. VAN DER LUGT, J.J., COETZER, J.A.W., SMIT, M.M.E. and CILLIERS, C. 1995. The diagnosis of Wesselsbron disease in a new-born lamb by immunohistochemical staining of viral antigen. *Onderstepoort J Vet Res*, 62:143-146.

5. WILLIAMS, R., SCHOEMAN, M., VAN WYK, A., ROOS, K., and JOSEMANS, E.J. 1997. Comparison of ELISA and HI for detection of antibodies against Wesselsbron disease virus. *Onderstepoort J Vet Res*, 64:245-250.

Corrie Brown, DVM, PhD, Department of Pathology, College of Veterinary Medicine, University of Georgia, Athens, GA 30602-7388, corbrown@uga.edu

IV

PHOTOGRAPHS

1. AFRICAN HORSE SICKNESS

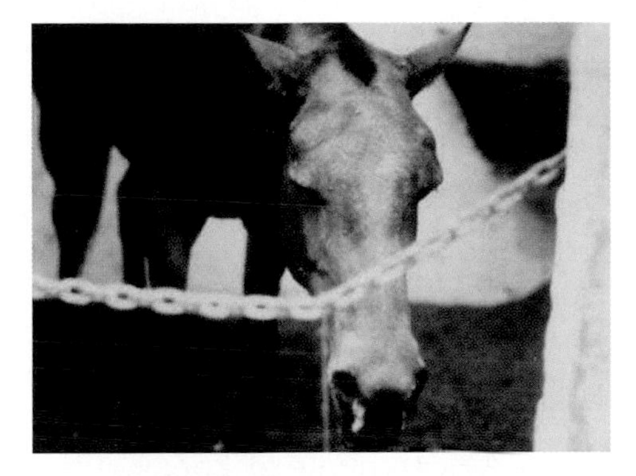

1A - African horse sickness - Horse showing depressions and marked bilateral supraorbital edema

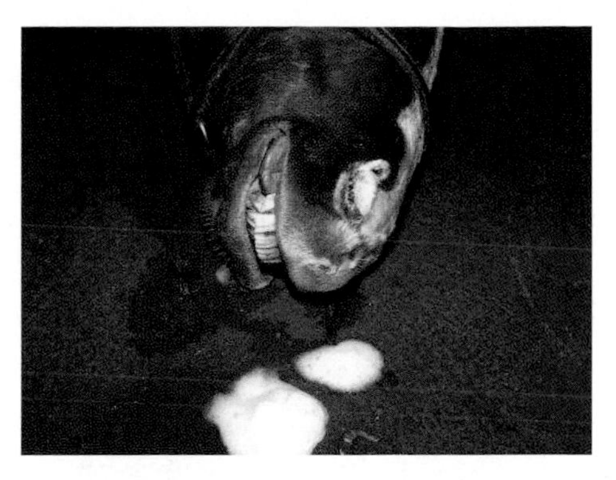

1B - African horse sickness - Foam from nares due to severe pulmonary edema

1C - African horse sickness - Lungs fill with edema, which is often first present at the ventral margins

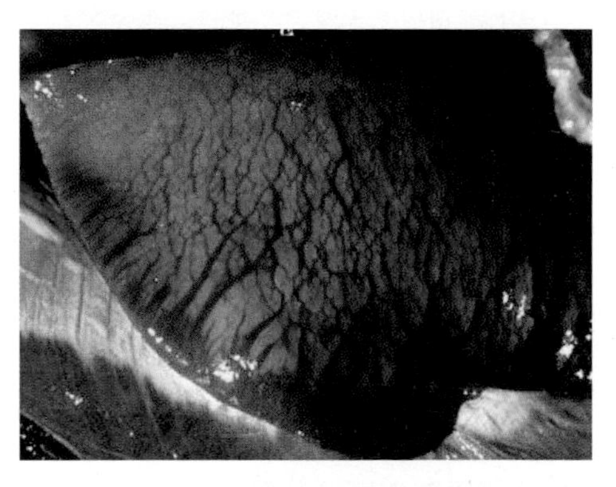

1D - African horse sickness - Overwhelming pulmonary edema and hydrothorax

(cont'd... 1. AFRICAN HORSE SICKNESS)

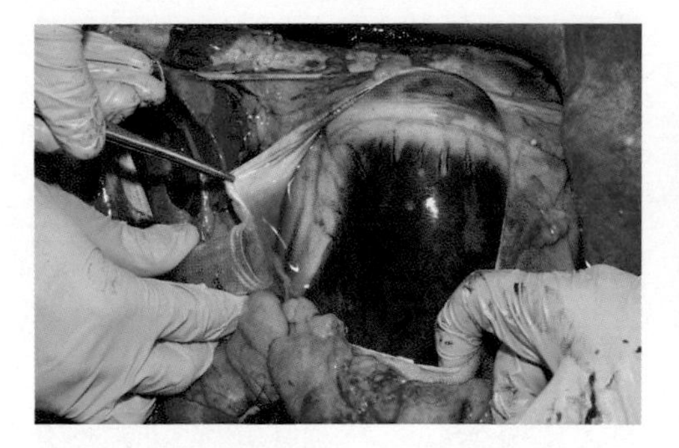

1E - African horse sickness - Fluid accumulates in the pericardial sac (hydropericardium)

1F - African horse sickness - Dramatic example of edema in the intermuscular fascia of the neck.

1G - African horse sickness - Petechiation on the serosal surface of intestine.

2. AFRICAN SWINE FEVER

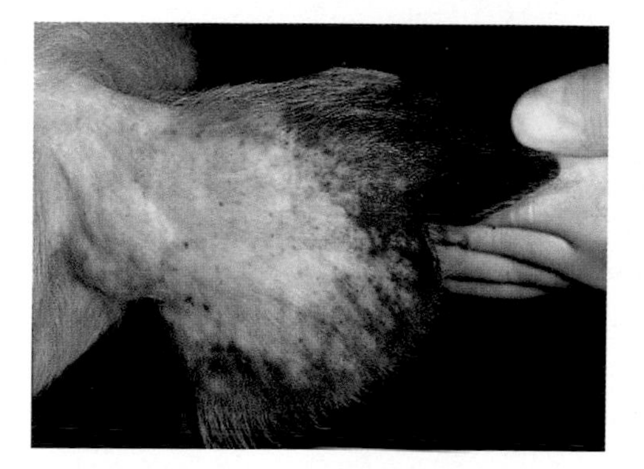

2A - African swine fever - Ecchymotic hemorrhages on the skin.

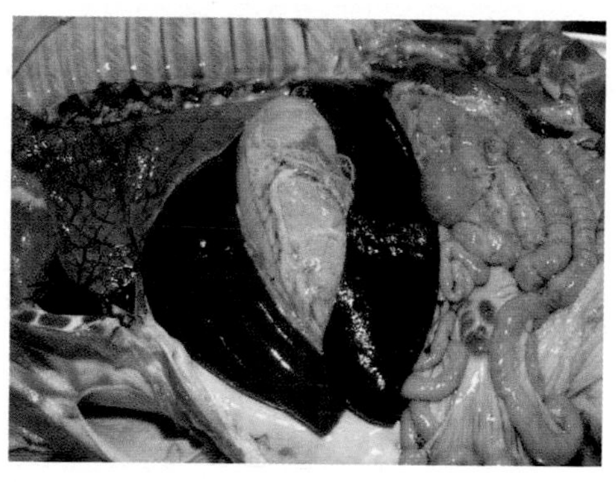

2B - African swine fever - A greatly enlarged dark red to black spleen from a pig infected with a highly virulent ASFV isolate.

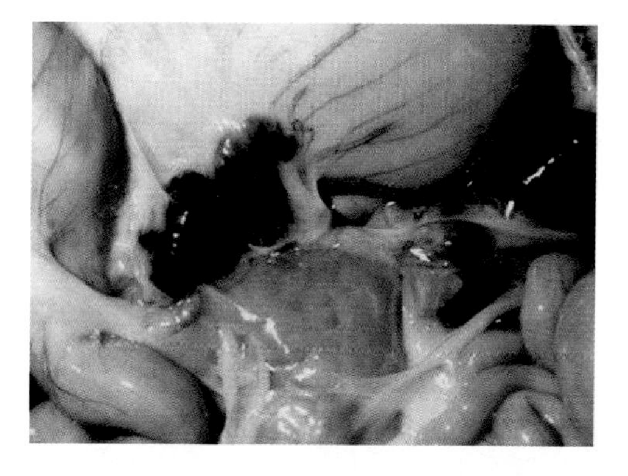

2C - African swine fever - Very enlarged dark red (hemorrhagic) gastrohepatic lymph nodes from a pig infected with a highly virulent isolate of ASFV.

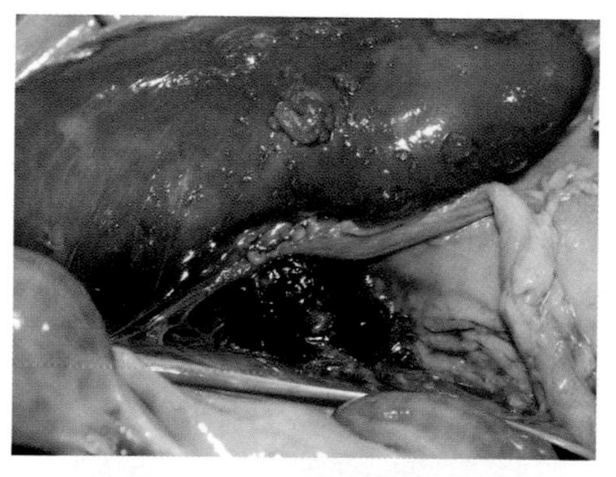

2D - African swine fever - Enlarged dark red renal lymph nodes.

(cont'd... 2. AFRICAN SWINE FEVER)

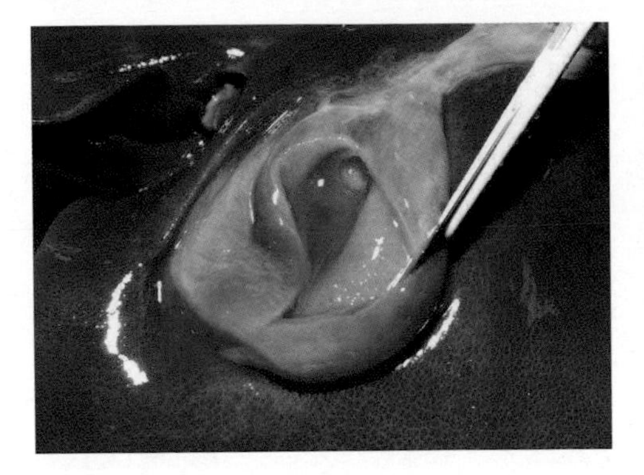

2E - African swine fever - Edema of the gall bladder.

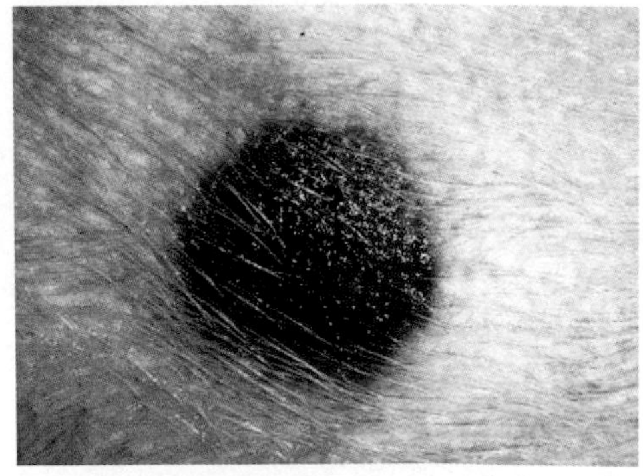

2F - African swine fever - Necrosis of the skin in chronic ASF can be focal; the areas begin as raised hyperemic areas and progress to areas of necrosis.

3. AKABANE DISEASE

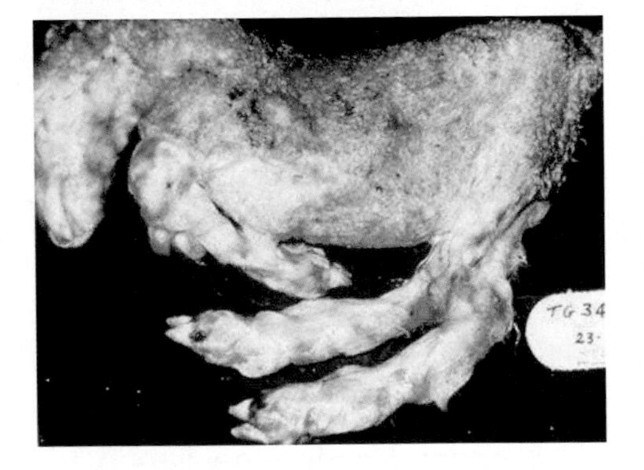

3A - Akabane disease - Arthrogryposis is the most frequently observed lesion.

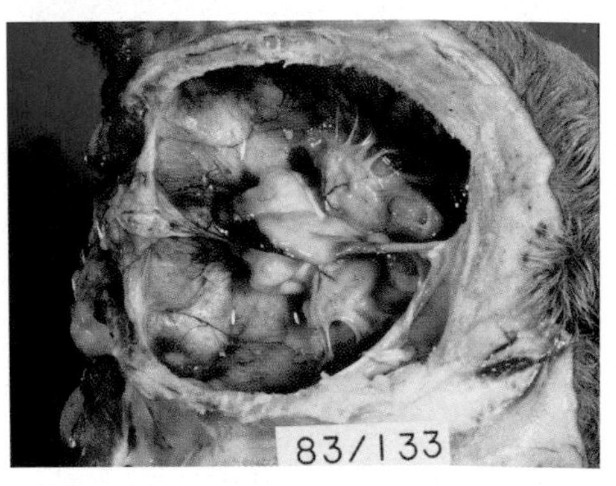

3B - Akabane disease - Hydranencephaly.

5. AVIAN INFLUENZA

5A - Highly pathogenic avian influenza - Cyanosis and edema of the comb and wattles.

5B - Highly pathogenic avian influenza - Cyanotic comb of an infected chicken on the left compared to a normal chicken on the right

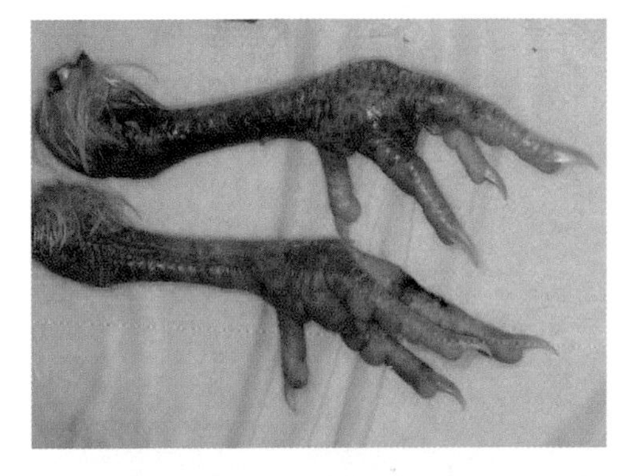

5C - Highly pathogenic avian influenza - Congestion and petechiae in the skin on the hocks and shanks.

5D - Highly pathogenic avian influenza - Severe subcutaneous edema, neck.

(cont'd… 5. AVIAN INFLUENZA)

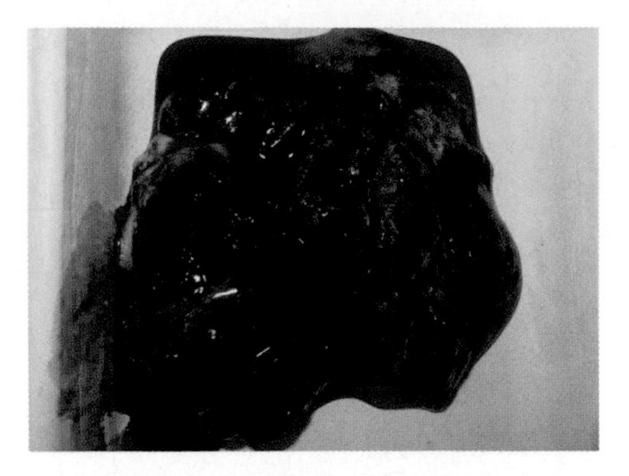

5E - Highly pathogenic avian influenza - Lungs can be very edematous and congested.

6. BABESIOSIS

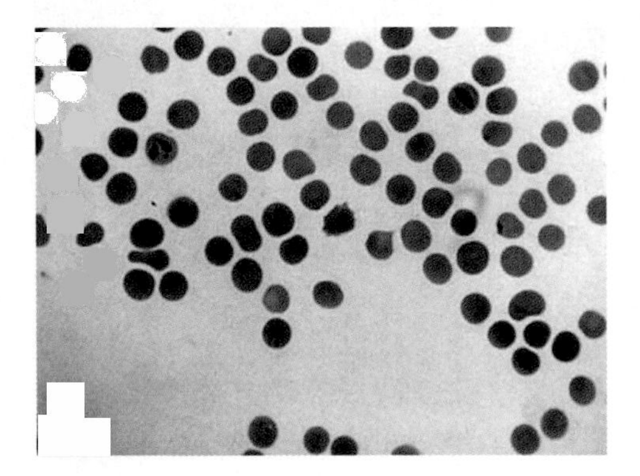

6A - Babesiosis - *Babesia bigemina* in erythrocytes.

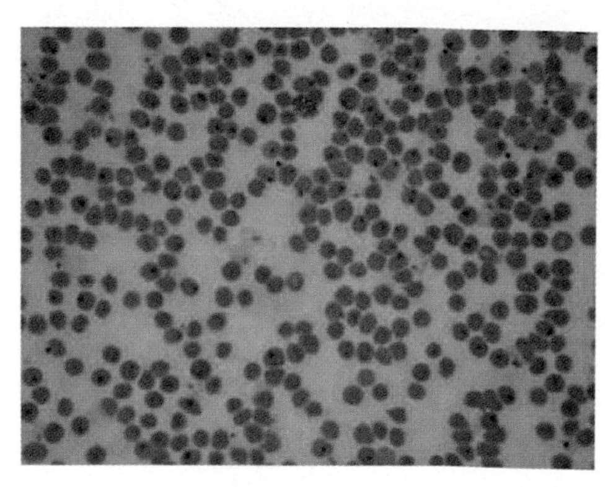

6B - Babesiosis - *Babesia bovis* in erythrocytes.

(cont'd… 6. BABESIOSIS)

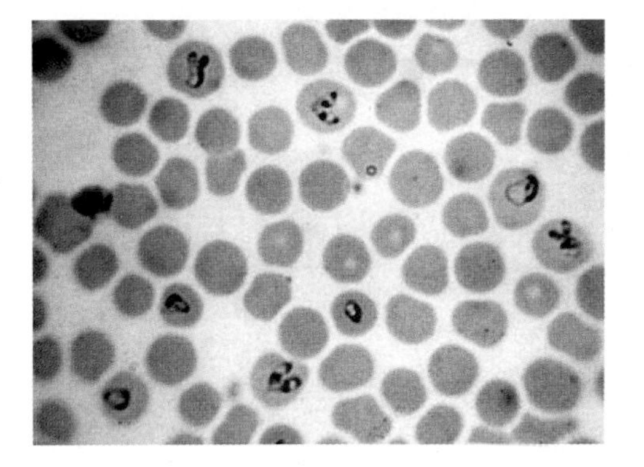

6C - Babesiosis - *Babesia caballi* in erythrocytes.

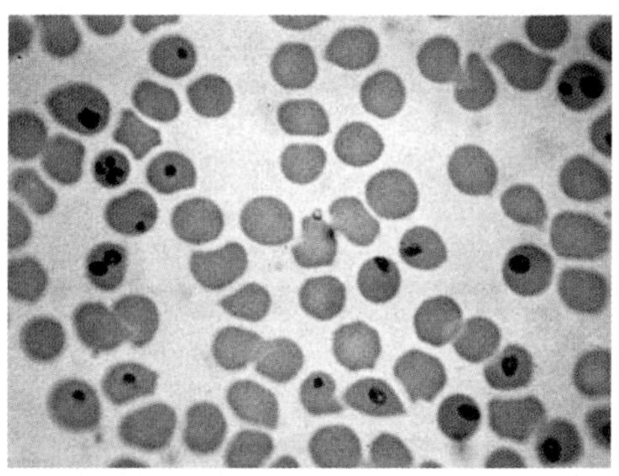

6E - Babesiosis - *Babesia equi* in erythrocytes.

6F - Babesiosis - Opisthotonus with cerebral babesiosis, *Babesia bovis* infection. (Photo courtesy of Dr. David Driemeier)

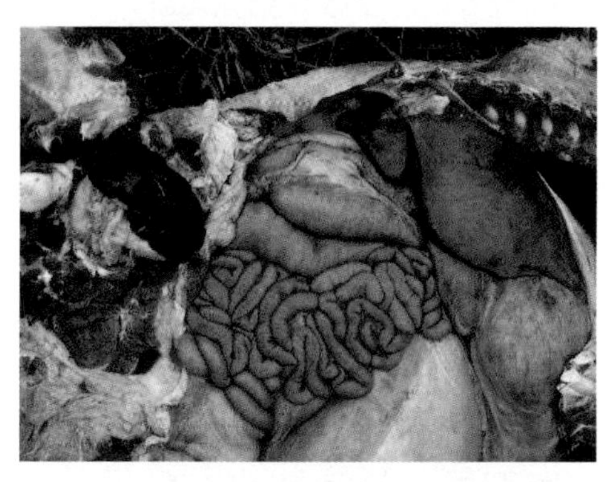

6G - Babesiosis - *Babesia bovis* infection - enlarged liver, serosa dark-pink due to hemogloblin imbibition, kidney with black discoloration due to hemoglobinuria, icterus of adipose tissue, and distension of the gall bladder.

(cont'd... 6. BABESIOSIS)

6H - Babesiosis - *Babesia bovis* infection - urinary bladder contains dark red urine due to hemoglobinuria (red water).

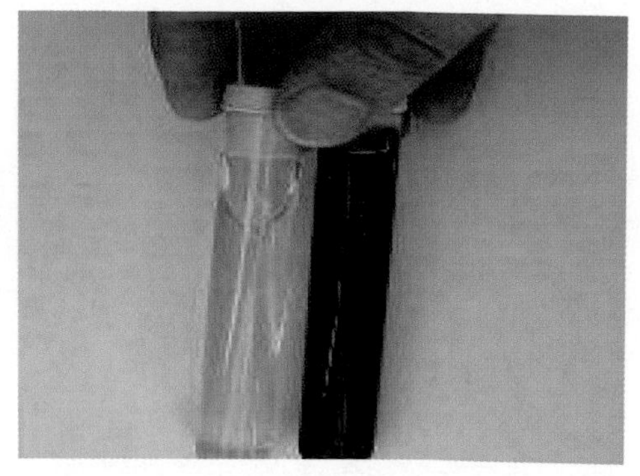

6I - Babesiosis - Babesiosis - Test tube at right with urine from an affected cow with babesiosis compared with normal urine in the tube at left.

6J - Babesiosis - splenomegaly and white pulp follicular hyperplasia.

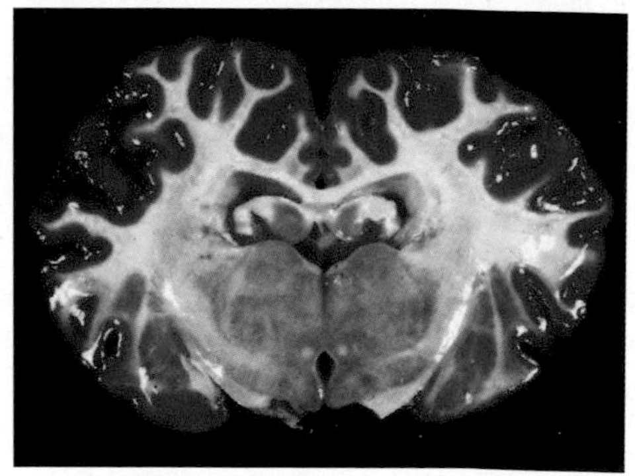

6K - Babesiosis - Cherry pink discoloration in the gray matter of the telencecephalic cortex. This gross lesion is a hallmark of cerebral babesiosis in cattle.

(cont'd… 6. BABESIOSIS)

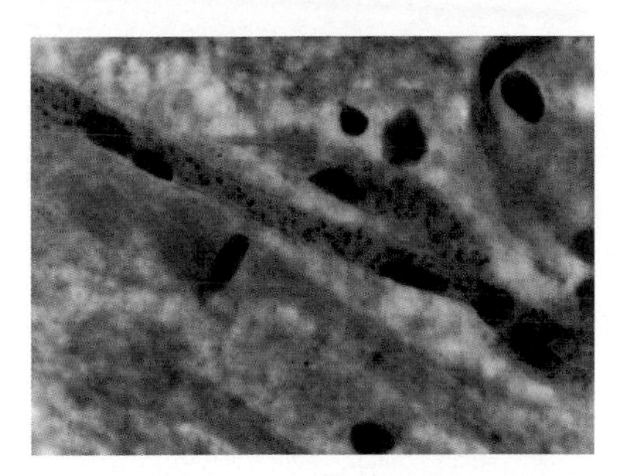

6L - Babesiosis - Field diagnosis - brain (telencephalic cortex) squash smear, showing capillaries distended with red blood cells parasitized by *B. bovis*.

7. BLUETONGUE

7A - Bluetongue - Ulceration at mucous membranes, secondary to vascular compromise (Photo courtesy of Dr. Mariano Domingo)

7B - Bluetongue - Petechial hemorrhages on hard palate (Photo courtesy of Dr. Mariano Domingo)

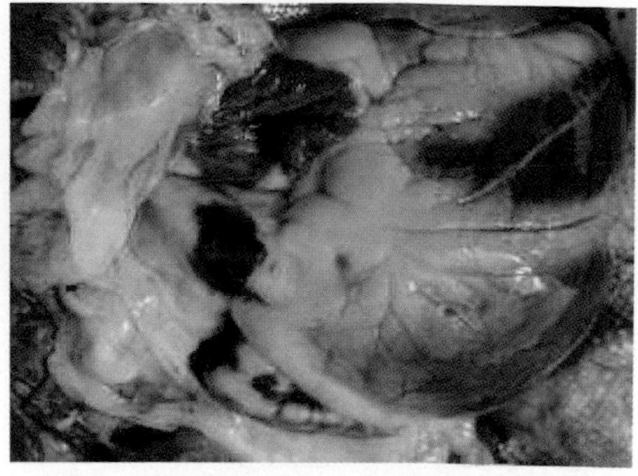

7C - Bluetongue - Highly characteristic lesion of hemorrhage in the pulmonary artery – the vasa vasorum vessels supplying the artery are compromised by the virus (Photo courtesy of Dr. Mariano Domingo)

8. BORNA

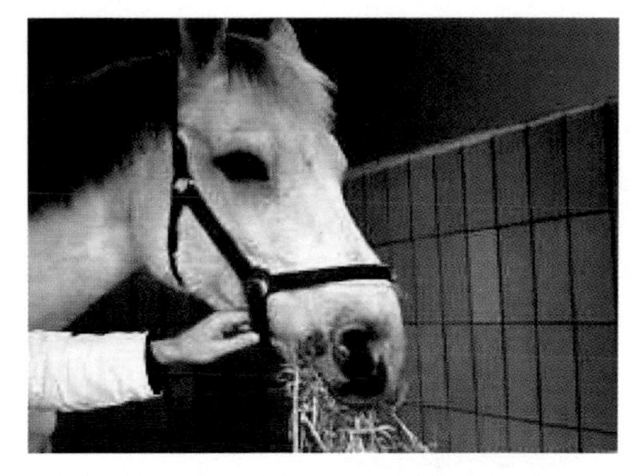

8A - Borna disease - arrested chewing movements are typical of borna disease

8B - Borna disease - ataxia is common

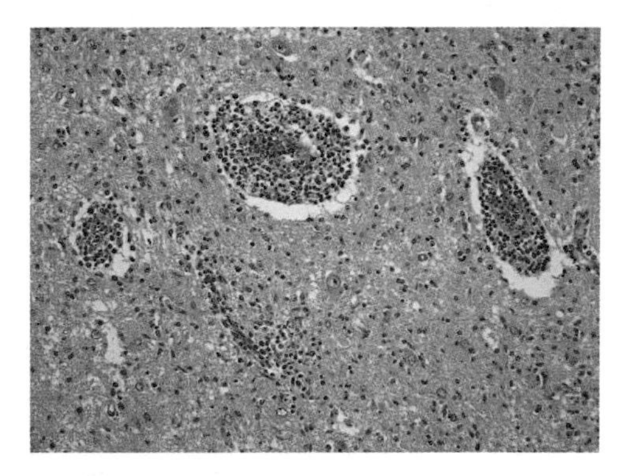

8C - Borna disease - severe nonsuppurative polioencephalomyelitis with massive pervascular cuffs is seen histologically

10. BOVINE SPONGIFORM ENCEPHALOPATHY

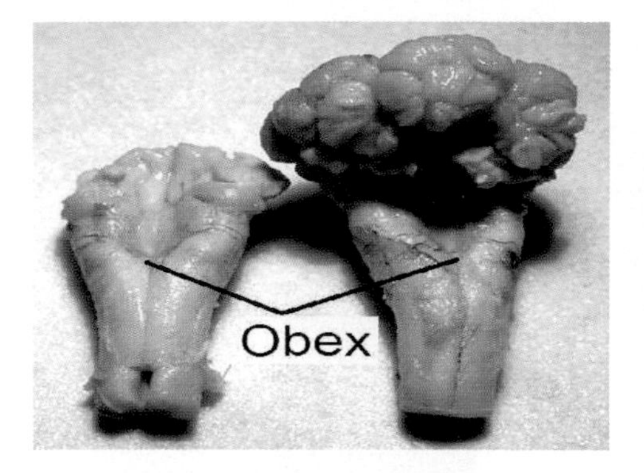

10A - Bovine spongiform encephalopathy - The obex is the preferred site for sampling for diagnosis of BSE. (Photo courtesy of Dr. Tim Baszler).

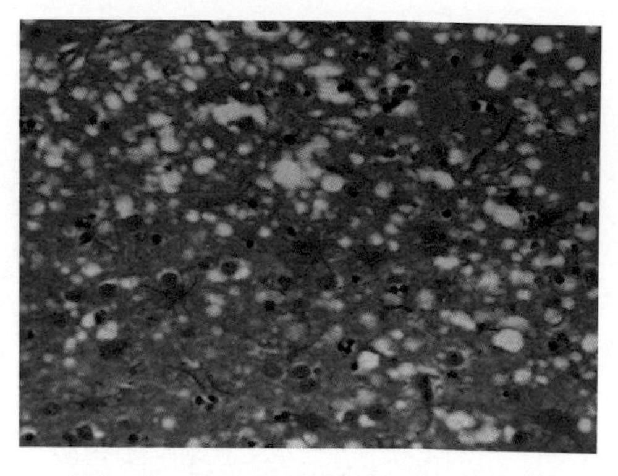

10B - Bovine spongiform encephalopathy - Vacuolation of neuropil, a characteristic feature of BSE. (Photo courtesy of the late Dr. Al Jenny).

11. CAPRIPOXVIRUS

11A - Capripoxvirus - A calf in Egypt infected with LSD, showing extensive dermal nodules.

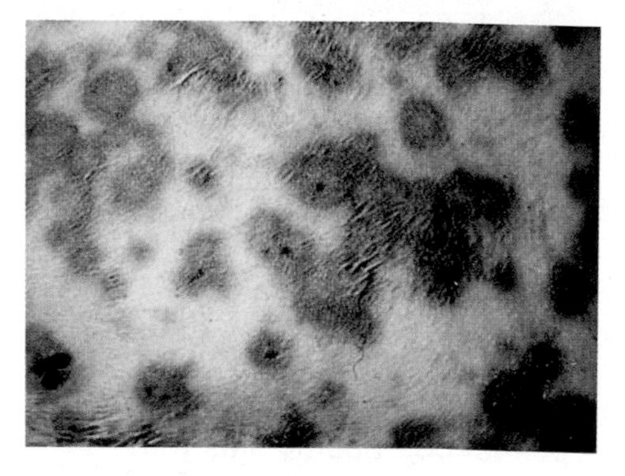

11B - Capripoxvirus - The dermal lesions appear as indurated, occasionally necrotic foci, scattered over the skin surface, sheep pox.

(cont'd... 11. CAPRIPOXVIRUS)

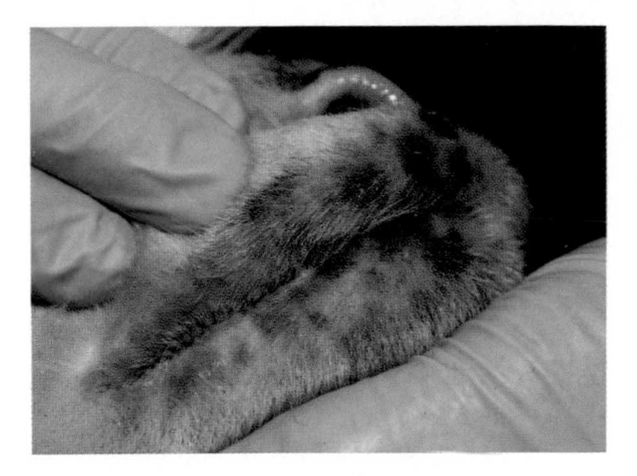

11C - Capripoxvirus - Capripoxvirus - Papules forming on the lips of a sheep infected with sheep pox.

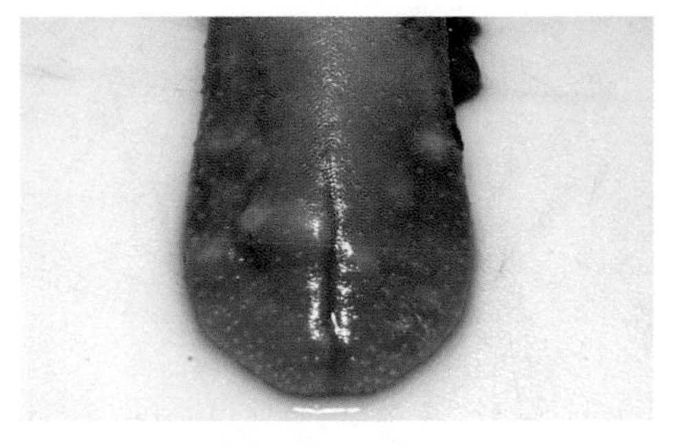

11D - Capripoxvirus - Tongue lesions forming in a sheep infected with sheep pox.

11E - Capripoxvirus - The pneumonia associated with sheep pox appears as embolically distributed firm edematous and hemorrhagic nodules.

11F - Capripoxvirus - Papules and nodules may form in visceral organs and be visible from either mucosal or serosal surfaces – here are some nodules in the reticulum mucosa.

12. CLASSICAL SWINE FEVER

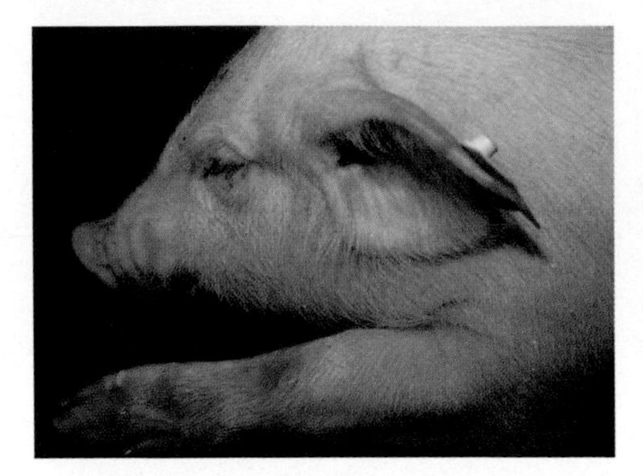

12A - Classical swine fever - Affected pigs are depressed, and there may be crusty, hemorrhagic exudates from the conjunctiva.

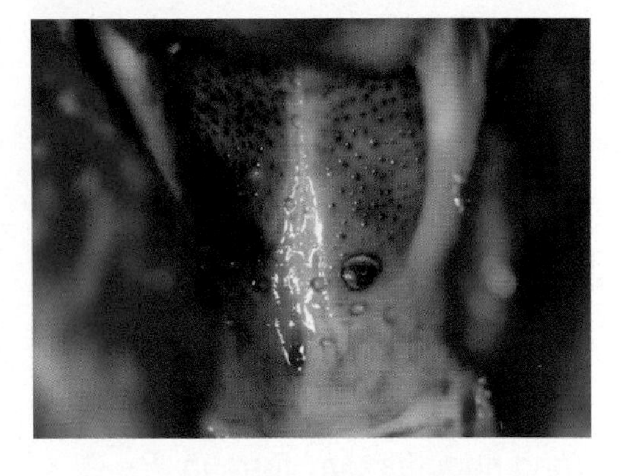

12B - Classical swine fever - Tonsillar necrosis is a typical feature of the disease.

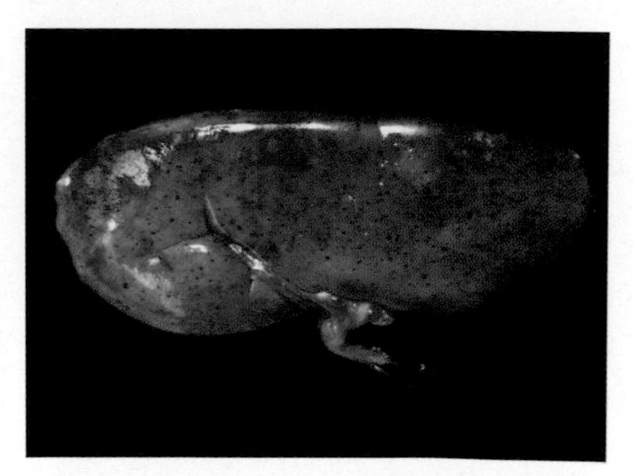

12C - Classical swine fever - Petechiation of the kidneys, also often called “turkey egg kidney”, is commonly seen in CSF, but is also seen in some other diseases, including African swine fever, erysipelas, and salmonellosis.

(cont'd… 12. CLASSICAL SWINE FEVER)

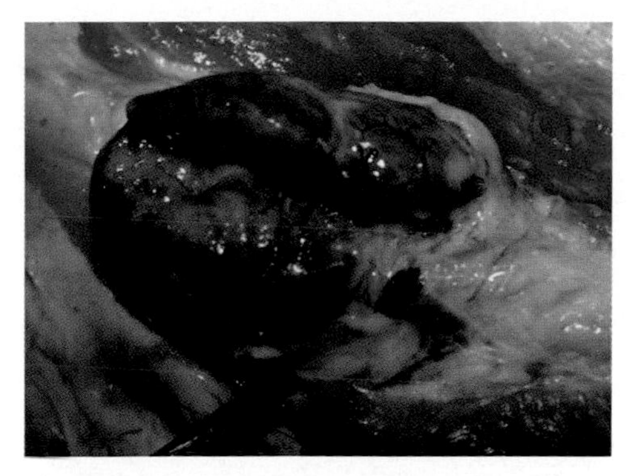

12D - Classical swine fever - A lymph node showing peripheral hemorrhage.

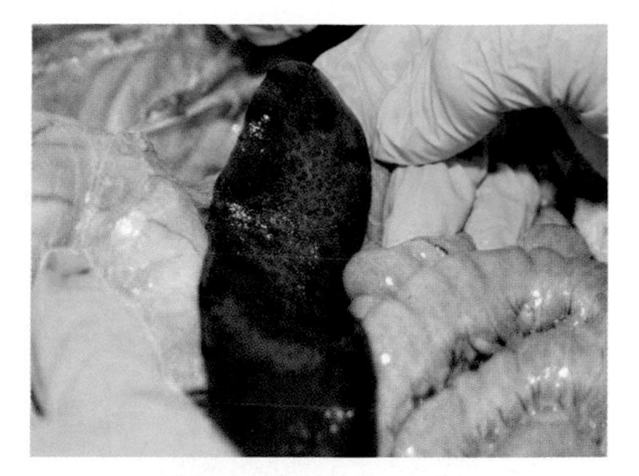

12E - Classical swine fever - Although infarcts in the spleen are heralded as being highly characteristic of CSF infection, the current strains circulating in the world rarely produce this lesion.

14. CONTAGIOUS BOVINE PLEUROPNEUMONIA

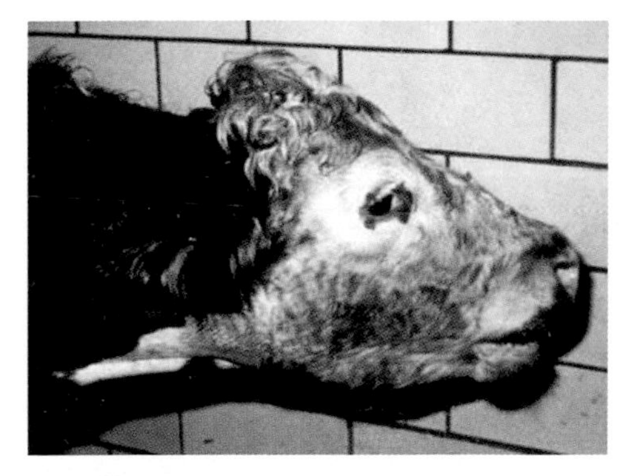

14A - CBPP - The extended neck and head is due to respiratory distress and coughing.

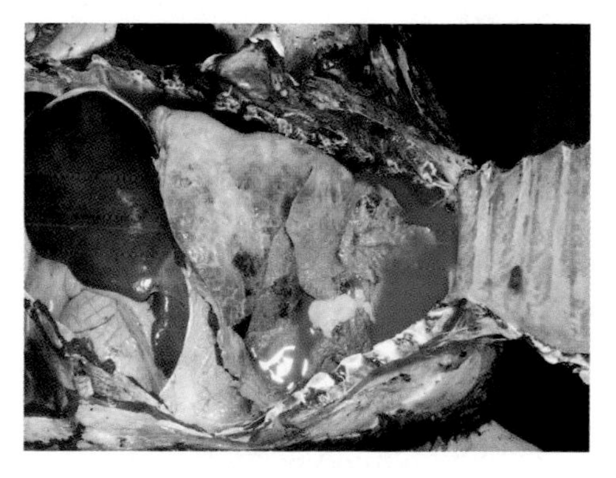

14B - Contagious bovine pleuropneumonia - Opened thoracic cavity showing extensive pulmonary and pleural involvement.

(cont'd...14. CONTAGIOUS BOVINE PLEUROPNEUMONIA)

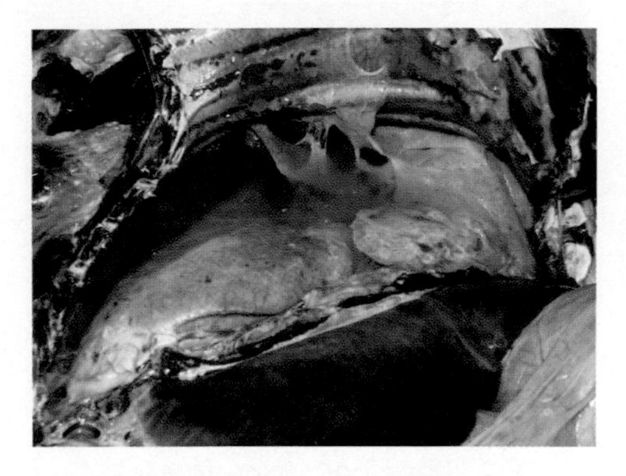

14C - Contagious bovine pleuropneumonia - Fibrin on the pleural surface is seen in almost all cases of CBPP.

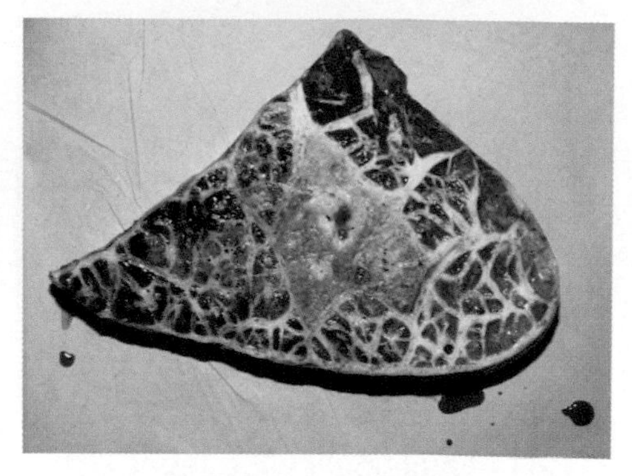

14D - Contagious bovine pleuropneumonia - This cross-section of lung demonstrates the classical prominent interlobular septal thickening or "marbling" as well as a lighter avascular zone in the center which is a sequestrum.

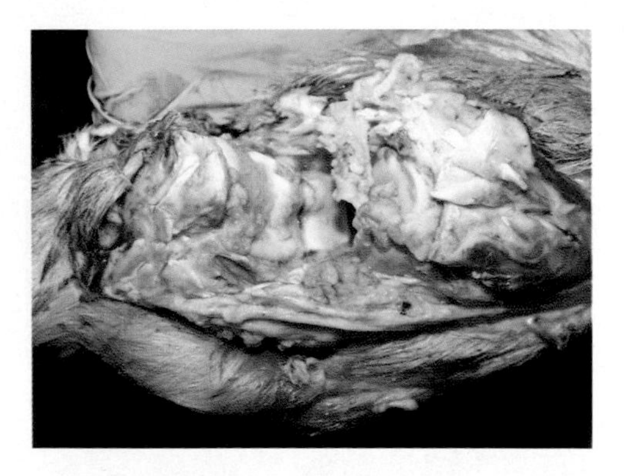

14E - Contagious bovine pleuropneumonia - In calves especially, a septicemia can result in fibrinous arthritis.

15. CONTAGIOUS CAPRINE PLEUROPNEUMONIA

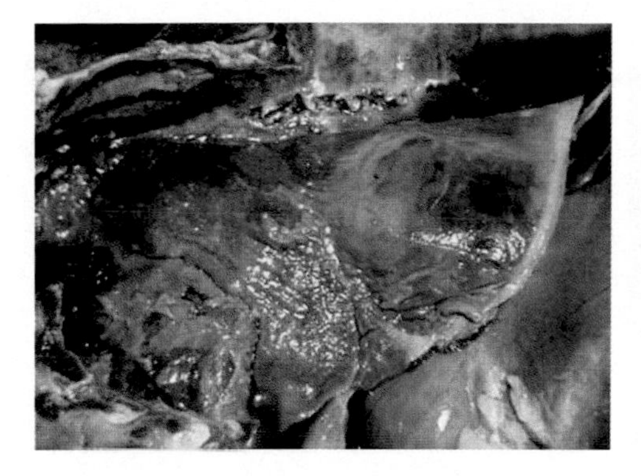

15A - Contagious caprine pleuropneumonia - There is usually both pulmonary and pleural involvement, often with abundant fibrin.

15B - Contagious caprine pleuropneumonia - The deposition of fibrin on the pleura can be extensive.

16. CONTAGIOUS EQUINE METRITIS

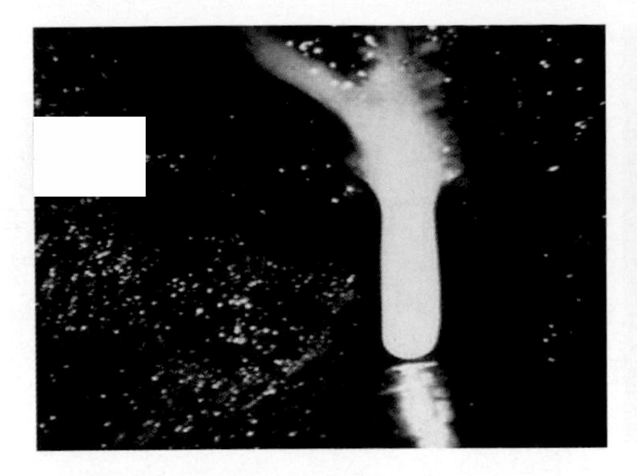

16A - Contagious equine metritis - Infected mares may have a copious mucopurulent vaginal discharge in the acute stage of the disease.

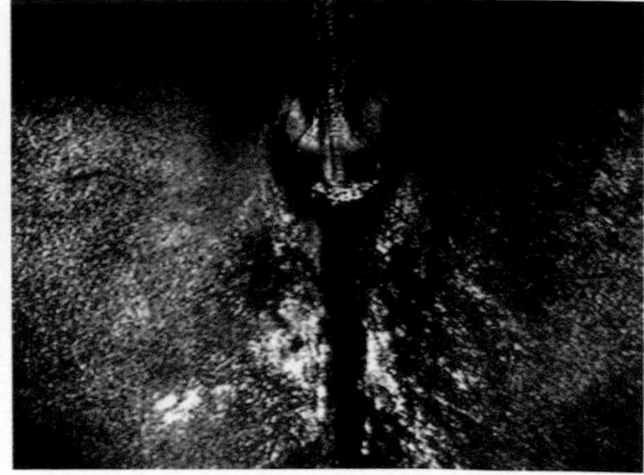

16B - Contagious equine metritis - Dried vaginal discharge just below the vulva.

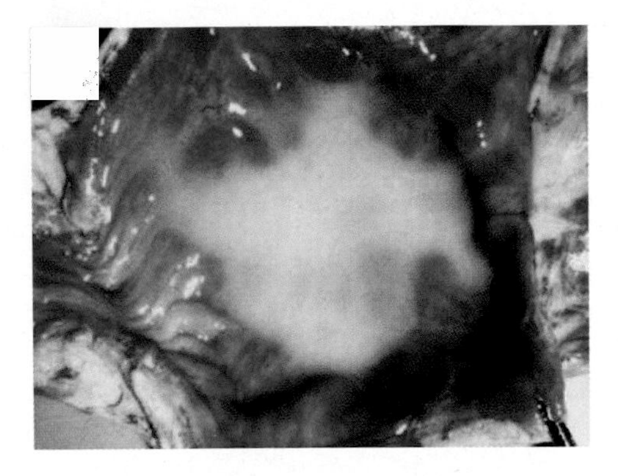

16C - Contagious equine metritis - There is a mucopurulent exudate in the lumen of the uterus in the acute stage of the disease.

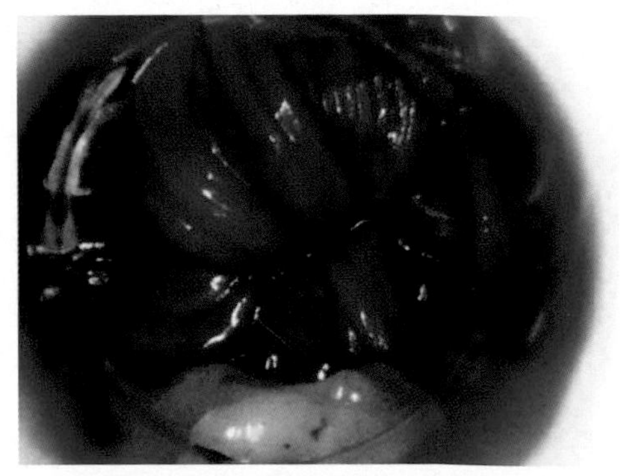

16D - Contagious equine metritis - There is a mucopurulent cerviitis in the acute stage of the disease.

17. DOURINE

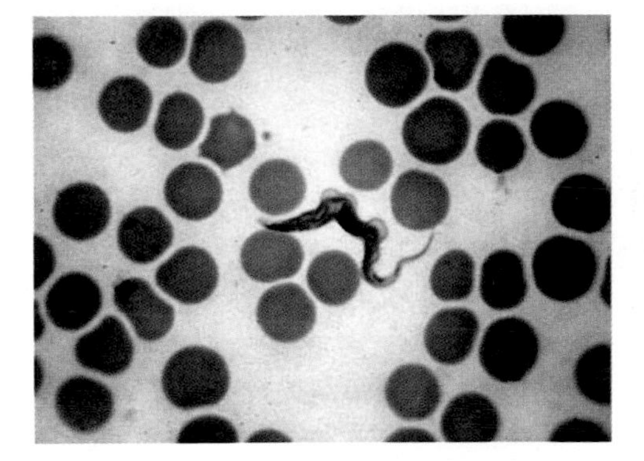

17A - Dourine - Trypanosomal organism as seen in a blood smear.

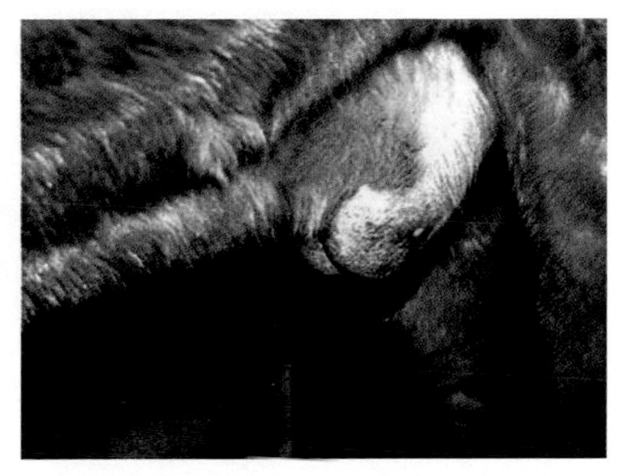

17B - Dourine - Swollen sheath can be seen in the disease.

18. DUCK VIRUS HEPATITIS

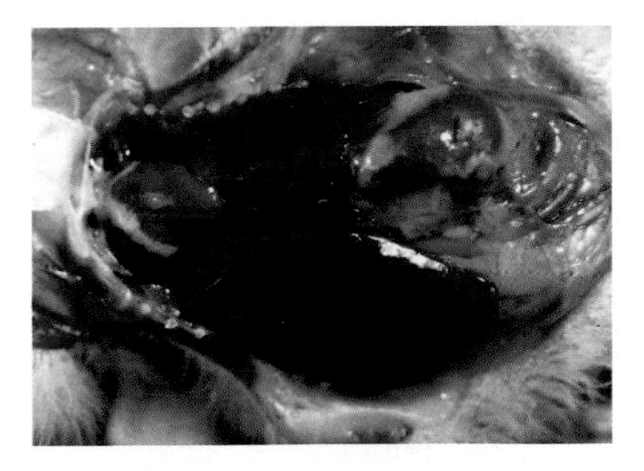

18 - Duck virus hepatitis - There is typically an enlarged liver with petechial to ecchmyotic hemorrhages.

19. EAST COAST FEVER

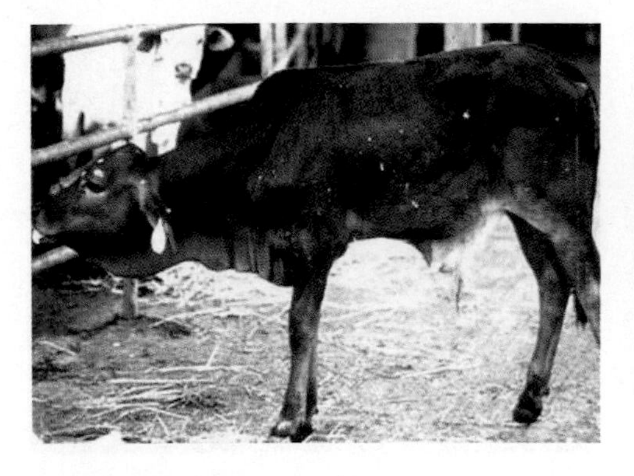

19A - East Coast fever - In ECF there is generalized lymphadenopathy; note the enlarged prescapular lymph node.

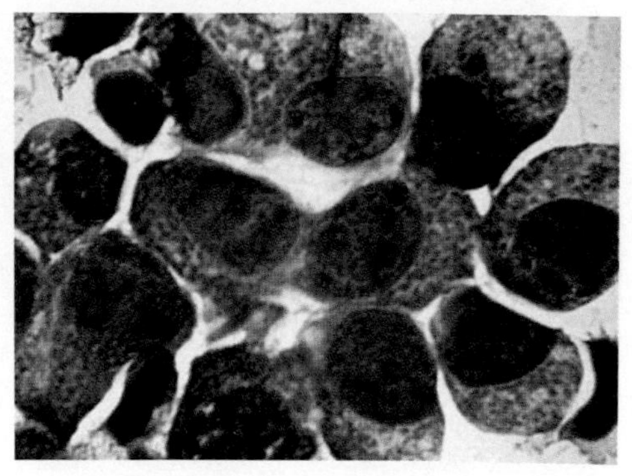

19B - East Coast fever - Lymphoblasts containing *Theileria* parasites.

22. FOOT-AND-MOUTH DISEASE

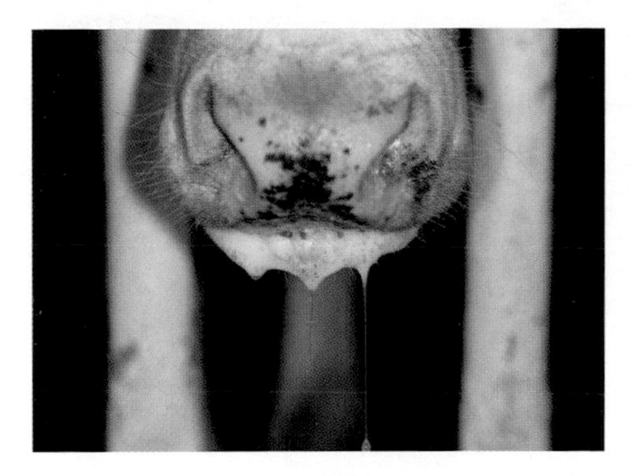

22A - Foot-and-mouth disease – Excess salivation is a characteristic feature in cattle infected with FMD, as animals are reluctant to swallow due to the painful developing lesions in the mouth.

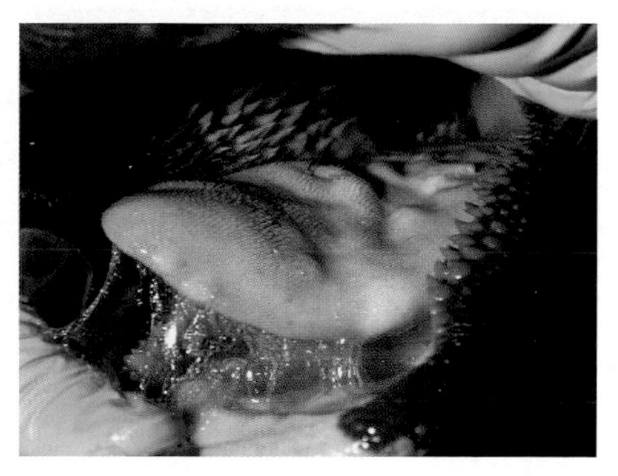

22B - Foot-and-mouth disease – Small vesicles in the tongue are enlarging and coalescing.

22C - Foot-and-mouth disease – Rupture of a vesicle leaves a raw, eroded area.

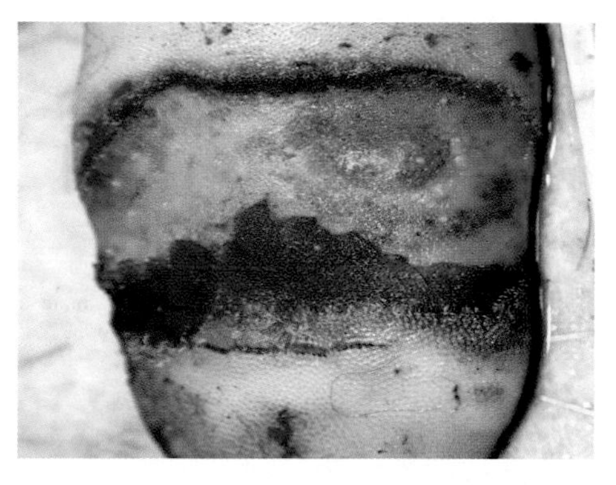

22D - Foot-and-mouth disease – Despite sometimes extensive ulcerative damage to the tongue, healing and re-epithelialization proceed rapidly, bovine.

(cont'd… 22. FOOT-AND-MOUTH DISEASE)

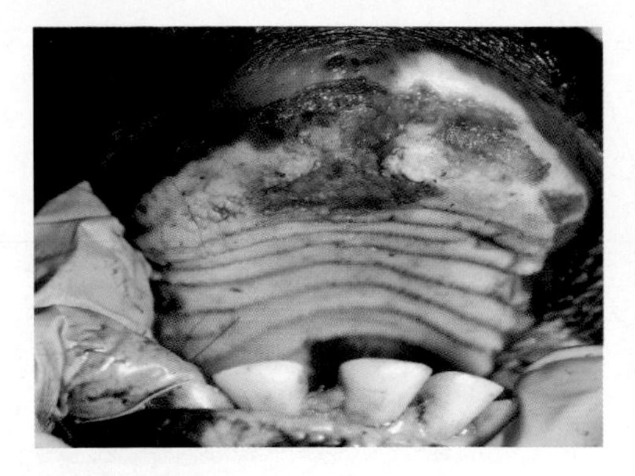

22E - Foot-and-mouth disease – Extensive erosion and ulceration of dental pad secondary to lesion development and rupture, bovine.

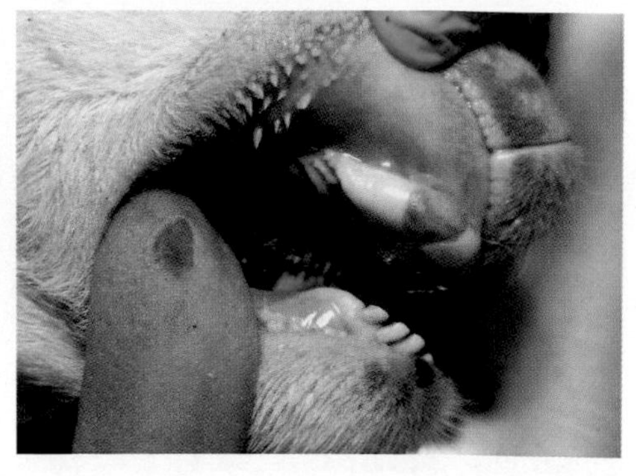

22F - Foot-and-mouth disease – Lesions on the dental pad and tongue of an infected sheep. Lesions in small ruminants are rarely as pronounced as in cattle.

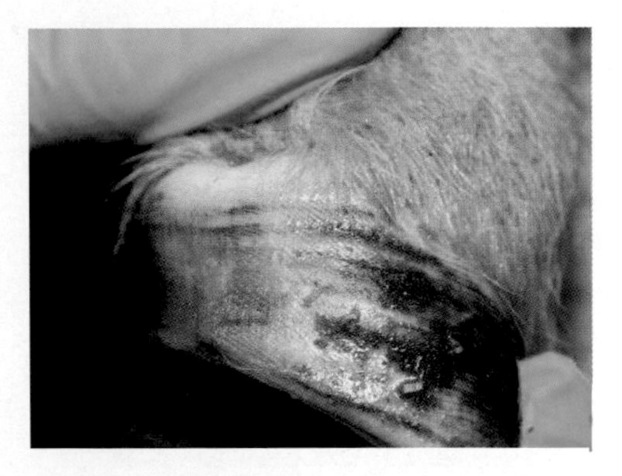

22G - Foot-and-mouth disease – Blanching of coronary band, a prequel to development of a vesicle, sheep.

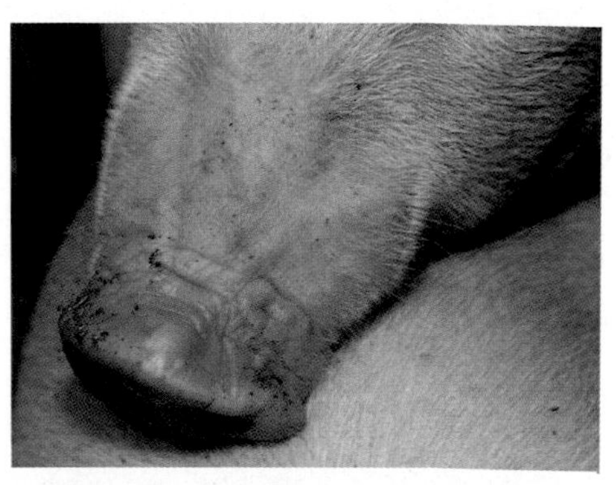

22H - Foot-and-mouth disease – Developing vesicle on pig snout.

(cont'd... 22. FOOT-AND-MOUTH DISEASE)

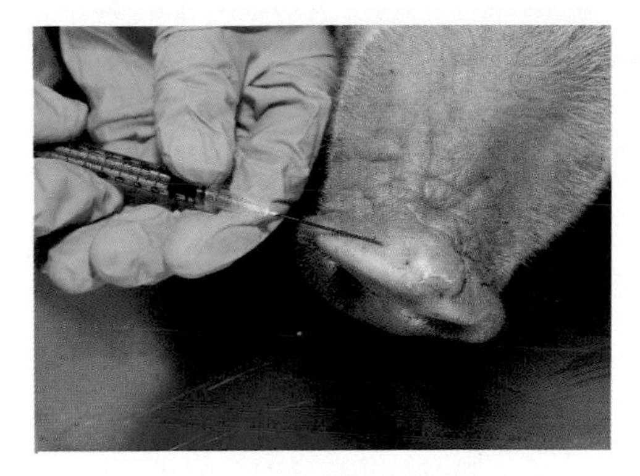

22I - Foot-and-mouth disease – Collecting vesicular fluid requires entering from beneath, so as to avoid spillage of diagnostic fluid.

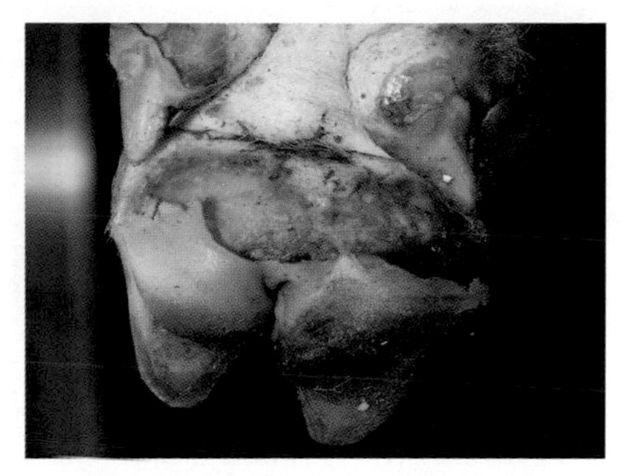

22J - Foot-and-mouth disease – Foot lesions in swine can be dramatic and lead to sloughing of the claw, rendering the animal permanently impaired.

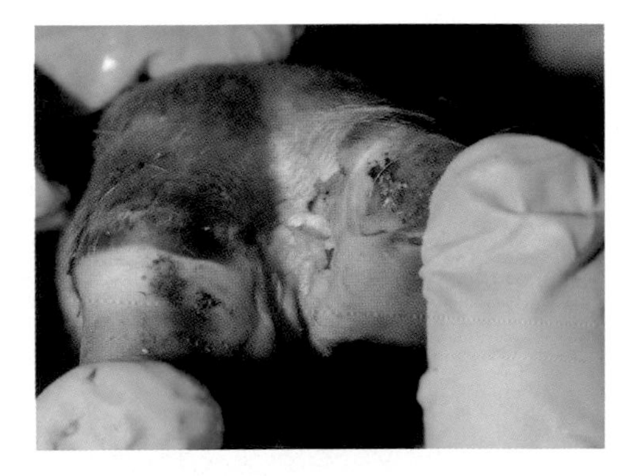

22K - Foot-and-mouth disease – Recently ruptured vesicle, interdigital cleft, sheep.

24. GLANDERS

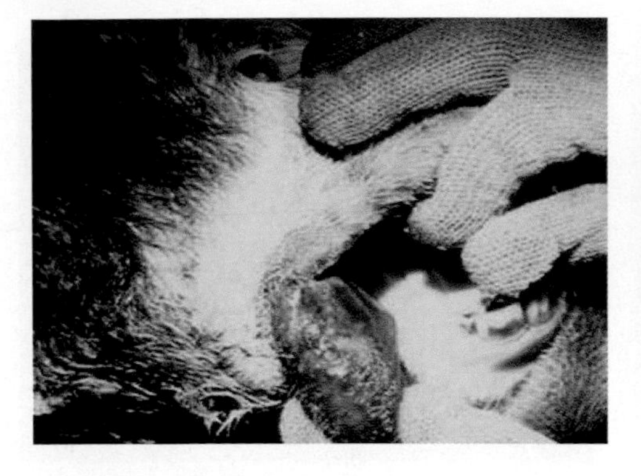

24A - Glanders - A granulomatous lesion in the lip of a donkey.

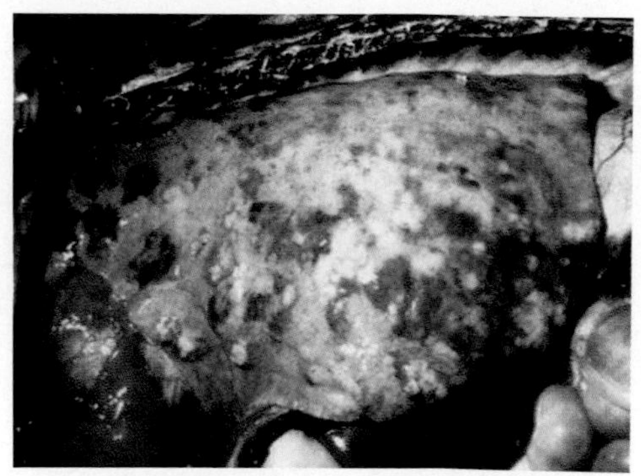

24B - Glanders - An extensive pyogenic granulomatous pneumonia in a donkey.

25. HEARTWATER

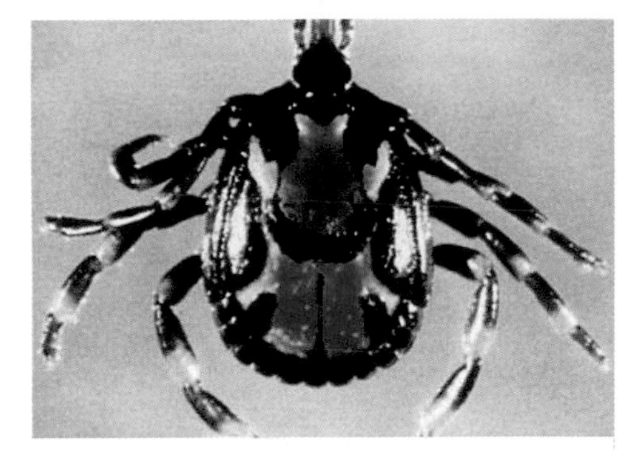

25A - Heartwater – Male tropical bont tick, *Amblyomma variegatum.*

25B - Heartwater - A heavy infestation of the bont tick (*Amblyomma variegatum*) on the dewlap of a cow.

25C - Heartwater – Hydrothorax seen in a sheep dying of HW

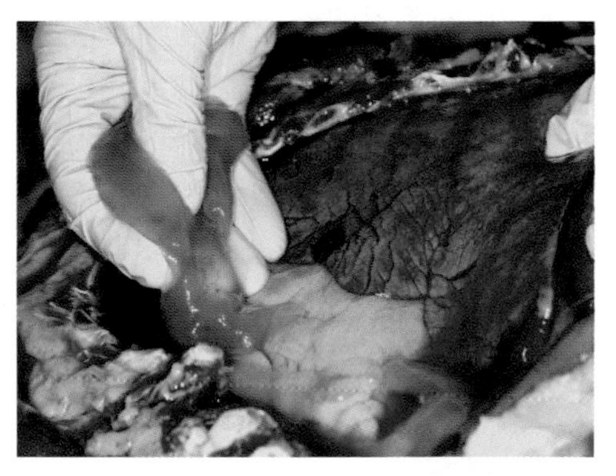

25D - Heartwater – Fibrin clots in the thoracic fluid demonstrate the protein-rich nature of the effusion.

(cont'd... 25. HEARTWATER)

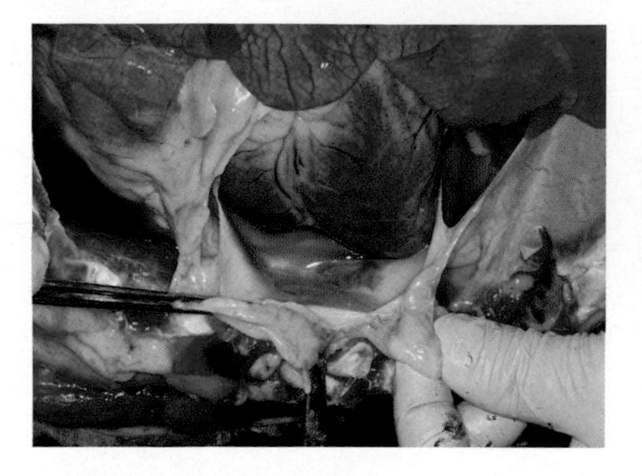

25E - Heartwater – Hydropericardium in an infected sheep. This is where the disease got its name (heartwater). Note also the extensive petechiation on the heart, a reflection of the vascular nature of the disease.

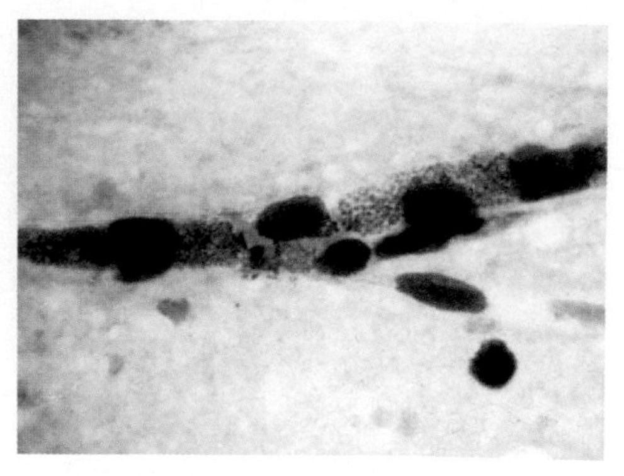

25F - Heartwater – Microscopic view of a brain smear from an infected animal. A line of endothelial cells shows prominent nuclei (large purple ovals) and cytoplasm filled with aggregates of *Ehrlichia* organisms, each represented as a single dot. These aggregates are referred to as "morulae."

26. HEMORRHAGIC SEPTICEMIA

26 - HS – Extensive edematous swelling of the head and neck.

27. HENDRA

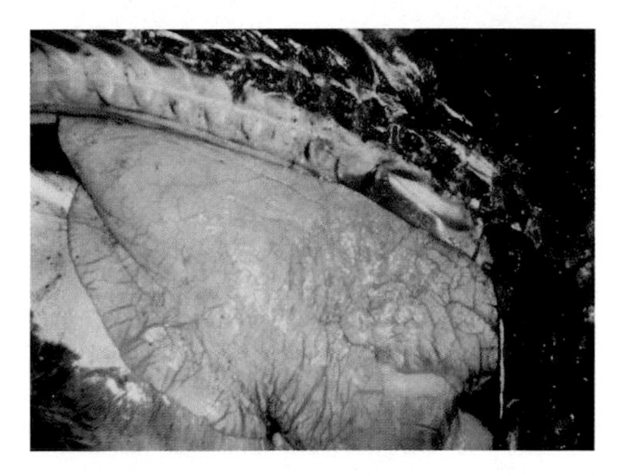

27 - Hendra virus – A characteristic feature is overwhelming pulmonary edema.

32. MALIGNANT CATARRHAL FEVER

32A - Malignant catarrhal fever – Cattle develop a bilateral mucoid nasal discharge that can become profuse.

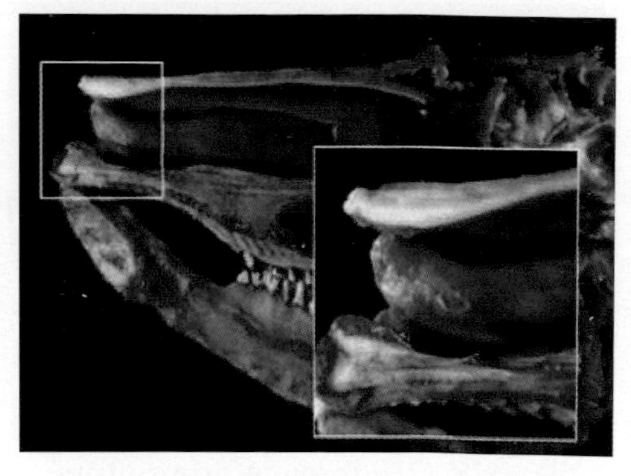

32B - Malignant catarrhal fever – An ulcerative rhinitis occurs in the cranial portion of the nasal cavity.

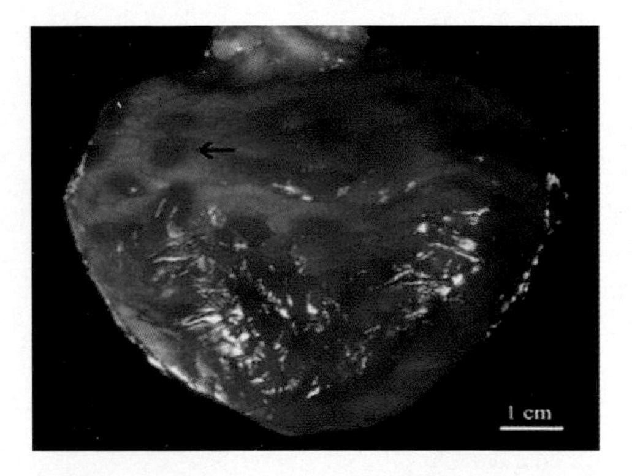

32C - Malignant catarrhal fever – Cystitis is seen in acute MCF of both cattle and bison.

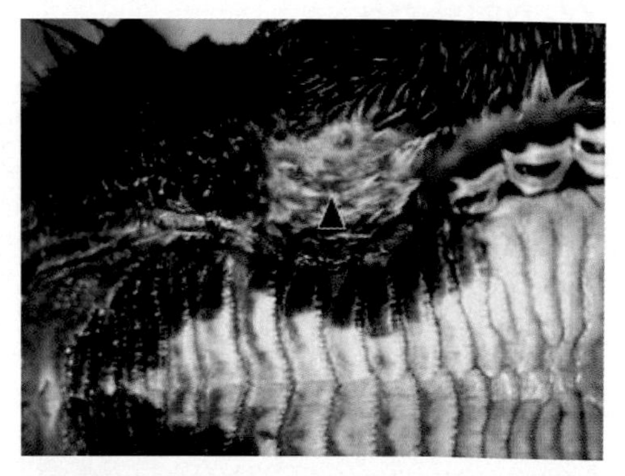

32D - Malignant catarrhal fever – The tips of the buccal papillae become reddened and then dark and necrotic; similarly, ulcers develop on the mucosa of the hard palate.

(cont'd... 32. MALIGNANT CATARRHAL FEVER)

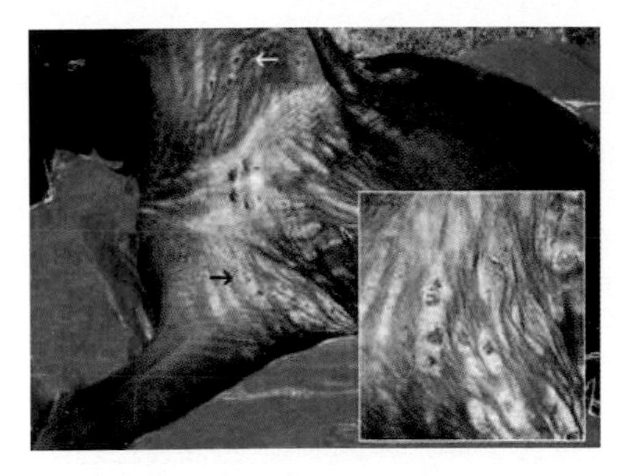

32E - Malignant catarrhal fever – Occasionally animals will have dermatitis, with ulcerative nodules present in the skin, especially evident in poorly haired regions.

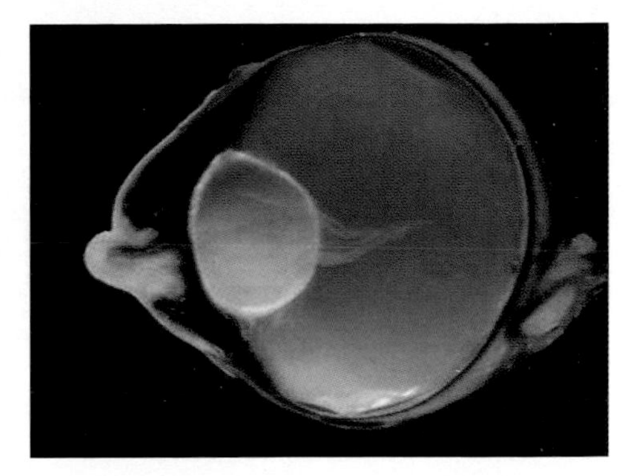

32F - Malignant catarrhal fever – Vascular compromise at the cornea can lead to edema and, in some cases, corneal necrosis and perforation, as seen here.

34. NEWCASTLE DISEASE

34A - Newcastle disease – Hemorrhage of the lower eyelid is often seen in Newcastle infection

34B - NDV – Newcastle disease - Splenic necrosis

(cont'd... 34. NEWCASTLE DISEASE)

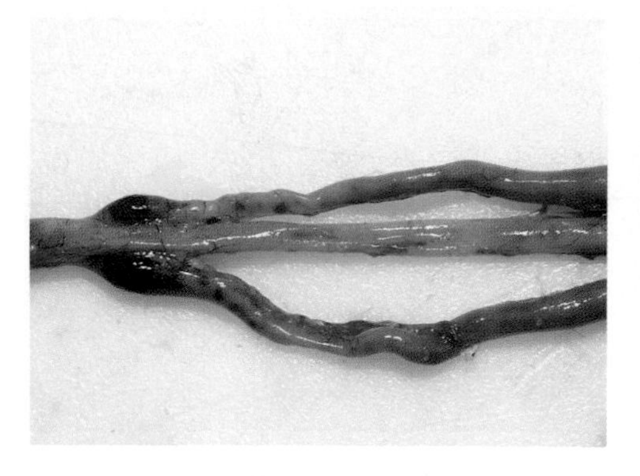

34C - Newcastle disease – Hemorrhagic necrosis occurs in the intestine in areas of lymphoid accumulations, seen here at the cecal tonsils.

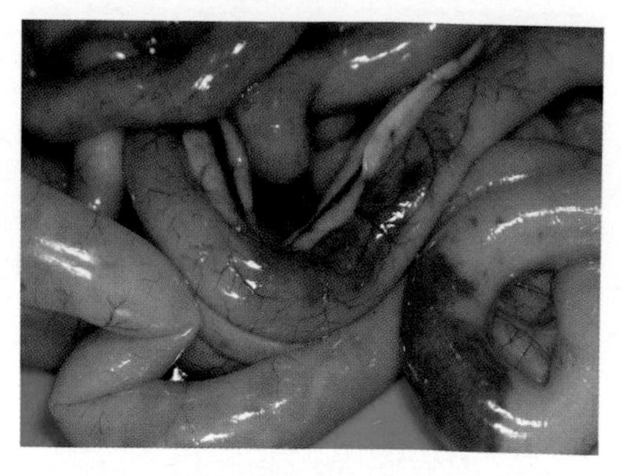

34D - Newcastle disease – Hemorrhagic necrosis can occur anywhere in the intestine where lymphoid tissue exists (Peyer's patches).

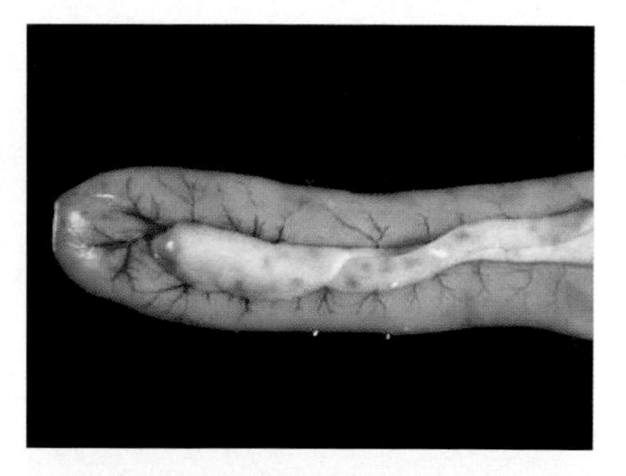

34E - Newcastle disease – Necrosis in the pancreas, seen as reddened foci.

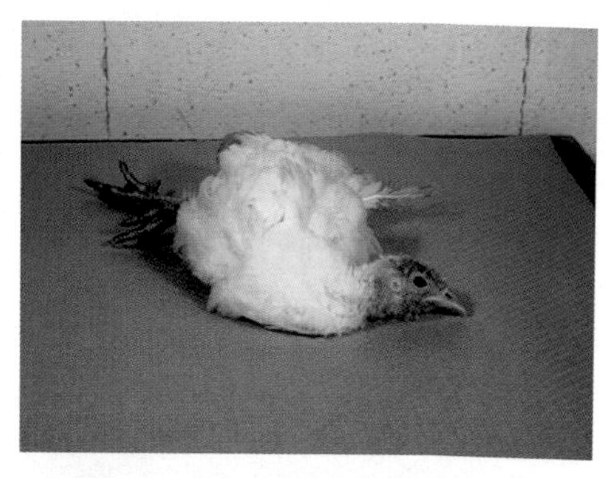

34F - Newcastle disease – Some strains of Newcastle will produce predominantly neurologic signs. Here an infected turkey is unable to right itself.

36. PESTE DES PETITS RUMINANTS

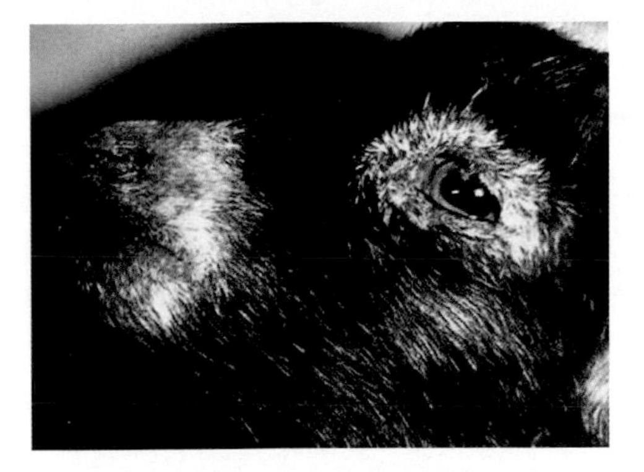

36A - Peste des petits ruminants - Dried exudate on the muzzle and around the eye resulting from rhinitis and conjunctivitis.

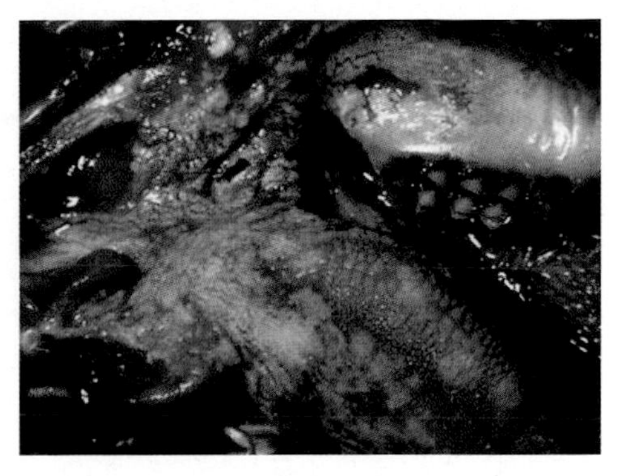

36B - Peste des petits ruminants - Necrosis (whitish areas) of the epithelium on the tongue and pharynx.

37. RABBIT HEMORRHAGIC DISEASE

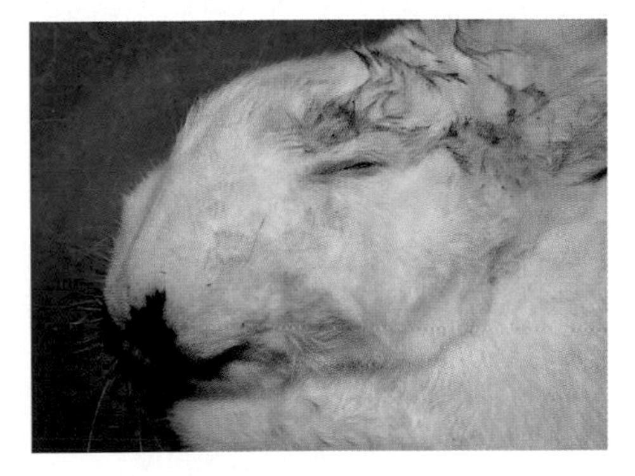

37A - Rabbit hemorrhagic disease – Terminal epistaxis which is often seen in rabbits dying of the disease.

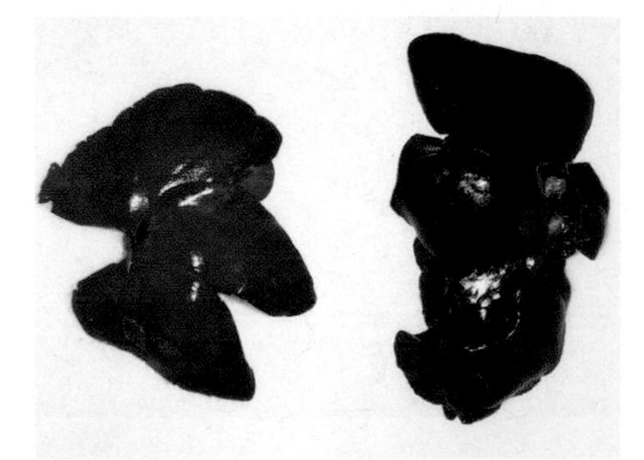

37B - Rabbit hemorrhagic disease – Livers from an affected (left) and nonaffected rabbit. The necrosis makes the liver appear lighter in color.

(cont'd... 37. RABBIT HEMORRHAGIC DISEASE)

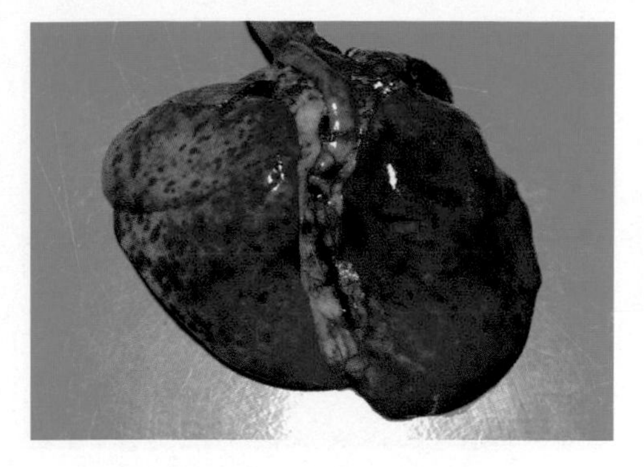

37C - Rabbit hemorrhagic disease – The lungs may be very congested or hemorrhagic.

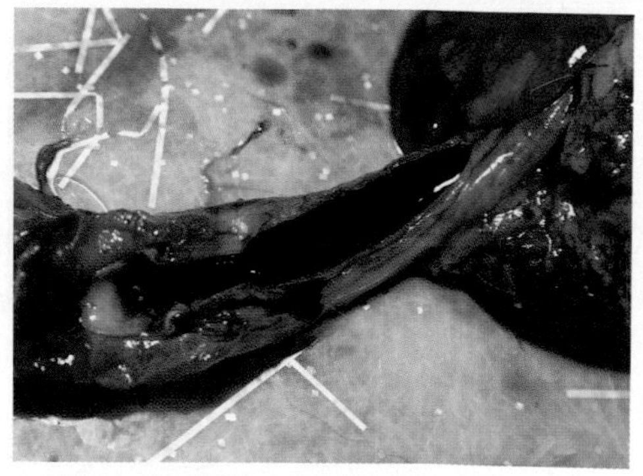

37D - Rabbit hemorrhagic disease - There may be hemorrhage in the tracheal mucosa.

39. RINDERPEST

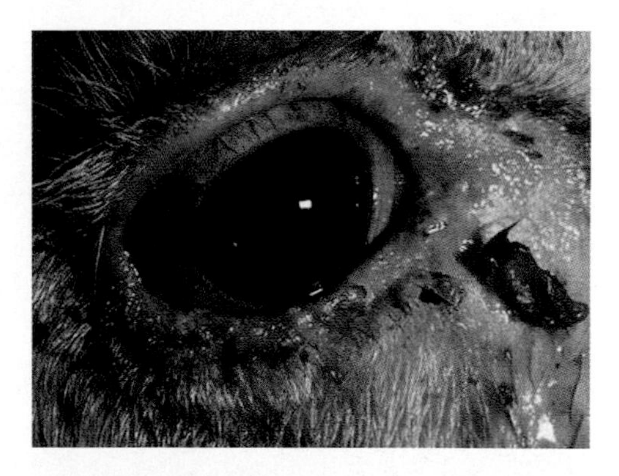

39A - Rinderpest – Conjunctivitis and mucopurulent exudate in the early stages of RP infection.

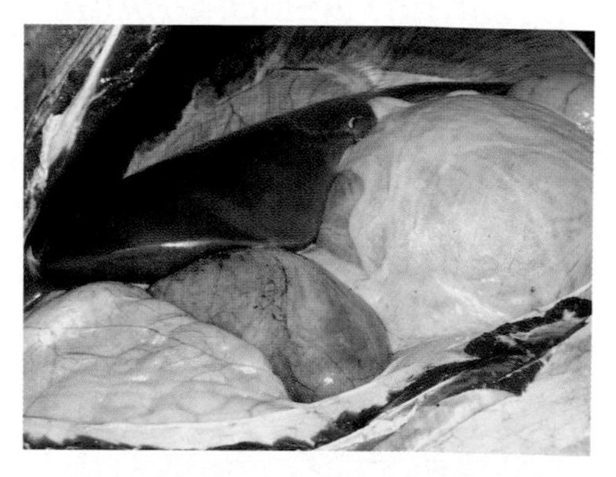

39B - Rinderpest – Greatly enlarged gall bladder, as a result of inanition.

(cont'd... 39. RINDERPEST)

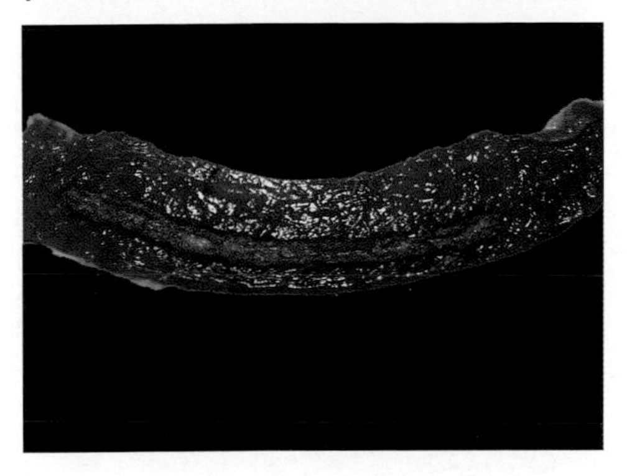

39C - Rinderpest – Ulceration and diphtheritic membrane over Peyer's patch, small intestine.

40. SCREWWORM MYIASIS

40A - Screwworm - Third instar larvae.

40B - Screwworm - Female fly.

44. AFRICAN ANIMAL TRYPANOSOMOSIS

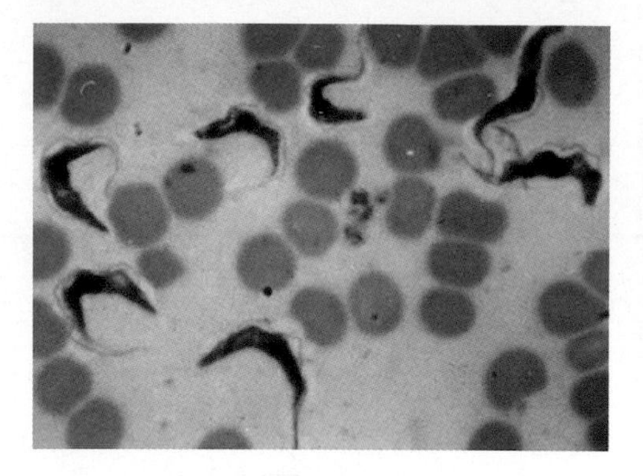

44A - Trypanosomiasis - Parasites in a blood smear.

44B - Trypanosomiasis - Severe emaciation due to one of the African trypanosomes, *T. congolense, T. vivax,* or *T. brucei brucei.*

47. VESICULAR STOMATITIS

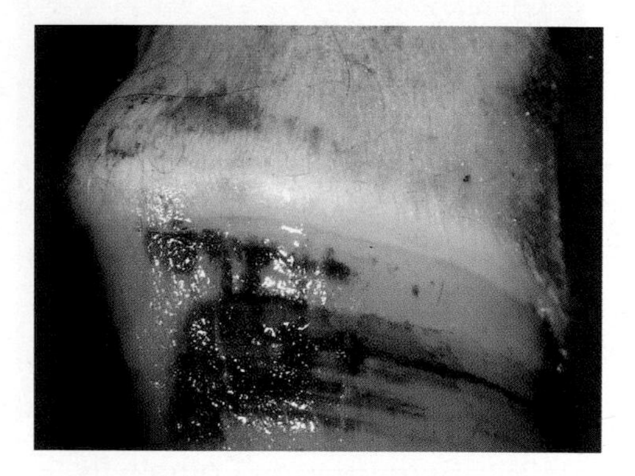

47A - Vesicular stomatitis - Blanching at the coronary band, just prior to the development of a vesicle.

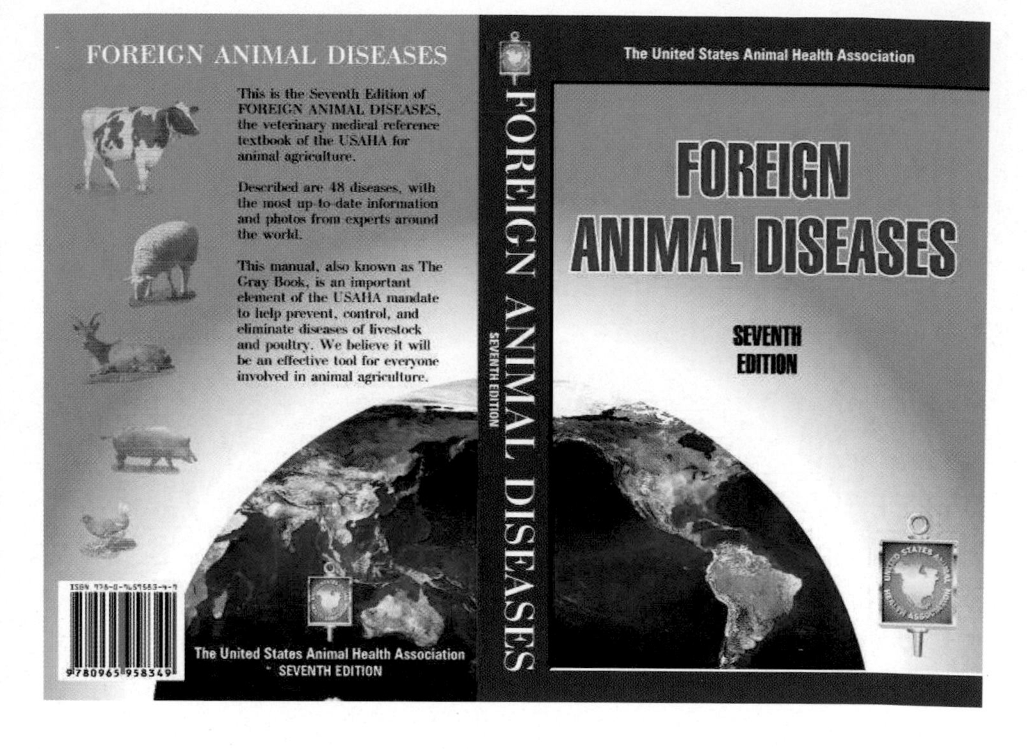

UNITED STATES ANIMAL HEALTH ASSOCIATION